Beneficial Fungi: Importance and their uses in Agriculture

Beneficial Fungi: Importance and their uses in Agriculture

Sunil Mekala

RANDOM

RANDOM PUBLICATIONS

NEW DELHI (INDIA)

Beneficial Fungi: Importance and their uses in Agriculture

ISBN 978-93-5111-770-4

Published in 2016 in India by

RANDOM PUBLICATIONS

4376-A/4B, Gali Murari Lal, Ansari Road
New Delhi-110 002
Phone : +9111-43580356, 011-23289044, 011-43142548
e-mail: sales@randompublications.com,
info@randompublications.com, randomexports@gmail.com

Reprinted 2026

Type Setting by : Friends Media, Delhi-110089

Preface

Fungi are one of the most important groups of organisms on the planet. This is easy to overlook, given their largely hidden, unseen actions and growth. They are important in an enormous variety of ways.

Fungi are also important directly as food for humans. Many mushrooms are edible and different species are cultivated for sale worldwide. While this is a very small proportion of the actual food that we eat, fungi are also widely used in the production of many foods and drinks. These include cheeses, beer and wine, bread, some cakes, and some soya bean products.

These organisms are responsible for the important process of decomposition in terrestrial ecosystems as they degrade and assimilate cellulose, the component of plant cell walls. Fungi are constituted by microscopic cells that usually grow as long threads or strands called hyphae of only a few micrometres in diameter but with the ability to span a length from a few cells to many metres

Although we often think of fungi as organisms that cause disease and rot food, fungi are important to human life on many levels. They influence the well-being of human populations on a large scale because they are part of the nutrient cycle in ecosystems. They also have other ecosystem uses, such as pesticides.

This book is benefit of unique uses of beneficial fungi. This include uses of fungal cellulases in biotechnology, utilization of fungi for the production of single cell proteins, and for the control of pests and exploiting plant diseases for human welfare.

– Author

Preface

Contents

1

Beneficial Pest Management of Crops

In addition to the soil improving benefits, cover crops can also enhance many pest management programmes. Ecologists tell us that stable natural systems are typically diverse, containing many different types of plants, arthropods, mammals, birds, and microorganisms. Growing cover crops adds diversity to a cropping system. In stable systems, serious pest outbreaks are rare, because natural controls exist to automatically bring populations back into balance.

Farmers and researchers in several locations have observed and documented increased beneficial insect numbers associated with cover crops. The cover crops provide pollen, nectar, and a physical location for beneficial insects to live while they search for pest insects. Conservation tillage proves a better option than tilling because it leaves more crop residue on the surface to harbour the beneficial insects. Strip tilling or no-tillage disturbs a minimum of the existing cover crop that harbours beneficial insects. Cover crops left on the surface may be living or in the process of dying. At either of these stages they protect beneficials. Once the main crop is growing, the beneficials move onto it.

By having the cover crop in place early in the growing season, the population of beneficials is much higher sooner in the growing season than would be the case if only the main crop were serving as habitat for the beneficials. Innovative farmers are paving the way by interplanting cover crops with the main crop and realizing pest management benefits as a result. Georgia cotton farmers Wayne Parramore and sons reduced their insecticide and fertilizer use by growing a lupine cover crop ahead of their spring-planted cotton. They started experimenting with lupines on 100 acres in 1993 and by 1995 were growing 1,100 acres of lupines. Ground preparation for cotton planting is begun about 10 days prior to planting by tilling 14-inch wide strips into the lupines. Herbicides are applied to the strips at that time and row middles remain untouched. The remaining lupines provide beneficial insect habitat and also serve as a smother crop to curtail weeds and grasses. The lupines in the row middles can be tilled in later in the season to release more legume nitrogen.

Dr. Sharad Phatak of the University of Georgia has been working with cotton growers in Georgia testing a strip cropping method using winter annual cover crops. Planting cotton into strip-killed crimson clover improves soil health, cuts tillage costs, and allows him to grow cotton without any insecticides and only 30 pounds of nitrogen fertilizer. Working with Phatak, farmer Benny Johnson reportedly saved at least $120/acre on his 16-acre test plot with the clover system. There were no insect problems in the test plot, while beet armyworms and whiteflies infested nearby cotton and required 8 to 12 sprays to control. Cotton intercropped with crimson clover yielded more than 3 bales per acre compared with 1.2 bales per acre in the rest of the field. Boll counts were 30 per plant with crimson clover and 11 without it. Phatak identified up to 15 different kinds of beneficial insects in these strip-planted plots.

Phatak finds that planting crimson clover seed at 15 pounds per acre in the fall produces around 60 pounds of nitrogen per acre by spring. By late spring, beneficial insects are active in the clover. At that time, 6- to 12-inch planting strips of clover are killed with Roundup herbicide. Fifteen to 20 days later the strips are lightly tilled and cotton is planted. The clover in the row-middles is left growing to maintain beneficial insect habitat. When the clover is past the bloom stage and less desirable as a preferred habitat, beneficials move onto the cotton. Even early-season thrips, which can be a problem following cover crops, are limited or prevented by beneficial insects in this system. Movement of the beneficials coincides with a period when cotton is most vulnerable to insect pests. Following cotton defoliation in the early fall, the beneficials hibernate in adjacent non-crop areas.

Phatak points out that switching to a whole-farm focus while reducing off-farm inputs is not simple. It requires planning, management, and several years to implement on a large scale. It is likewise important to increase and maintain organic matter, which stimulates beneficial soil microorganisms. Eventually a "living soil" will help keep harmful nematodes and soil-borne fungi under control.

COVER CROPS FOR CORN BELT GRAIN AND OILSEED PRODUCTION

In addition to providing winter cover and building soil structure, nitrogen (N) management will probably be a major factor in your cover crop decisions for the corn>soybean rotation. A fall- planted grass or small grain will scavenge leftover N from the previous corn or soybean crop. Legumes are much less efficient at scavenging N, but will add N to the system for the following crop. Legume/grass mixtures are quite good at both.

CROP ROTATION WITH COVER CROPS

One of the biggest challenges of cover cropping is to fit cover crops into your current rotations, or to develop new rotations that take full advantage of their benefits. Some of the systems used successfully by farmers in different regions of the U.S., One might be easily adapted to fit your existing crops, equipment and management. Other examples may point out ways that you can modify your rotation to make the addition of cover crops more profitable and practical.

Whether you add cover crops to your existing rotations or totally revamp your farming system, you should devote as much planning and attention to your cover crops as you do to your cash crops. Failure to do so can lead to failure of the cover crop and cause problems in other parts of your system. Also remember that there is likely no single cover crop that is right for your farm. Before you start:

- Review Benefits of Cover Crops and Selecting the Best Cover Crops for Your Farm.
- Decide which benefits are most important to you.
- Then consider how these cover crop rotations might be adapted to your particular conditions.
- Talk to your neighbours and the other "experts" in your area, including the contact people listed in Regional Experts.
- Start small on an easily accessible plot that you will see often.

Be an opportunist—and an optimist. If your cropping plans for a field are disrupted by weather or other conditions outside of your control, this may be the ideal window for establishing a cover crop. Consider using an early-maturing cash crop to allow for timely planting of the cover crop. Cover crops can be used for feed. Consider harvesting or grazing for forage or alternative livestock such as sheep and goats.

Corn>Soybean Systems

Keep in mind that corn is a heavy N feeder, soybeans benefit little, if at all, from cover crop N and that you have a shorter time for spring cover crop growth before corn than before soybeans.

Precaution. If you use herbicides, be sure to check labels for plant-back or rotation intervals to ensure that your cover crop isn't adversely affected.

Cover crop features: rye provides winter cover, scavenges N after corn, becomes a long-lasting (6 to 12 week) residue to hold moisture and suppress weeds for your soybeans; hairy vetch provides spring ground cover, abundant N and a moderate-term (4 to 8 week) mulch for a no-till corn crop; field peas are similar to vetch, but residue breaks down faster; red clover is also similar, but produces slightly less N and has less vigourous spring regrowth; berseem clover grows quickly to provide several cuttings for high-N green manure. Here are some options to consider adapting to your system:

Corn>Rye>Soybeans>Hairy Vetch

In Zone 7 and warmer, you can grow a cover crop every year between your corn and full-season beans. Also, you can use wheat or another small grain to replace the cover crop before beans, in a three-crop, two-year rotation (corn>wheat>doublecrop beans). In all cases, another legume or a grass/legume mixture can be used instead of a single species cover crop. Where it is adapted, you can use crimson clover or a crimson/grass mixture instead of vetch.

In cooler areas, plant rye as soon as possible after corn harvest. If you need more time in the fall, try overseeding in rowed beans at drydown "yellow leaf" stage in early fall, or in early summer at the last cultivation of corn. Seeding options include aerial application where the service is economical, using a specialty high-body tractor with narrow tires, or attaching a broadcast seeder, air seeder or seed boxes to a cultivator.

Precaution. Broadcast seeding of cover crops into standing crops is less dependable than other seeding methods. Success will depend on many factors, including adequate rainfall amount and distribution after seeding.

Kill the rye once it is about knee-high, or let it go a bit longer, killing it a couple of weeks before planting beans. Killing the rye with herbicides and no-tilling beans in narrow rows allows more time for cover crop growth, since you don't have to work the ground. If soil moisture is low, consider killing the rye earlier. Follow the beans with hairy vetch or a vetch/small grain mixture.

Legumes must be seeded at least 6 weeks before hard frost to ensure winter survival. Seed by drilling after soybean harvest, or by overseeding before leaf drop. Allow the vetch (or mixture) to grow as long as possible in spring for maximum N fixation.

Harvesting sweet corn, seed corn or silage corn opens a window for timely cover crop planting. Harvest or graze the small grain or legume/small grain mixture in spring if needed for feed.

In Pennsylvania, Ed Quigley seeds rye or spring oats after corn silage harvest. The oats can be cut for silage in fall if planted by early September. Rye can be made into rylage or sprayed before no-till corn the following spring.

Worried about planting your corn a bit late because you're waiting for your cover crop to mature? Research in Maryland, Illinois and elsewhere suggests that no-tilling corn towards the end of the usual window when using a legume cover crop has its rewards. The delay can result in greater yields than earlier planting, due to greater moisture conservation and more N produced by the cover crop, or due to the timing of summer drought. In Pennsylvania, however, delayed planting sometimes reduced corn yields following rye.

Check your state variety trial data for a shorter season corn hybrid that yields nearly as well as slightly longer season corn. The cover crop benefit should overcome many yield differences. Worried about soil moisture? There's no question that growing cover crops may consume soil moisture needed by the next crop. In humid regions, this is a problem only in an unusually dry spring. Time permitting, allow 2 to 3 weeks after killing the cover crop to alleviate this problem. While spring rainfall may compensate for the moisture demand of most cover crops by normal planting dates, cover crops can quickly dry out a field. Later in the season, killed cover crop residues in conservation tillage systems can conserve moisture and increase yields.

In any system where you are using accumulated soil moisture to grow your cash crop, you need to be extra careful. However, farmers and researchers are finding that water-thrifty cover crops may be able to replace even a fallow year without adversely affecting the cash crop.

Corn>Rye>Soybeans>Small Grain>Hairy Vetch. This rotation is similar to the corn>rye>soybeans rotation, except you add a year of small grains following the beans. In crop rotation research from different areas, many benefits accrue as the rotation becomes longer. This is because weed, disease and insect pest problems generally decrease with an increase in years between repeat plantings of the same crop.

Residue from small grains provides good organic matter for soil building, and in the case of winter grains, the plants help to prevent erosion over winter after soybeans loosen up the soil. The length of the growing season will determine how you fit in cover crops after full-season soybeans in the rotation. Consider using a short- season bean if needed in order to achieve timely planting after soybean harvest. Calculate whether cover crop benefits will compensate for a possible yield loss on the shorter season beans. If there is not enough time to seed a legume after harvest, use a small grain rather than no cover crop at all.

The small grain scavenges leftover N following beans. Legume cover crops reduce fertilizer N needed by corn, a heavy N feeder. If you cannot seed the legume at least six weeks before a hard frost, consider overseeding before leaf drop or at last cultivation.

Precaution. Because hairy vetch is hard seeded, it will volunteer in subsequent small grain crops.

An alternate rotation for the lower mid-South is corn>crimson clover (allowed to go to seed) > soybeans > crimson clover (reseeded) > corn. Allow the crimson clover to go to seed before planting beans. The clover germinates in late summer under the beans.

Kill the cover crop before corn the next spring. If possible, choose a different cover crop following the corn this time to avoid potential pest and disease problems with the crimson clover.

Precaution. In selecting a cover crop to interseed, do not jeopardize your cash crop if soil moisture is usually limiting during the rest of the corn season! Banding cover crop seed in row middles by using insecticide boxes or other devices can reduce cover crop competition with the cash crop.

3 Year: Corn>Soybean>Wheat/Red Clover. This well-tested Wisconsin sequence provides N for corn as well as weed suppression and natural control of disease and insect pests. It was more profitable in recent years as the cost of synthetic N increased. Corn benefits from legume-fixed N, and from the improved cation exchange capacity in the soil that comes with increasing organic matter levels.

Growers in the upper Midwest can add a small grain to their corn>bean rotation. The small grain, seeded after soybeans, can be used as a cover crop, or it can be grown to maturity for grain. When growing wheat or oats for grain, frost-seed red clover or sweet-clover in March, harvest the grain, then let the clover grow until it goes dormant in late fall. Follow with corn the next spring. Some secondary tillage can be done in the fall, if conditions allow. One option is to attach sweeps to your chisel plow and run them about 2 inches deep, cutting the clover crowns.

Alternatively, grow the small grain to maturity, harvest, then immediately plant a legume cover crop such as hairy vetch or berseem clover in July or August. Soil moisture is critical for quick germination and good growth before frost. For much of the northern U.S., there is not time to plant a legume after soybean harvest, unless it can be seeded aerially or at the last cultivation. If growing spring grains, seed red clover or sweet-clover directly with the small grain.

An Iowa study compared no-till and conventional tillage corn>soybean>wheat/clover rotation with annual applications of composted swine manure. Berseem clover or red clover was frostseeded into wheat in March. Corn and soybean yields were lower in no-till plots the first year, while wheat yield was not affected by tillage. With yearly application of composted swine manure, however, yield of both corn and soybean were the same for both systems beginning in year two of the 4-year study.

Adding a small grain to the corn>soybean rotation helps control white mold on soybeans, since two years out of beans are needed to reduce pathogen populations. Using a grain/legume mix will scavenge available N from the bean crop, hold soil over winter and begin fixing N for the corn. Clovers or vetch can be harvested for seed, and red or yellow clover can be left for the second year as a green manure crop.

Using a spring seeding of oats and berseem clover has proved effective on Iowa farms that also have livestock. The mix tends to favour oat grain production in dry years and berseem production in wetter years. Either way the mixture provides biomass to increase organic matter and build soil. You can clip the berseem several times before flowering for green manure. Precaution.

Planting hairy vetch with small grains may make it difficult to harvest a clean grain crop. Instead, seed vetch after small grain harvest. Be sure to watch for volunteer vetch in subsequent small grain crops. It is easily controlled with herbicides but will result in significant penalties if found in wheat grain at the elevator.

THE FOUNDER OF THE STUDY OF CROP EVOLUTION

Alphonse de Candolle, the founder of the study of crop evolution, was primarily interested in biogeography, the study of distribution of plants in relation to their environment. In his opinion, among the factors influencing this distribution are 'historic' factors such as glaciation events that are not related to the intrinsic adaptation characteristics of the species. When he considered crop plants, he posited that one particular historic event - domestication - had had a paramount importance in determining their adaptation and, hence, distribution, as well as their genetic characteristics Four types of evidence could be utilized to determine the origin of domestication of crop plants These were archaeological (or more specifically archaeobotany), botanical (*i.e.* the distribution of the wild, ancestral relative), historical (or the existence of a written record documenting the existence or importance of the crop) and linguistic evidence (*i.e.* the existence of words designating the crop or objects or concepts related to the crop in native languages). Of these four types, the first two provide the most abundant and reliable evidence. Since the time of de Candolle, however, additional scientific methods have increased considerably our power to determine centers of crop domestication based on archaeological and botanical arguments

Although common bean is not a species originating in the Near East, the focus of the present symposium, it presents nevertheless an interesting domestication history, which is only recently being elucidated. The first botanically accurate description of common bean in European herbalsWhereas barely 50 years had passed after the conquest of the Americas, the actual origin of the crop had already been lost to the writer of the herbal and later European botanists. Linnaeus assigned the origin of common bean to India; de Candolle himself expressed doubts that common bean had been domesticated in the Americas. It was only during and after the Second World War that it became generally admitted that common bean originated in the Americas. The conclusion was based on the discovery of wild common bean in Argentina and Guatemala and archaeological remains in the Americas

Based on the extensive distribution of the wild relative speculated that common bean originated through multiple domestications as did other crops originating in the Americas, such as cotton, pepper and amaranth. observed that there was a Mesoamerican cultivar group with smaller seeds and an Andean group with larger seeds.

None of these observations, however, provided substantial experimental evidence and hence a clearer picture of the actual process of domestication for this species. For example, one could ask whether the two cultivated groups identified resulted from distinct domestications or a single domestication event followed by divergence into a small-seeded and a large-seeded group. One could also ask where these single or multiple domestications took place. The archaeological record of the origins of agriculture in the Americas is rather limited. In Mesoamerica, This important type of information therefore pales in comparison with what is available in the Near East In part, this may be due to the relatively humid environment required by many crop relatives in the Americas, which render long-term conservation of crop remains less likely. Therefore, in relative terms compared with archaeological information, there is a relatively higher emphasis on botanical evidence to decipher crop evolution in the Americas than in the Near East.

The use of molecular marker information has increased our understanding of the domestication process in common bean a crop originating in the Americas. In addition, I hope to demonstrate how molecular analyses can bridge the gap that has separated botanical and archaeological evidence, the twin pillars on which rests the study of crop evolution. The power of molecular biology to change scientific paradigms is by no means unique to this field but extends to many other fields as well.

GENERAL METHODOLOGICAL PRINCIPLES

Our current understanding of the origins of domestication in common bean is based on a two-pronged approach relying on the study of wild-growing *P. vulgaris* beans, on one hand, and the use of biochemical or molecular markers, on the other. The use of wild beans is predicated on the fact that their dispersal occurs over much smaller distances (at most a few meters after each vegetation cycle, if at all) than their domesticated descendants. The latter are subject to long-distance exchange, trade, gift or human migration, all of which tend to obscure the original geographic distribution patterns of genetic diversity. Recent explorations have provided a more complete representation of the actual distribution of wild *P.* beans are distributed from approximately 30°N Lat. to 30°S Lat. at mid-elevations from 1000-2500 m asl in regions with moderate rainfall. A common characteristic of the environments in which wild beans are found is the existence of a marked dry season, an important feature needed for seed dispersal.

Genetic relatedness is traditionally assessed with morphological phenotypic traits. Plants with a similar phenotype are inferred to be genetically related either through common ancestry or through geneflow (hybridization). Occasionally, however, inferences based on phenotypic similarity are erroneous. Plants can arise from different selection episodes leading to the same phenotype (convergent evolution). In addition, similar phenotypes can have a different genetic basis (genetic heterogeneity). In both cases, an inference of genetic relatedness would be unwarranted. In contrast, biochemical, and even more so, molecular markers provide information on genetic relatedness that is generally (although not always) devoid of selective influences. All biochemical and molecular markers were created equal! Differences have been observed with regard to the ease of analysis, reproducibility, level of polymorphism, number and genome distribution of loci Of particular note is the molecular basis of polymorphisms. The more complex the basis, the more desirable the marker, because complexity will reduce the probability of repeat mutations, especially in conspecific materials, which are, by definition, closely related. For example, polymorphisms based on DNA nucleotide substitutions are simple changes that could presumably occur repeatedly.

Analyses based on single or few RAPD or RFLP markers are therefore unreliable in determining genetic relatedness that sample a large number of changes at the molecular level, because they include either a large number of markers or complex markers such as actual DNA sequences or multigene families, are more robust with regard to repeated changes. For example, banding patterns of seed proteins such as phaseolin in common bean, when analyzed in denaturing SDS-polyacrylamide gel electrophoresis, represent a compomy image resulting from changes at the DNA nucleotide level (duplication of genes, within-gene duplications, nucleotide substitutions) and at the translation level (co - and post-translational modifications). Therefore each pattern is unlikely to have arisen more than once. As a consequence, genotypes that exhibit the same protein pattern are likely to share a common ancestor

One of the major experimental difficulties in crop evolutionary studies is the distinction one has to make between ancestry (domestication) and geneflow as a cause of genetic relatedness. This is particularly the case given that wild ancestors and domesticated descendants belong to the same biological species and can therefore freely hybridize. Hence, similarity between ancestors and descendants even at the molecular level cannot be assumed to be due to domestication For example, DNA sequence data and the utilization of an outgroup (*e.g.* closely related species) can shed additional light on this problem because they can provide information on the nature of the polymorphism and the direction of the change, *i.e.* the ancestral vs. descendant states of the molecular polymorphism.

Domestication Pattern

The origin of the *Phaseolus vulgaris* progenitor in its wild state can be traced to the Pacific slope of Ecuador and northern Peru, based on the distribution pattern of direct repeats in phaseolin seed protein genes Phaseolin is coded by a small multigene family inherited as a single Mendelian unit and located on linkage group B7 The phaseolin locus is at most 190 Kb long and contains 6 to 9 genes Differences among these phaseolin sequences include nucleotide substitutions and the presence or absence of tandem direct repeats. Three repeats have been identified so far: a 21 bp repeat in the third intron, a 15 bp repeat in the fourth exon and a 27 bp repeat in the sixth exon. The existence of these tandem repeats provides an interesting landmark to follow the evolution of phaseolin genes. Indeed, the generation of a repeat is a more likely event than the precise loss of such a repeat, which would be needed to restore a sequence without repeats. Hence, phaseolin genes without repeats represent probably a more ancestral state than phaseolin genes with repeats. This was verified by analyzing the presence of tandem repeats with a polymerase chain reaction (PCR) test of phaseolin sequences in two subspecies of a related species, *P. coccineus* subsp. *polyanthus* and *P. coccineus* subsp. *coccineus.*

In these two taxa, none of the three repeats could be identified, suggesting that the repeatless state is indeed the ancestral state. A survey was then conducted among wild *P. vulgaris* to determine if any phaseolin haplotypes lacked genes with tandem repeats. Only wild bean populations from Ecuador and northern Peru on the Pacific slope of the Andes appeared to be devoid of phaseolin genes with tandem repeats It was inferred that these wild populations therefore represent the presumed ancestor of the species. This conclusion is somewhat unexpected as most wild *Phaseolus* species are distributed in Mexico.

One could have thought that *P. vulgaris* would have originated in that region as well and that other wild *P. vulgaris* populations distributed elsewhere were ultimately derived from Mexican ancestors. It may be, however, that some wild *Phaseolus* ancestors ('proto-vulgaris' and other species) were dispersed from Mexico to Central America and the Andes to give rise to locally distributed wild *Phaseolus* species, such as *P. costaricensis* From such a local distribution, a species such as *P. vulgaris* could have acquired a secondary, broader distribution by dispersal towards the north and south. It is worth remembering that lima bean (*P. lunatus*) also shows an extensive distribution of its wild relative from Mexico to Argentina and that the species consists of two genepools, an Andean genepool originating on the Pacific side of the Andes in Ecuador and northern Peru (although at lower altitudes than wild *P. vulgaris*) and a 'Mesoamerican' genepool distributed from Mexico to the southern Andes. Although the domestication region of the 'Mesoamerican' genepool is unknown so far, it is likely to be located somewhere in Mesoamerica, given that most Mesoamerican cultivars are located there

From the small nuclear area in Ecuador and northern Peru, wild *P. vulgaris* were dispersed towards both the north (Colombia and Venezuela, Central America and Mexico) and the south (southern Peru, Bolivia and Argentina) to achieve their current distribution In a subsequent step, multiple domestications took place in the Mesoamerican and Andean regions. At least one domestication may have taken place in west-central Mexico (Jalisco), a conclusion based on phaseolin data. Most Mesoamerican cultivars exhibit the same phaseolin type ('S'-type). Wild beans with the same phaseolin type and without morphological signs of past hybridizations to domesticated beans are concentrated in the Mexican state of Jalisco, Interestingly, Doebley *et al.* suggested a similar domestication area for maize.

A second major domestication took place in the southern Andes although a more specific region awaits confirmatory. Phaseolin suggest possible multiple domestications within the Andean region because at least four phaseolin types have been identified among Andean beans. Alternatively, the original populations that were domesticated were polymorphic for phaseolin. Further are needed to distinguish between these possibilities. A third, likely minor domestication may have taken place in Colombia, although regions in Central America deserve further study because samples of wild *P. vulgaris* from that area were unavailable.

CROP EVOLUTION AND HUMAN CIVILIZATION

Agriculture evolved independently in many places on this earth, but the earliest evidence of farming dates 10,000 years ago in present day Iraq For much of the 200,000 or so years prior to agriculture, humans lived as nomadic hunters, gatherers, and scavengers surviving solely on wild plants and animals. Subsequent domestication of these wild plants and animals from their natural habitats launched agriculture, thus radically transforming human societies. This occurred initially in the Fertile Crescent, the Andean region in South America, Mexico, and parts of Asia, but diffused throughout much of the globe. A change from the nomadic lifestyle to farming led us to become community dwellers, eventually spawning the development of languages, literature, science, and technology as people were freed from the continuous daily task of finding food.

Some regions caught on much faster than others, by margins of thousands of years Plants have also evolved or, more accurately, they have been changed rapidly by human intervention Every crop plant grown today is related to a wild species occurring naturally in its centre of origin, and progenitors of many of our crops are still found in the wild. Early humans must have tried eating thousands of feral plant species from a pool of a quarter of a million flowering plants before settling down on less than one thousand such species, which were subsequently tamed and adapted to farming. A little over 100 crop species are now grown intensively around the world, with only a handful of them supplying us with most of what we now eat.

Through a process of gradual selection, our ancestors chose a very tiny section of the wild plant community and transformed it into cultivated crops. Some profound alterations in the plant phenotype occurred during such selection, and these include determinate growth habit; elimination of grain shattering; synchronous ripening; shorter maturity; reduction of bitterness and harmful toxins; reduced seed dispersal, sprouting and dormancy; greater productivity, including bigger seed or fruit size; and even an elimination of seeds, such as in banana. These changes reduced the survivability of crops in the wild, and thus a feature that transcends all of our crops is the reduction of weedy traits from wild plants. Present crops are thus totally dependent upon human care for their survival, and modern crop varieties would persist in the wild "no longer than a Chihuahua would last in the company of wolves"

Most crops that supply our food were thus obtained at the end of the Stone Age, often from a relatively narrow pool of extant wild genetic diversity. Additional diversity arose within such cultivated crops through new mutations and natural hybridization, and through judicious selection and perpetuation by farmers who maintained them as land races. Varied uses and preferences brought forth further diversification such as in corn (popcorn, sweet corn, dent corn, broom corn, and flour corn for tortilla and corn bread) or the derivatives of ancestral cabbage (kale, kohl rabi, brussels sprouts, cabbage, cauliflower, and broccoli). With the advent of transoceanic navigation and the "discovery" of the New World, crops were moved around the world rapidly, often achieving prominence in adopted homes far beyond their natural centers of origin or domestication.

For instance, the United States is the leading producer of corn and soybean in the word, yet these crops are native to Mexico and China, respectively. The world's largest traded commodity, coffee, had a humble origin in Ethiopia, but now much of it is produced in Latin America and Asia. Florida oranges have their roots in India, while sugarcane arose in Papua New Guinea. Food crops that are now so integral to the culture or diet in the Old World, such as the potato in Europe, chili pepper in India, cassava in Africa, and sweet potato in Japan, were introduced from South America. For that matter, every crop in North America other than the blueberry, Jerusalem artichoke, sunflower, and squash are borrowed from elsewhere! A few sources of our food are also recent domesticates.

Chinese gooseberry occurs wildly in China and is not edible. But careful breeding made it palatable, and it was re-christened "Kiwi fruit" in New Zealand after its introduction there early in the 20th century. The modern strawberry with big fruits is a product of the accidental crossing of two wild species from Virginia (United States) and Chile in France in the mid-18th century. Rapeseed, grown in India for centuries, was altered recently through classical breeding to eliminate the toxic erucic acid and smelly glucosinolates to result in canola—Canadian oil.

Triticale, a completely new crop, was artificially sythesized a few decades ago by combining the genomes of wheat and rye (two distinct genera that do not interbreed in nature). It is now grown on over three million acres worldwide. Modern bread wheat itself is also a fairly recent crop in the evolutionary time scale, having arisen only about 4,000 years ago through hybridization of tetraploid (pasta or durum) wheat with inedible goat grass.

FROM MESOPOTAMIA TO MENDEL

While humans have always molded the evolution of crop plants, such changes imposed by farmers occurred over several millennia, leading to rich crop diversity—especially in traits related to their planting or consumption. At the same time, global population grew very slowly until the mid-19th century. It took 1,800 years for the global population to climb from an estimated 300 million around the time when Christianity began, to reach its first billion. But it took only 12 years to add the last billion, rising from five billion people in 1987 to six billion two years ago.

Fortunately, parallel scientific developments in agriculture ensured that food production kept pace with the population explosion of the past century Beginning with Mendel's study of peas, knowledge of genetics helped usher in scientific crop development, resulting in high-yielding varieties. Food production increased in every part of the world in the past few decades, including in Africa. Per capita food consumption has also increased steadily everywhere except in parts of sub-Saharan Africa. In the United States and Canada, where such scientific developments and their applications were most intense, one average farmer now produces enough to feed nearly 150 people! In crops subject to intensive scientific attention—corn, wheat, and rice—the productivity levels increased severalfold. For example, U.S., corn growers averaged 26 bushels of corn per acre in 1928 and 134 bushels per acre in 1998

Such a prodigious increase in agricultural production was underpinned by scientific crop improvement methods along with other developments, including the use of irrigation, improved soil fertility management, mechanization, and control of diseases and pests To develop better crop varieties, scientists have used an array of tools. Artificial crossing, or hybridization, helped us assimilate desirable traits from several varieties into elite cultivars. When desired characteristics were unavailable in the cultivated plants, genes were liberally borrowed from wild relatives and introduced into crop plants.

When a crop variety refused to mate with the wild species, various tricks were employed to force them to intermingle, such as the use of the carcinogenic chemical colchicine or by rescuing the hybrid embryos with tissue culture methods. Hybrid vigour was exploited in crops such as corn and cotton to boost productivity. When existing genetic variation within the cultivated germplasm was not adequate, breeders created new variants using ionizing irradiation (gamma ray, x-ray, neutron), mutagenic chemicals (ethyl methane sulfate, mustard gas), or through somaclonal variation

Most people who are concerned about modern biotechnology have little or no knowledge of the processes that have been used to transform crops in the past. Nor are they likely aware that crops have been continually altered over time or that, without human care, they would cease to exist. Using a variety of tools over the past few decades, plant breeders have radically transformed our crop plants by altering their architecture (such as the development of dwarf wheat and rice), shortening growing seasons, developing greater resistance to diseases and pests (all crops), and developing bigger seeds and fruits These crops are also more responsive to management and better adapted to diverse ecological conditions. Improved food quality also resulted through fewer toxins (canola), better digestibility (beans), increased nutrition (high-protein corn), better taste, longer shelf life (thus withstanding long transportation and storage), and enhanced freshness in many vegetables and fruits. A 1,000-fold increase in the marble-sized wild *Lycopersicon* resulted in the modern tomato that can now weigh as much as a kilogram. Modern farming has steadily increased the supply of relatively safe, affordable, and abundant food not only in the developed world, but also in most developing countries.

An average American family now spends only 11 per cent of its income on food and yet has access to better food choices with more variety and nutrition than ever before. Without scientific developments in agriculture, we would otherwise be farming on every square inch of arable land to produce the same amount of food! Using gene transfer techniques to develop GM crops thus can be seen as a logical extension of the continuum of devices we have used to amend our crop plants for millennia. When compared to the gross genetic alterations using wide-species hybridization or the use of mutagenic irradiation, direct introduction of one or a few genes into crops results in subtle and less disruptive changes that are relatively specific and predictable. The process is also clearly more expeditious, as the development of new cultivars by classical breeding typically takes from 10 to 15 years. The primary attraction of the gene transfer methods to the plant breeder, however, is the opportunity to tap into a wide gene pool to borrow traits, obviating the constraints of cross-compatible crop species.

ADDRESSING PUBLIC CONCERNS

While direct gene transfer is still a relatively new approach, many concerns arising from its use may be addressed with the "benchmark" of conventionally bred varieties, as we have the accumulated experience and knowledge with the latter for more than a century. While it seems logical to express a concern such as "I don't know what I am eating with GM foods!" it must be remembered that we really never had that information before with classically bred crops. With GM crops, at least we know what new genetic material is being introduced, so we can test for predictable and even many unpredictable effects. Consider, for example, how conventional plant breeders would develop a disease-resistant tomato.

They would introduce chromosome fragments from its wild relative to add a gene for disease resistance. In the process, hundreds of unknown and unwanted genes would also be introduced, with the risk that some of them could encode toxins or allergens, armaments that wild plants deploy to survive. Yet we never routinely tested most conventionally bred varieties for food safety or environmental risk factors, and they were not subject to any regulatory oversight. We have always lived with food risks, but in the last few decades we have become increasingly more adept at asking questions.

To address the concern about long-term health consequences of GM foods, it is instructive to recognize that we worried little about such impacts when massive amounts of new proteins (and unfamiliar chemicals) were introduced into our foods from wild species or when unknown changes were created through mutation breeding. When new foods from exotic crops are introduced, we often assimilate them easily into our diets. What's more we rarely, if ever, before asked the same questions that we now pose about GM crops. Many so-called functional foods, health foods, and nutraceuticals have been entering into the mainstream American diet lately, with little or no regulation or testing. We do not question the long-term health implications of these food supplements, even though they involve relatively large changes in our food intake.

In contrast, the GM foods currently on the market have been tested extensively and judged to be substantially equivalent to their conventional counterparts, with just one or two additional proteins present in miniscule amounts (introduced into a background of thousands of proteins). And, those proteins are broken down either during processing or digestion, with little long-term consequence. In food products such as oil, starch, and sugar, such proteins are not even found. A nagging potential problem with a new protein in food is that it could be a potential allergen. As most food allergens are now well studied, we know that they are found in few defined sources (peanut and other grain legumes, shellfish, tree nuts, and a handful of other foods) and share many similar structural features. Moreover, they must be present in huge proportions in our food, and we must be sensitized to them over time for them to cause any adverse effects. Thus, it is highly unlikely for new allergens to be introduced into our food supply from GM plants.

HISTORICAL ABSENCE

There is no such thing as safe food, and there never has been! That is not to suggest that all of our foods are dangerous, only an acknowledgment that trace levels of such contaminants as toxins and carcinogens are present in everything we eat. But a primary rule of toxicology, articulated over 400 years ago by Paracelsus, refers to the importance of dosage: "Every substance is a poison, but it is the dosage that makes it poisonous" While not alarming, our daily food naturally contains thousands of chemicals, and many of them are shown to be carcinogenic or hazardous in lab animal studies with huge doses.

We consume roughly 5,000 to 10,000 natural toxins daily, as plants have evolved to produce an array of chemicals to protect themselves against pests, diseases, and herbivores 7 have been tested and 19 of them found to be rodent carcinogens The fat-soluble neurotoxins solanine and chaconine are present in potatoes and can be detected in the bloodstream of all potato eaters transferring genes through conventional methods, the resistance is often accompanied by an increase in such toxic compounds. Thus, it is not true that we never had problems with conventionally bred varieties.

Any crop variety found to pose a real health risk was promptly removed from the market, but those varieties (in contrast to GM crops) were never routinely tested. One pest-resistant celery variety produced rashes in agricultural workers and subsequently was found to contain 6,200 ppb of carcinogenic psoralens compared to 800 ppb in the control celery This celery was removed from cultivation and that was also the case with the potato variety Lenape, which contained very high levels of toxic solanine. We have always learned from trial and error with all innovations. Similarly, crop improvement practices evolved over time with continued refinement. It is common, though, for human nature to generate an exaggerated fear of new innovations while perceiving older or "natural" products as always more benign. Huber discusses this double standard in the larger context of risk regulation. We have always been lenient towards existing known and greater hazards, even as we create "gatekeepers" to minimize new risks. Thus, we fail to recognize and "exorcise" much larger older risks.

While most food hazards arise from pathogens such as *Escherichia coli* 0:157, *Listeria monocytogenes*, and *Salmonella enterica* along with mycotoxins produced by fungi (and thus a function of food storage and handling), certain foods containing toxic compounds are known to produce adverse health consequences over time. Cassava, eaten by a large population in Africa, contains cyanogenic glucosides, which cause limb paralysis if consumed before extensive processing.

Solanin in tomato and potato is known to cause spina bifida. Vetch pea, a common legume known for its hardiness—and thus popular in India among poor farmers—contains highly dangerous neurotoxins that cause untold misery. Phytohemagglutinin, found in undercooked kidney beans, is toxic. And peach seeds are extremely rich in cyanogenic glucosides. None of these were subject to any mandatory testing before they were introduced into the food chain, nor are they subject to any regulation now. But if the current regulatory standards imposed on GM crops were to be invoked for traditional crops, most of them would fail to meet their requirements.

Humans have built-in natural defences that protect us against normal exposure to toxins. But, according to Ames and Gold we have not evolved to achieve "toxic harmony" with everything we eat, because natural selection occurs much too slowly and because much of what is in our diet today was not eaten at all when we were hunter-gatherers. A balanced mixture of foods normally provides adequate nutrition. However, none of the crops grown today were selected with our nutritional requirements in mind. Instead they were chosen intuitively, by our ancestors, from among the edibles that could be found around them. Thus, the most important food crop in the developing world—rice—has no provitamin A and little iron in its endosperm. This has led to horrific problems, such as blindness among millions of children due to vitamin A deficiency, and iron-deficiency anemia in nearly a billion women dependent on a rice diet. Biotechnology research, far from causing any new food safety problems, has already demonstrated its potential in enhancing the nutritional quality of our food and is also being employed to reduce harmful toxic compounds that exist in our food.

CONSUMPTION OF AGRICULTURAL CROPS FOR FOOD

Global demand and consumption of agricultural crops for food, feed, and fuel is increasing at a rapid pace. This demand for plant materials has been expanding for many years. However, recent increases in meat consumption in emerging economies together with accelerating use of grain for biofuel production in developed countries have placed new pressures on global grain supplies. To satisfy the growing, worldwide demand for grain, two broad options are available:

1. The area under production can be increased or
2. Productivity can be improved on existing farmland.

These two options are not mutually exclusive and both will be employed to produce the additional 200 million tonnes/year of corn (Zea mays) and wheat (Triticum aestivum) estimated to be needed by 2017. Both options will alter the environmental footprint of farming. Of the two options, increasing productivity on existing agricultural land is preferable as it avoids greenhouse gas emissions and the large-scale disruption of existing ecosystems associated with bringing new land into production.

In the United States, breeders, agronomists, and farmers have a documented history of increasing yield. U.S., average corn yields have increased from approximately 1.6 tonnes/ha in the first third of the 20th century to today's approximately 9.5 tonnes/ha. This dramatic yield improvement is due to the development and widespread use of new farming technologies such as hybrid corn, synthetic fertilizers, and farm machinery. The introduction of biotechnology traits and development of new breeding methodology using DNA-based markers are further improving yields. Outside the United States, similar farming practices have been adopted in some agricultural nations, but in many major grain-producing countries, yields still lag well behind world averages. By continuing to develop new farming technologies and deploying of them on a global basis, demand for feed, fuel, and food can be met without the commitment of large land areas to new production.

VIGOUR OF CROPS

Weeds, pests and diseases all compete with or directly reduce the vigour of crops. Many pests and diseases are soil-borne. Weed life-cycles depend on replenishment of the soil seed bank and survival of the seeds against natural decay, predation by soil animals and depletion by human management, particularly cultivations. Ecological weed control thus aims to minimize recruitment of new seed into the soil as a long-term strategy as well as trying to reduce artificially the size of the weed seed bank in the soil.Weed life cycles and strategic weed control are well understood for temperate wet-and-dry climates. The life cycle of weed populations and gives informative data for weed populations in north India (2500 mm AAR of which 70-90 per cent falls in 5 months). The weed populations were measured under the contrasting conditions of slash-and-burn and terrace cultivation. Seed recruitment to the soil seed bank was higher under terrace cropping than under slash and burn but fewer survived.

It is generally accepted that there are links between weed populations, their competitiveness, and soil fertility. A good example of this is the various species of witch weed. These species infect the roots of millet, sorghum, Maize and sugar cane. *S. generioides* also infects legumes and tuberous crops. They are a serious nuisance and are widespread in semi-arid Africa. It is estimated that they currently infect 21 million ha and a further 44 million ha are at risk. Their spread and increased aggressiveness is considered to be due largely to declining fertility and soil carbon contents.

The incidence of insect pests and crop diseases depends on the complexity of the cropping system and the types of crops grown. It depends too on management factors, such as the timing and type of tillage and the treatment of the crop residues. These aspects are so complex they warrant a Bulletin to themselves, so they are only touched upon here.

The incidence of pests and diseases is generally believed to be less frequent and/or intense in diverse cropping systems. Data on insects are relatively scarce but Andow is frequently quoted as showing a positive relationship between insect pest populations and the practice of monoculture. His experience, however, is mainly with cotton, a unique crop, and wheat, though he gives a few examples from Maize and rice.

Power shows that the incidence of virus diseases is less in complex cropping systems. He suggests that disease transmitted by insects is less common in multiple-cropping than in monocultures because aphids and some other insects settle on plants in response to visual cues that do not enable them to distinguish between host and non-host plants. In this way polyphagous insects can acquire a pathogen from an infected plant and transfer it when feeding on a non-host plant. If the insect is preferentially attracted to the non-host crop, it may be speculated that this crop acts as a 'trap' or 'protection' crop. Power reviews situations where intercropping or the inclusion of a second, barrier species has led to lower disease levels. In contrast, bacteria and fungi in the root region and rhizosphere are reported to be slightly fewer in a coconut monocrop than in a multi-storied mixed cropping system. Differences in populations, and in their varied (positive or negative) effects on the food crop, are no doubt best understood through analysis of the feeding habits and predation of the organisms themselves, rather than by making superficial generalisations about their numbers. Fick and Power and Hearn and Fitt review pest management generally and pests and diseases in cotton crops specifically.

SOIL BIOLOGY AND NUTRITION

The aim of management should be to create balanced organic matter and mineral budgets. It should ensure that, over several years (a complete crop rotation), soil organic matter is not depleted and that nutrients added equal or exceed those removed by cropping or lost in various ways.

Various approaches to good management. These complement the aspects and the advice on saline water management given by Kandiah and that on water harvesting offered by Critchley and Siegert. When managing organic matter farmers should recognise that the effects of animal and human manure, sewage sludge and plant residues last longer than those of green manure crops. Green manures, though valuable, usually last only one or two seasons because they are incorporated before they are mature and lignified. The longer-term beneficial effects of manures on soil organisms.

Juo and Payne advise: 'In spite of the many proven benefits of soil organic matter, its management and recycling in an intensified, modem agro-ecosystem must necessarily revolve around two fundamental characteristics of the system, namely, the availability of organic material at the farm level, and the economic incentive for conserving and recycling organic matter'. To this may be added consideration of the benefits and costs of treating and transporting human wastes from cities. On-site organic matter, organic material brought in from outside and industrially-produced fertilizers will be used according to their benefit cost ratios and the attitudes of governments and crop managers.

Tillage reduces the frequency of VAM, at least in invertebrates. Tillage reduces earthworm populations 10-30-fold, both killing them directly and destroying their burrows. Haines and Uren show that differences in earthworm populations between tillage treatments and stubble burning are reflected by 50 per cent differences in numbers of soil pores.

Tillage also effects the location, numbers and activity of micro-organisms within the soil, depending on the type of tillage. For example a mouldboard plough that inverts topsoil has more effect than minimal tillage. Doran found microbial biomass and potentially-mineralisable nitrogen to be 54 per cent and 37 per cent higher respectively in the top 7.5 cm of no-till than ploughed soil. These would be critical differences if translated to the semi-arid tropics.

Microbial and fungal biomasses may be 20-70 per cent higher after 1 and 2 years no-till. However, conventional tillage, by burying topsoil, may cause microbial populations and their activities to be higher than under no-till at depths of say 7.5 to 15 cm. Differences in soil organisms (*e.g.,* microbial biomass) under differing tillage treatments are seasonal, generally being greatest in the period when the weather is most favourable for soil organism activity and plant growth. Rasmussen and Collins summarise many data on the impact of tillage on soil properties and conclude that stubble-mulching with zero tillage conserves up to 2 per cent more organic matter per year than ploughing. Cropping patterns and rotations affect soil nutrition. Diversity of cropping increases the number and variety of soil organisms and reduces pests and diseases.

Soil acidification and salination are extreme cases of nutrient imbalance and, unlike other deficiencies, cannot be corrected simply by adding mineral fertilizers. The management techniques to reduce acidification are:

- To reduce the net proton production of the system by minimising nitrate leaching (a special problem in seasonally wet-and-dry climates);
- To avoid using ammonium fertilizers and reduce the accumulation of organic matter;
- To lime the soil.

Liming is a high cost option while adoption of perennial species (to avoid seasonal accumulation of organic matter and nitrate leaching) is less costly. By contrast, salinity is usually managed by strategic leaching. This is done with the sparing use of low-salt water when both soil moisture.

CRUCIAL DIFFERENCE BETWEEN CULTIVATED CEREAL CROPS AND THEIR WILD RELATIVES

The crucial difference between cultivated cereal crops and their wild relatives is in ear fragility at maturity. Easy shattering of spikes into spikelets upon maturity is essential for seed dispersal and survival in the wild, whereas forms with non-brittle ears survive only under cultivation. It is generally assumed that most Triticeae crops have been domesticated from their wild relatives by selection of non-shattering individuals which sporadically appear in wild populations as rare mutants However, there are no documented cases of the appearance of non-brittle mutants in wild populations, except observations of introduction from cultivars into weedy forms. The appearance of non-brittle mutants seems to be a rare event which may be induced under specific conditions.

Interspecific hybridization may activate transposition of mobile genetic elements which may be one potential source of increased mutation rates. A hypothesis was proposed that rye and wheat forms with varying ear fragility may have arisen as a result of interspecific hybridization processes between different wild species (Jaaska 1975). Taking into account that non-brittle mutants will persist in the wild only for a limited number of generations after their appearance, it follows that rye and barley should have been domesticated *in statu nascendi* from the ancient hybrid populations of their wild relatives.

RYE

Despite considerable disagreement among taxonomists about the delimitation of rye species and their intraspecific taxa the rye genus *Secale* L. may be treated as consisting of three valid biological species

- The Outcrossing annual *Secale cereale* L. s.l, including the European cultivated rye (subsp. *cereale),* the Transcaucasian cultivated and weedy rye with a tough rachis (subsp. *segetale* Zhuk.), and various weedy and wild forms of the annual self-incompatible rye with a brittle rachis such as subsp. *ancestrale* Zhuk. s.l, including weedy subsp, *ancestrale* Zhuk. s.str, weedy subsp. *afghanicum* (Vav.) Hammer, weedy subsp. *dighoricum* Vav., and wild subsp. *vavilovii* (Grossh.) Kobyl. s.str., non s.l. of Kobyljanskij
- The Outcrossing perennial *Secale strictum* Presl., syn. *S. montanum* Guss., with several subspecies: subsp. *strictum* (= *S. dalmaticum* Vis.), subsp. *anatolicum* (Boiss.) Hammer, subsp. *kuprijanovii* (Grossh.) Hammer, subsp. *ciliatoglume* (Boiss.) Hammer, and subsp. *africanum* (Stapf) Hammer.
- *Secale Sylvestre* Host. - a cleistogamous annual of sandy dunes and coasts.

A cleistogamous annual, originally collected by Kuckuck in Iran and attributed to *S. vavilovii* may deserve a specific rank as *S. iranicum* Kobyl., syn. *S. vavilovii* Grossh. sensu Khush, non- sensu Grossheim or sensu lato of Gandilyan owing to its reproductive isolation as a result of its cleistogamous breeding system. Likewise, South African endemic perennial rye may be recognized as *S. africanum* Stapf because of its autogamous breeding system.

Grossheim described *S. vavilovii* as a short-statured (stems 20-35 cm) wild annual with short (4-8 cm) fragile spikes which he collected in Nakhitshevan from a dry habitat on volcanic ash soil. Populations of short wild annual rye were later found to occur in different dry regions of Armenia by Gandilyan who attributed the short stature to a phenotypic adaptation to a dry and nutrient-poor habitat.

On those grounds he proposed to treat all wild and weedy annual ryes with a brittle rachis as *S. vavilovii* s.l. separately from the cultivated *S. cereale* s.str. To give these forms a specific rank is questionable because they are not reproductively isolated from non-brittle weedy *S. cereale* subsp. *segetale.* Their inclusion under *S. cereale* subsp. *vavilovii* s.l. as proposed by Kobyljanskij contradicts the botanical nomenclature rules because of the priority of subsp. *ancestrale* Zhuk. At the subspecies level Vavilov expressed a view that cultivated rye evolved from a wild perennial rye through the appearance of annual weedy forms which in turn were domesticated by unconscious selection of non-brittle forms under cultivation in barley and wheat fields. Later he reported finding a weedy rye in Afghanistan with fragile ears which he attributed to *S. cereale* var. *afghanicum* and proposed that it might be an initial form from which semi-brittle and non-brittle forms of weedy rye were spontaneously selected under cultivation in wheat and barley fields in northern areas and higher altitudes He also further argued that *S. cereale* may have evolved from a perennial *S. montanum* with wild *S. vavilovii* Grossh. acting as a link between them. rye was domesticated in the Near East region as a secondary weedy crop (subsp. *segetale*) and only thereafter spread to European countries and was used as a distinct crop.

Alternatively, the present-day weedy ryes with brittle and semi-brittle rachis are products of introgressive hybridization of the cultivated non-brittle subsp. *segetale* with wild rye subsp. *vavilovii* s.str. Segregants from this hybridization acquired a weedy trait from subsp. *segetale* and became field weeds. Artificial hybrids between wild, weedy and cultivated forms of *S. cereale* are completely interfertile and show regular meiosis (that they comprise a single biological species. It is generally accepted that all annual rye species have evolved from the perennial *S. strictum*. *Secale cereale* and *S. strictum* differ by two large translocations between three of the seven chromosomes, but intraspecific taxa have the same chromosome arrangement in both species Polymorphism for reciprocal translocations has been described in populations of *S. cereale* The evolution of *S. cereale* from *S. strictum* should thus involve fixation of two reciprocal translocations which would be possible only by the involvement of an autogamous intermediate. Stutz (suggested that cleistogamous annuals *S. sylvestre* and *S. vavilovii* sensu Khush (= *S. iranicum*) might represent intermediate stages in the evolution of *S. cereale* from *S. strictum.*

Secale sylvestre has the same chromosomal arrangement as *S. strictum,* but a low crossability with it, or differs in only a small chromosomal translocation, whereas *S. iranicum* was found to have the same chromosomal arrangement as *S. cereale* (Nürnberg-Krüger Stutz suggested that cleistogamous annual *S. sylvestre* has been derived from *S. strictum* and has then given rise to a similarly cleistogamous annual *S. iranicum* by fixation of two chromosomal translocations, while *S. cereale* was assumed to emerge from *S. iranicum* by acquisition of the outcrossing breeding system by introgression from *S. strictum.*

Variability among the rye species established that *S. cereale* and *S. strictum* display partially homologous polymorphism of many isozymes with most allozymes shared between them. that *S. iranicum* is fixed for one of the allozymes found in *S. cereale* and *S. strictum,* whereas *S. sylvestre* has unique allozymes of acid phosphatase ACP-B, anodal peroxidase PRX-C, and aromatic alcohol dehydrogenases AAD-A and AAD-E In the it has been concluded that

S. sylvestre is a lateral branch of evolutionary divergence form S. *strictum* and not an intermediate leading to *S. iranicum* and *S. cereale.*

Morever, *S. strictum* and *S. sylvestre* are well isolated in nature by growing in different ecological habitats and do not hybridize freely. Instead, I suggest that some unknown inbreeding form of subsp. *vavilovii* related to *S. iranicum* might be an intermediate from which both brittle- and tough-rachis forms of *S. cereale* evolved by introgression of self-incompatibility from *S. strictum* It was also assumed that non-brittle rye *S. cereale* subsp. *segetale* might be domesticated *in statu nascendi,* directly from some such hybrid populations currently extinct, while fragile subsp. *vavilovii* s.str. has persisted in the wild. Later, *S. sylvestre* was found to differ from the other rye species also in allozymes of cathodal peroxidases CPX-4 and CPX-5 and chloroplast DNA whereas *S. cereale* and *S. strictum* revealed homologous polymorphism and could not be definitely distinguished by allozymes and chloroplast DNA

Zhukovsky (1pointed out that there have been no reports of spontaneous appearance of non-brittle forms for more than 100 years of cultivation of wild and weedy ryes in different botanical gardens and collections. The reason may be in the recessive nature of the non-brittle mutation combined with the outcrossing breeding system of rye The rare non-brittle mutants will remain phenotypically unexpressed in outcrossing rye populations in heterozygous genotypes until two parents with recessive non-brittle alleles cross and segregate into homozygotes in a progeny.

The probability for this depends on the frequency of non-brittle mutations which may be extremely low. Therefore, the involvement of an inbreeding intermediate would be an important premise for the appearance of recessive non-brittle mutants phenotypically expressed in homozygotes.

Zhukovsky also has proposed a hypothesis that weedy subsp. *segetale* might have arisen from interspecific hybridization between wild perennial and annual ryes with the appearance of non-brittle forms in the hybrid progeny. He assumed that interspecific hybridization might be the crucial mutagenic factor which has caused the appearance of semi- and non-brittle forms of rye. Stutz "on the slopes of Mt. Ararat, *S. vavilovii* makes contact and hybridizes rather freely with *S. montanum.*"

The seed obtained from crosses between *S. cereale* and *S. strictum* were found to germinate only when *S. cereale* was the mother parent indicating that the introgression is possible from the perennial rye. Sympatric populations of wild perennial and annual ryes in the Near East region should be examined for the fragility, chromosome rearrangements and molecular characters in order to have more information about the origin of wild, weedy and cultivated forms of *S. cereale.* Electrophoretic variants of aliphatic and aromatic alcohol dehydrogenase (ADH and AAD), aspartate aminotransferase (AAT), acid phosphatase (ACP), esterase (EST) and anodal peroxidase (APX) isoenzymes in rye species: major electrophoretic variants (allozymes) are numbered in the order of their decreasing mobilities and listed in the order of decreasing occurrence; rare allozymes are labelled by letter 'r'; unique allozymes are in italics.

Species	ADH -A	AAD -A	AAD -E	AAT -B	AAT -C	ACP- B	EST A	EST B	APX C
S. strictum	2;1r	1	2;1;3r	1	2;1	4;4f;5;3	2f;2;3	3;1;2	1
S. cereale									
subsp. *cereale*	2	1	2;1	1	2;3	4;1;3;2	2	1	1
subsp. *segetale*	2	1	2,1	1	2;3	4;1;3;2	2	1	1
subsp. *ancestral*	2	1	2;1	1;2r	2;3	4;3;1;2	2;2f;3r	1;3;2	1
S. iranicum	2	1	1	1	2	4	2	1	1
S. sylvestre	2	2	*1f*	1	2	*4f*	2;0	1	2

BARLEY

The cultivated barley *Hordeum vulgare* L., incl. *H. distichon* L. and *H. hexastichon* L., and its closest wild relative *H. spontaneum* C. Koch are autogamous annuals which with the allogamous perennial *H. bulbosum* L. share basic genome a number of allozymes Their hybrids are completely interfertile and have normal chromosome pairing in meiosis. Cultivated barley and its closest wild relative, *H. spontaneum,* are now commonly recognized as subspecies of *H. vulgare* s.l.

However, taking into account their autogamous breeding system, which itself provides a reproductive barrier between them, and the fact that the biological species concept is not strictly applicable to uniparentals, they could formally be accepted at a species level as separate phylogenetic lineages. There is a general agreement among investigators that cultivated barley has originated from its closest wild relative *H. spontaneum* through the latter's domestication Different views, however, have been expressed with respect to the place, time and mechanisms of the origin of different forms of cultivated barley, particularly six-row forms

Monophyletic origin of the cultivated barley as a two-row form by domestication of *H. spontaneum* and the secondary origin of the six-row barley from a cultivated two-row form is now generally accepted thanks to experimental evidence on the secondary origin of the Tibetan semi-brittle six-row barley by hybridization between the cultivated six-row barley and wild *H. spontaneum.* The semi-brittle forms in the present-day mixed populations of weedy and cultivated barley are evidently derived by introgression from the cultivated barley. Genetic studies of hybrid progeny between *H. vulgare* and *H. spontaneum* have shown that ear fragility is controlled by two closely linked genes (*Bt* and *Bt2*) with a recessive mutation at either loci giving rise to non-brittle ears In the case of a simple *in statu nascendi* domestication of barley by selecting out non-brittle mutants, the cultivated barley has evidently been domesticated repeatedly from different genotypes of wild barley, *i.e.* it has a polytopic origin. This, however, does not explain a strikingly wider morphological variability among cultivated barley than in other cereals. Indeed, morphological variation within *H. spontaneum* is limited to only three botanical varieties which differ in the apex shape and awn length of lateral spikelets. At the same time, the cultivated barley is remarkably more variable with over 100 botanical varieties described among primitive barley landraces Although a remarkable morphometric variation was observed among a set of 77 accessions of wild barley under cultivation around Moscow (Russia) under unusual climatic conditions, it still remained relatively limited and no non-brittle cultivar-like forms were observed Characters of the wild parent dominate in hybrids with cultivated barley and segregation appears in later generations.

No morphologically new types were found among the progeny of hybrid generations between various morphological types of *H. spontaneum.* Therefore, the question as to why cultivated barley is so morphologically variable still remains unanswered.

Comparative studies of isoenzyme variation among a set of morphologically different accessions of *Hordeum vulgare* s.str. and *H. spontaneum* showed that both-had the same allozymes of essentially monomorphic isoenzymes and displayed variation of 10 polymorphic esterase isozymes.

This shows that the isozyme loci which were monomorphic in wild barley have remained also in the cultivated barley, *i.e.* cultivation has not induced new variation at the isozyme loci in contrast to those controlling variable morphological characters. The same also has been demonstrated with respect to wild and cultivated wheats and The cultivated barley and its wild progenitor could not definitely be distinguished by any of the isoenzymes studied except by a frequency of allozymes of some isoesterases.

Hordeum bulbosum displays extensive intrapopulation polymorphism of several isozymes which were largely monomorphic in *H. vulgare* and *H. spontaneum.*

The allozyme genepool of the two annuals was found to be a subset of most frequent allozymes of *H. bulbosum* This result is consistent with a view that *H. spontaneum* might have evolved from *H. bulbosum* by fixation of genes controlling self-compatibility and annual habit, polytopic origin of *H. spontaneum,* as well as against later, repeated introgression from the bulbous barley. Moreover, crossing experiments between *H. vulgare* s.l. and *H. bulbosum* have shown that hybrids could be obtained only by the embryo rescue technique, had disturbed meiosis, were sterile and frequently haploid owing to spontaneous elimination of *bulbosum* chromosomes

Electrophoretic variants of aliphatic and aromatic alcohol dehydrogenase (ADH and AAD), aspartate aminotransferase (AAT) and esterase (EST) isoenzymes in cultivated barley (*vulgare*) and its wild relative species (*spontaneum, bulbosum*): major electrophoretic variants (allozymes) are numbered in the order of their decreasing mobilities and listed in the order of decreasing occurrence; rare variants are labelled by letter 'r'.

Isozyme	Vulgare	Spontaneum	Bulbosum	Vulgare	Spontaneum
ADH-A		1	1	1;2r	
ADH-B	1	1	1		
ADH-D	3	3	3;2;1		
AAD-A	1	3	3;4;2;1		
AAD-B	2;0	2;0	2;2s;1;0		
AAD-E	2	2	2;1;3		
AAT-A	1	1;2	1		
AAT-B	1	1	1;2;3;0		
AAT-C	2	2	2;3;1		
EST-A				1	1
EST-B				1	1
EST-D				3;2;4;0	2;3;4;1;0
EST-E				2	2;1r;3r
EST-F				1;2;0	2;1;3;4
EST-H				2;0	2;1;0;3
EST-I				2;0;1;3	2;0;3;1
EST-J				1;0	1;0
EST-K				3;0;2	0;1

Hordeum spontaneum and *H. bulbosum* are reproductively isolated by strong sterility barriers. Observations that crossability and chromosome elimination varies depending on the genotype of parental species suggested that introgression from the bulbous barley in some genotype combinations may be possible.

2

Soil Microbes and Pesticides

INTRODUCTION

The extensive use of pesticides in agriculture to eliminate pest, such as grasshoppers, has resulted in the accumulation of pesticide residues in soil and has altered the soil microbial communities by favouring the growth of those pesticide degrading organisms. Due to the xenobiotic features of pesticides, pesticides in soil can be persistent in the environment and eventually enter the food chains, which cause reproductive failure in birds and even cancer in humans (Arias-Estevez *et al.* 2008). Pesticides can inhibit the growth of certain microorganisms by interfering with enzymatic activity. For example, one type of organophosphate pesticide can inhibit nitrogenase activity involved in nitrogen fixation (Hussain *et al.* 2009). The inhibition of nitrogenase activity can reduce the amount of nitrogen available to plants, thus reducing crop yields. The abundance of pesticide applied to agricultural soils (5.1 billion pounds/ annually in US) has created selective pressures for microorganisms that have the ability to use pesticides in soil as sources of energy and nutrients. This can be useful in bioremediation to detoxify contaminated soil.

OVERVIEW OF SOIL MICROBES

Soil looks lifeless but a quarter of a teaspoon of soil (1 gram) contains many microbes - 100 million bacteria, 1 million actinomycetes and 5 metres of fungal filaments. Soil is certainly living! In pastures grazed by sheep near Armidale, NSW, Australia, we have measured the weight of microbes in soil and calculated that there is the equivalent weight of soil microbes of 88 sheep/ha.

Since the pastures were grazed with only 20 sheep/ha above-ground, there was the equivalent weight of 4 sheep (as microbes) in the soil to every real sheep grazing above. Scientists in England and New Zealand have calculated that they have equivalent weights of 110 sheep/ha in microbes in pasture soil. These English and New Zealand soils would be moister and have more organic matter food for microbes than our Australian pasture soil which has periodic droughts and where organic soils are rare.

Ways of Feeding

Bacteria and fungi have similar ways of feeding. Their main food is organic detritus in leaf litter, soil, dung and carrion. In contrast to invertebrate animals which engulf food and digest it within their bodies, microbes digest their food outside themselves by secreting digestive enzymes over their food. Under the action of the enzymes, organic matter breaks down into simpler molecules external to the microbe, which then re-absorbs the simpler products of digestion through its cell wall. Microbes have a vast number of enzymes to do this job – 50 to 60 different ones have been discovered. Few invertebrates possess this many digestive enzymes.

PHYSICAL ENVIRONMENT

Factors Affecting the Bioavailability of Pesticides

The bioavailability of pesticides to microorganisms is determined by several environmental factors and chemical properties of pesticides. Volatilization, leaching, plant uptake, run-off and sorption of pesticides can influence the transport of pesticides from soil to air, from soil to groundwater, or within soil, thus affecting the availability of pesticides for degradation by soil microbes (Arias-Estevez *et al.* 2008). Sorption of pesticides is the most important factor that can be influenced by soil pH and organic matter and clay contents in soil (Arias-Estevez *et al.* 2008). For example, higher pH indicates that the soil is more acidic, thus attracting basic and neutral pesticides. Organic matter and clay possess weak negative charges and, therefore, attract weakly acidic and neutral pesticides. Although increase in sorption can limit the amount of pesticides available to soil microbes, the degradation of pesticides by soil microbes does not decrease dramatically (Arias-Estevez *et al.* 2008).

MICROBE PROCESS

Microbes

Bacteria and fungi help in the formation of water-stable soil aggregates. Fungal hyphae grow around and between soil mineral and organic particles and physically bind them together. Both bacteria and fungi secrete polysaccharide mucilages which are sticky and glue the soil particles together into aggregates. These aggregates can be stable to the action of water for several months and help prevent slaking and dispersion of the soil.

Soil organisms clear away and degrade organic debris such as dead plants, animals and dung and use it as a source of food and nutrients, and in the process release chemical elements (nutrients) into the soil solution so that living plants may re-use them. This process sustains soil chemical fertility. But soil organisms are also responsible for soil physical fertility as well. They help the formation of good soil structure. In addition, soil organisms degrade chemicals and pollutants that enter soil. Also, some soil organisms live in mutually beneficial relationships with plants, enhancing the plant's nutrition by increasing the nitrogen uptake from the atmosphere (Rhizobium) or phosphorus uptake from the soil by soil (mycorrhizae).

Role of Invertebrate Animals

Soil animals and soil microbes have different ways of decomposing organic detritus. Those soil animals which have mouthparts (*e.g.* arthropods) bite off bits of organic matter and fragment it into small pieces. Some earthworms select organic particles that are larger than their mouths, and can split them into smaller fragments. On passage through the invertebrate guts, further disintegration of the organic matter occurs.

Watch as the springtail eats dead plant litter and passes a faecal pellet out the other end. The springtail is head down, tail up! Springtails can deposit faecal pellets at the rate of once every 5 minutes. You can see how a large pile of dung has built up over the course of the 30 minutes as I watched it feed. The plant leaf is being fragmented as the springtail munches.

Role of Microbes

Microbes (bacteria and fungi) digest organic material externally by exuding a huge variety of digestive enzymes (>50 different ones) onto their food. These enzymes help chemically degrade organic detritus, breaking up large organic molecules (*e.g.* proteins) in which nutrient elements are bonded. Eventually, simpler molecules (*e.g.* amino acids) are split off and passed back through the microbial cell walls to be absorbed into their tissues and used for growth and maintenance. The simple organic molecules are eventually reduced to an inorganic form (*e.g.* NH4+) which plants can re-use.

Hexachlorocyclohexane Degradation

Recently, a few bacterial strains have been observed to have the ability to degrade one of the very toxic pesticides, organochlorine. Organochlorine pesticides are lipophilic compounds that tend to bio-magnify along the food chains through accumulation in body fats (Cruz *et al.* 2003). Hexachlorocyclohexane (HCH), one type of organochlorines, has four isomers, α, β, γ, and δ (Gupta *et al.* 2000).

The insecticide, γ- HCH, has been mostly used among the four isomers due to its fast degradation rate under aerobic condition (Chaudhary *et al.* 2006). Under aerobic condition, Sphingomonas paucimobilis has all the enzymes necessary to degrade HCH to acetyl-CoA (Camacho-Pérez *et al.* 2012). Therefore, S. paucimobilis is able to use Hexachlorocyclohexane as a source of energy and nutrients.

Parathion Degradation

Parathion is one of the most studied organophosphorous pesticides (Singh and Walker 2006). Organophophorous pesticides can inhibit the activity of enzyme acetylcholine esterase, involved in nervous system that disrupts the transmission of nerve impulse (Kanekar *et al.* 2004).

Pseudomonas sp. is proposed to hydrolyze parathion to yield p-nitrophenol and diethyl thiophosphate (Singh and Walker 2006). Furthermore, these microorganisms can degrade p-nitrophenol to p-benzoquinone by using monooxygenase (Singh and Walker 2006). Further degradation of p-benzoquinone produces unidentified intermediates that can enter TCA cycle; therefore, parathion can act as a source of carbon and nitrogen for Pseudomonas sp. (Singh and Walker 2006). Pseudomonas stutzeri is able to degrade parathion co-metabolically to diethyl thiophosphate and p-nitrophenol (Kanekar *et al.* 2004); however, it cannot use p-nitrophenol as a source of energy. Interestingly, Pseudomonas aeruginosa can use p-nitrophenol made by P. stutzeri as a source of carbon and energy (Kanekar *et al.* 2004).

KEY ORGANISMS

Hexachlorocyclohexane Degrading Bacteria

Several bacteria that are able to degrade hexachlorocyclohexane under aerobic condition have been isolated and characterized from the HCH contaminated soil (Gupta *et al.* 2000). Only some organisms have fully proposed mechanisms that describe the process of HCH degradation (Camacho-Pérez *et al.* 2012).

Parathion Degrading Bacteria

Pseudomonas sp. is one major organism found to degrade and use parathion as a source of energy and nutrients (Singh and Walker 2006).

- Pseudomonas stutzeri
- Pseudomonas aeruginosa

CURRENT RESEARCH

Conditions for degrading parathion

Different conditions are examined to determine the best environment for increasing the efficiency of p-nitrophenol degradation by a common parathion degrader, Pseudomonas aeruginosa (Zheng *et al.* 2009).

The addition of ammonium chloride as a nitrogen source increases the p-nitrophenol degradation, but the addition of glucose as a carbon source decreases the p-nitrophenol degradation (Zheng *et al.* 2009).

Effect of Pesticide on Microbial Community Composition

Different pesticides have been shown to promote or inhibit the growth of certain soil microorganisms in either aerobic or anaerobic conditions (Lo 2010). This may be helpful to identify possible pesticide degraders and the efficiency of the degradation of pesticides.

SOIL MICROORGANISM —BACTERIA

Amongst the different microorganisms inhabiting in the soil, bacteria are the most abundant and predominant organisms. These are primitive, prokaryotic, microscopic and unicellular microorganisms without chlorophyll.

Morphologically, soil bacteria are divided into three groups viz.:

1. Bacilli
2. Cocci (round/spherical), (rod-shaped)
3. Spirilla/Spirllum (cells with long wavy chains).

Bacilli are most numerous followed by Cocci and Spirilla in soil. The most common method used for isolation of soil bacteria is the "dilution plate count" method which allows the enumeration of only viable/living cells in the soil.

The size of soil bacteria varies from 0.5 to 1.0 micron in diameter and 1.0 to 10.0 microns in length. They are motile with locomotory organs flagella. Bacterial population is one-half of the total microbial biomass in the soil ranging from 1, 00000 to several hundred millions per gram of soil, depending upon the physical, chemical and biological conditions of the soil.

Winogradsky, on the basis of ecological characteristics classified soil microorganisms in general and bacteria in particular into two broad categories *i.e.* Autochnotus (Indigenous species) and the Zymogenous (fermentative). Autochnotus bacterial population is uniform and constant in soil, since their nutrition is derived from native soil organic matter (*e.g.* Arthrobacter and Nocardia whereas Zymogenous bacterial population in soil is low, as they require an external source of energy, *e.g.* Pseudomonas and Bacillus.

The population of Zymogenous bacteria increases gradually when a specific substrate is added to the soil. To this category belong the cellulose decomposers, nitrogen utilizing bacteria and ammonifiers.

As per the system proposed in the Bergey's Manual of Systematic Bacteriology, most of the bacteria which are predominantlyencountered in soil are taxonomically included in the three orders, Pseudomonadales, Eubacteriales and Actinomycetales of the class Schizomycetes.

The most common soil bacteria belong to the genera Pseudomonas, Arthrobacter, Clostridium Achromobacter, Sarcina, Enterobacter etc. The another group of bacteria common in soils is the Myxobacteria belonging to the genera Micrococcus, Chondrococcus, Archangium, Polyangium, Cyptophaga.

Bacteria are also classified on the basis of physiological activity or mode of nutrition, especially the manner in which they obtain their carbon, nitrogen, energy and other nutrient requirements.

They are broadly divided into two groups:

1. Autotrophs
2. Heterotrophs

Autotrophs

Autotrophic bacteria are capable synthesizing their food from simple inorganic nutrients, while heterotrophic bacteria depend on pre-formed food for nutrition. All autotrophic bacteria utilize CO_2 (from atmosphere) as carbon source and derive energy either from sunlight (photoautotrophs, *e.g.* Chromatrum. Chlorobium). Rhadopseudomonas or from the oxidation of simple inorganic substances present in soil (chemoautotrophs *e.g.* Nitrobacter, Nitrosomonas, Thiaobacillus)

ACTINOMYCETES

These are the organisms with characteristics common to both bacteria and fungi but yet possessing distinctive features to delimit them into a distinct category. In the strict taxonomicsense, actinomycetes are clubbed with bacteria the same class of Schizomycetes and confined to the order Actinomycetales. They are unicellular like bacteria, but produce a mycelium which is non-septate (coenocytic) and more slender, tike true bacteria they do not have distinct cell-wall and their cell wall is without chit in and cellulose (commonly found in the cell wall of fungi). On culture media unlike slimy distinct colonies of true bacteria which grow quickly, actinomycetes colonies grow slowly, show powdery consistency and stick firmly to agar surface. They produce hyphae and conidia/ sporangia like fungi. Certain actinomycetes whose hyphae undergo segmentation resemble bacteria, both morphologically and physiologically. Actinomycetes are numerous and widely distributed in soil and are next to bacteria in abundance.

They are widely distributed in the soil, compost etc. Plate count estimates give values ranging from 10^4 to 10^8 per gram of soil. They are sensitive to acidity/low PH (optimum PH range 6.5 to 8.0) and water logged soil conditions. The population of actinomycetes increases with depth of soil even up to horizon 'C' of a soil profiler They are heterotrophic, aerobic and mesophilic (25-30 °C) organisms and some species are commonly present in compost and manures are the rmophilic growing at 55-65 °C temperature (eg. Thermoatinomycetes, Streptomyces). Actinomycetes belonging to the order of Actinomycetales are grouped under four families *viz.* Mycobacteriaceae, Actinomycetaceae, Streptomycetaceae and Actinoplanaceae. Actinomycetous genera which are agriculturally and industrially important are present in only two families of

Actinomycetaceae and Strepotmycetaceae. In the order of abundance in soils, the common genera of actinomycetes are Streptomyces (nearly 70 per cent), Nocardia and Micromonospora although Actinomycetes, Actinoplanes, Micromonospora and Streptosporangium are also generally encountered.

Functions/ Role of actinomycetes:

- Degrade/decompose all sorts of organic substances like cellulose, polysaccharides, protein fats, organic-acids etc.
- Organic residues/substances added soil are first attacked by bacteria and fungi and later by actinomycetes, because they are slow in activity and growth than bacteria and fungi.
- They decompose/degrade the more resistant and indecomposable organic substance/matter and produce a number of dark black to brown pigments which contribute to the dark colour of soil humus.
- They are also responsible for subsequent further decomposition of humus (resistant material) in soil.
- They are responsible for earthy/musty odour/smell of freshly ploughed soils.
- Many genera species and strains (*e.g.* Streptomyces if actinomycetes produce/synthesize number of antibiotics like Streptomycin, Terramycin, Aureomycin etc.
- One of the species of actinomycetes Streptomyces scabies causes disease "Potato scab" in potato.

FUNGI

Fungi in soil are present as mycelial bits, rhizomorph or as different spores. Their number varies from a few thousand to a few -million per gram of soil. Soil fungi possess filamentous mycelium composed of individual hyphae. The fungal hyphae maybe as eptate/coenocytic (Mastigomycotina and Zygomycotina) or septate (Ascomycotina, Basidiomycotina and Deuteromycotina).

As observed by C.K. Jackson, most commonly encountered genera of fungi in soil are: Alternaria, Aspergillus, Cladosporium,Cephalosporium Botrytis, Chaetomium, Fusarium, Mucor,Penicillium, Verticillium, Trichoderma, Rhizopus, Gliocladium, Monilia, Pythium, etc. Most of these fungal genera belong to the subdivision Deuteromycotina/ Fungi imperfeacta which lacks sexual mode of reproduction.

As these soil fungi are aerobic and heterotrophic, they require abundant supply of oxygen and organic matter in soil. Fungi are dominant in acid soils, because acidic environment is not conducive/suitable for the existence of either bacteria oractinomycetes.

The optimum PH range for fungi lies-between 4.5 to 6.5. They are also present in neutral and alkaline soils and some can even tolerate pH beyond 9.0.

Functions/Role of Fungi:

- Fungi plays significant role in soils and plant nutrition.
- They plays important role in the degradation/ decomposition of cellulose, hemi cellulose, starch, pectin, lignin in the organic matter added to the soil.
- Lignin which is resistant to decomposition by bacteria is mainly decomposed by fungi.
- They also serve as food for bacteria.
- Certain fungi belonging to sub-division Zygomycotina and Deuteromycotina are predaceous in nature and attack on protozoa and nematodes in soil and thus, maintain biologica lequilibrium in soil.
- They also plays important role in soil aggregation and in the formation of humus.
- Some soil fungi are parasitic and causes number of plant diseases such as wilts, root rots, damping-off and seedling blightseg. Pythium, Phyiophlhora, Fusarium, Verticillium etc.
- Number of soil fungi forms mycorrhizal association with the roots of higher plants (symbiotic association of a fungus with the roots of a higher plant) and helps in mobilization of soil phosphorus and nitrogen *e.g.* Glomus, Gigaspora, Aculospora, (Endomycorrhiza) and Amanita, Boletus, Entoloma, Lactarius (Ectomycorrhiza).

PROTOZOA

These are unicellular, eukaryotic, colourless, and animal like organisms (Animal kingdom). They are larger than bacteria and size varying from few microns to a few centimeters. Their population in arable soil ranges from 10, 000 to 1, 00,000 per gram of soil and are abundant in surface soil. They can withstand adverse soil conditions as they are characterized by "cyst stage" in their life cycle. Except few genera which reproduce sexually by fusion of cells, rest of them reproduces asexually by fission/binary fission. Most of the soil protozoa are motile by flagella orcilia or pseudopodia as locomotors organs.

Depending upon the type of appendages provided for locomotion, protozoa are:

- Rhizopoda (Sarcondia)
- Mastigophora
- Ciliophora (Ciliata)
- Sporophora (not common Inhabitants of soil)

Class-Rhizopoda

It consists protozoa without appendages usually have naked protoplasm without cell-wall, pseudopodia astemporary locomotory organs are present some times. Important genera are Amoeba, Biomyxa, Euglypha, etc.

Class Mastigophora

It belongs flagellated protozoa, which are predominant in soil. Important genera are: Allention, Bodo, Cercobodo, Cercomonas, Entosiphon Spiromonas, Spongomions, and Testramitus. Many members are saprophytic and some posses chlorophyll and are autotrophic in nature. In this respect, they resemble unicellular algae and hence are known as "Phytoflagellates".

Ciliophora

The soil protozoa belonging to the class ciliate/ciliophora are characterized by the presence of cilia (short hair-like appendages) around their body, which helps in locomotion. The important soil inhabitants of this class are Colpidium, Colpoda, Balantiophorus, Gastrostyla, Halteria, Uroleptus, Vortiicella, Pleurotricha etc.

Sporophora

Protozoa are abundant in the upper layer (15 cm) of soil. Organic manures protozoa. Soil moisture, aeration, temperature and pH are the important factors affecting soil protozoa.

Function/Role of Protozoa:

- Most of protozoans derive their nutrition by feeding oringesting soil bacteria belonging to the genera Enterobacter, Agrobacterium, Bacillus, Escherichia, Micrococcus, and Pseudomonas and thus, they play important role in maintainingmicrobial/ bacterial equilibrium in the soil.
- Some protozoa have been recently used as biological controlagents against phytopathogens.
- Species of the bacterial genera *viz.* Enterobacter and Aerobacter are commonly used as the food base for isolation andenumeration of soil protozoans.
- Several soil protozoa cause diseases in human beings whichare carried through water and other vectors, eg. Amoebicdysentery caused by Entomobea histolytica.

ALGAE

Algae are present in most of the soils where moisture and sunlight are available. Their number in soil usually ranges from 100 to 10,000 per gram of soil. They are photoautotrophic, aerobic organisms and obtain CO_2 from atmosphere and energy from sunlight and synthesize their own food. They are unicellular, filamentous or colonial.

Soil algae are divided in to four main classes or phyla as follows:

1. Cyanophyta (Blue-green algae)
2. Chlorophyta (Grass-green algae)

3. Xanthophyta (Yellow-green algae)
4. Bacillariophyta (diatoms or golden-brown algae)

Out of these four classes/phyla, blue-green algae and grass-green algae are more abundant in soil. The green-grass algae and diatoms are dominant in the soils of temperate region while blue-green algae predominate in tropical soils. Green-algae prefer acid soils while blue green algae are commonly found in neutral and alkaline soils. The most common genera of green algae found in soil are Chlorella, Chlamydomonas, Chlorococcum, Protosiphon etc. and that of diatoms are Navicula, Pinnularia, Synedra, Frangilaria.

Blue green algae are unicellular, photoautotrophic prokaryotes containing Phycocyanin pigment in addition to chlorophyll. They do not posses flagella and do not reproduce sexually. They are common in neutral to alkaline soils. The dominant genera of BGAin soil are: Chrococcus, Phormidium, Anabaena, Aphanocapra, Oscillatoria etc. Some BGA posses specialized cells known as "Heterocyst" which is the sites of nitrogen fixation. BGA fixes nitrogen (non-symbiotically) in puddle paddy/water logged paddy fields (20-30 kg/ha/season). There are certain BGA which possess the character of symbiotic nitrogen fixation in association with other organisms like fungi, mosses, liverworts and aquatic ferns Azolla, *e.g.* Anabaena-Azolla association fix nitrogen symbiotically in rice fields.

FUNCTIONS/ROLE OF ALGAE OR BGA

- Plays important role in the maintenance of soil fertility especially in tropical soils.
- Add organic matter to soil when die and thus increase the amount of organic carbon in soil.
- Most of soil algae (especially BGA) act as cementing agent in binding soil particles and thereby reduce/prevent soil erosion.
- Mucilage secreted by the BGA is hygroscopic in nature and thus helps in increasing water retention capacity of soil for longer time/period.
- Soil algae through the process of photosynthesis liberate large quantity of oxygen in the soil environment and thus facilitate the aeration in submerged soils or oxygenate the soil environment.
- They help in checking the loss of nitrates through leaching and drainage especially in un-cropped soils.
- They help in weathering of rocks and building up of soil structure.

SOIL MICROORGANISMS IN BIODEGRADATION OF PESTICIDES AND HERBICIDES

Pesticides are the chemical substances that kill pests and herbicides are the chemicals that kill weeds. In the context of soil, pests are fungi, bacteria insects, worms, and nematodes etc. that cause damage to field crops. Thus, in broad sense pesticides are insecticides, fungicides, bactericides, herbicides and nematicides that are used to control or inhibit plant diseases and insect pests. Although wide-scale application of pesticides and herbicides is an essential part of augmenting crop yields; excessive use of these chemicals leads to the microbial imbalance, environmental pollution and health hazards. An ideal pesticide should have the ability to destroy target pest quickly and should be able to degrade non-toxic substances as quickly as possible. The ultimate "sink" of the pesticides applied in agriculture and public health care is soil. Soil being the storehouse of multitudes of microbes, in quantity and quality, receives the chemicals invarious forms and acts as a scavenger of harmful substances. The efficiency and the competence to handle the chemicals vary with the soil and its physical, chemical and biological characteristics.

Effects of Pesticides

Pesticides reaching the soil in significant quantities have direct effect on soil microbiological aspects, which in turn influence plant growth.

Some of the most important effects caused by pesticides are:

- Alterations hi ecological balance of the soil micro flora,
- Continued application of large quantities of pesticides may cause everlasting changes in the soil micro flora,
- Adverse effect on soil fertility and crop productivity,
- Inhibition of N2 fixing soil microorganisms such as Rhizobium, Azotobacter, Azospirillum etc. and cellulolytic and phosphate solubilizing microorganisms,
- Suppression of nitrifying bacteria, Nitrosomonas and Nitrobacter by soil fumigants ethylene bromide, Telone, and vapam have also been reported,
- Alterations in nitrogen balance of the soil,
- Interference with ammonification in soil,
- Adverse effect on mycorrhizal symbioses in plants and nodulation in legumes, and
- Alterations in the rhizosphere microflora, both quantitatively and qualitatively.

Persistence of Pesticides in Soil

How long an insecticide, fungicide, or herbicide persists in soil is of great importance in relation to pest management and environmental pollution.

Persistence of pesticides in soil for longer period is undesirable because of the reasons:

- Accumulation of the chemicals in soil to highly toxic levels,
- May be assimilated by the plants and get accumulated in edible plant products,

- Accumulation in the edible portions of the root crops,
- To be get eroded with soil particles and may enter into the water streams, and finally leading to the soil, water and air pollutions.

The effective persistence of pesticides in soil varies from a week to several years depending upon structure and properties of the constituents in the pesticide and availability of moisture in soil. For instance, the highly toxic phosphates do not persist for more than three months while chlorinated hydrocarbon insecticides (*e.g.*, DOT, aldrin, chlordane etc.) are known to persist at least for 4-5 years and some times more than 15 years.

From the agricultural point of view, longer persistence of pesticides leading to accumulation of residues in soil may result into the increased absorption of such toxic chemicals by plants to the level at which the consumption of plant products may prove deleterious/ hazardous to human beings as well as livestock's.

There is a chronic problem of agricultural chemicals, having entered in food chain at highly inadmissible levels in India, Pakistan, Bangladesh and several other developing countries in the world. For example, intensive use of DDT to control insect pests and mercurial fungicides to control diseases in agriculture had been known to persist for longer period and thereby got accumulated in the food chain leading to food contamination and health hazards. Therefore, DDT and mercurial fungicides has been, banned to use in agriculture as well as in public health department.

BIODEGRADATION OF PESTICIDES IN SOIL

Pesticides reaching to the soil are acted upon by several physical, chemical, and biological forces. However, physical and chemical forces are acting upon/ degrading the pesticides to some extent, microorganism's plays major role in the degradation of pesticides. Many soil microorganisms have the ability to act upon pesticides and convert them into simpler non-toxic compounds. This process of degradation of pesticides and conversion into non-toxic compounds by microorganisms is known as "biodegradation". Not all pesticides reaching to the soil are biodegradable and such chemicals that show complete resistance to biodegradation are called "recalcitrant".

The chemical reactions leading to biodegradation of pesticides fall into several broad categories which are discussed in brief in the following paragraphs.

- *Detoxification*: Conversion of the pesticide molecule to anon-toxic compound. Detoxification is not synonymous with degradation. Since a single chance in the side chain of a complex molecule may render the chemical non-toxic.
- *Degradation*: The breaking down/ transformation of acomplex substrate into simpler products leading finally to mineralization. Degradation is often considered to be synonymous with mineralization, *e.g.* Thirum

(fungicide) is degraded by a strain of Pseudomonas and the degradation products are dimethlamine, proteins, sulpholipaids, etc.

CONJUGATION (COMPLEX FORMATION OR ADDITION REACTION)

In which an organism make the substrate more complex or combines the pesticide with cell metabolites. Conjugation or the formation of addition product is accomplished by those organisms catalyzing the reaction of addition of an amino acid, organic acid or methyl crown to the substrate, for *e.g.*, in the microbial metabolism of sodium dimethly dithiocarbamate, the organism combines the fungicide with an amino acid molecule normally present in the cell and thereby inactivate the pesticides/chemical.

Activation

It is the conversion of non-toxic substrate into atoxic molecule, for eg. Herbicide, 4-butyric acid (2, 4-D B) and the insecticide Phorate are transformed and activated microbiologically in soil to give metabolites that are toxic to weeds and insects.

Changing the Spectrum of Toxicity

Some fungicides/pesticides are designed to control one particular group of organisms/ pests, but they are metabolized to yield products inhibitory to entirely dissimilar groups of organisms, for *e.g.* the fungicide PCNB fungicide is converted in soil to chlorinated benzoic acids that kill plants. Biodegradation of pesticides/ herbicides is greatly influenced by the soil factors like moisture, temperature, PH and organic matter content, in addition to microbial population and pesticide solubility. Optimum temperature, moisture and organic matter in soil provide congenial environment for the break down or retention of any pesticide added in the soil. Most of the organic pesticides degrade within a short period (3-6 months) under tropical conditions. Metabolic activities of bacteria, fungi and actinomycetes have the significant role in the degradation of pesticides.

CRITERIA FOR BIOREMEDIATION/BIODEGRADATION

For successful biodegradation of pesticide in soil, following aspects must be taken into consideration:

- Organisms must have necessary catabolic activity required for degradation of contaminant at fast rate to bring down the concentration of contaminant,
- The target contaminant must be bioavailability,
- Soil conditions must be congenial for microbial/plant growth and enzymatic activity and
- Cost of bioremediation must be less than other technologies of removal of contaminants.

According to Gales (1952) principal of microbial infallibility, for every naturally occurring organic compound there is a microbe/enzyme system capable its degradation.

STRATEGIES FOR BIOREMEDIATION

For the successful biodegradation/ bioremediation of a given contaminant following strategies are needed:

- *Passive/intrinsic Bioremediation*: It is the natural bioremediation of contaminant by tile indigenous microorganisms and the rate of degradation is very slow.
- *Biostimulation*: Practice of addition of nitrogen and phosphorus to stimulate indigenous microorganisms in soil.
- *Bioventing*: Process/way of Biostimulation by which gases stimulants like oxygen and methane are added or forced into soil to stimulate microbial activity.
- *Bio-augmentation*: It is the inoculation/introduction of microorganisms in the contaminated site/soil to facilitate biodegradation.
- *Composting*: Piles of contaminated soils are constructed and treated with aerobic thermophilic microorganisms to degrade contaminants. Periodic physical mixing and moistening of piles are done to promote microbial activity.
- *Phytoremediation*: Can be achieved directly by planting plants which hyper accumulate heavy metals or indirectly by plants stimulating microorganisms in the rhizosphere.
- *Bioremediation*: Process of detoxification of toxic/unwanted chemicals/ contaminants in the soil and other environment by using microorganisms.
- *Mineralization*: Complete conversion of an organic contaminant to its inorganic constituent by a species or group of microorganisms.

SOIL MICROORGANISMS IN BIODEGRADATION OF PESTICIDES AND HERBICIDES

Pesticides are the chemical substances that kill pests and herbicides are the chemicals that kill weeds. In the context of soil, pests are fungi, bacteria insects, worms, and nematodes etc. that cause damage to field crops. Thus, in broad sense pesticides are insecticides, fungicides, bactericides, herbicides and nematicides that are used to control or inhibit plant diseases and insect pests. Although wide-scale application of pesticides and herbicides is an essential part of augmenting crop yields; excessive use of these chemicals leads to the microbial imbalance, environmental pollution and health hazards. An ideal pesticide should have the ability to destroy target pest quickly and should be able to degrade non-toxic substances as quickly as possible.

The ultimate "sink" of the pesticides applied in agriculture and public health care is soil. Soil being the storehouse of multitudes of microbes, in quantity and quality, receives the chemicals in various forms and acts as a scavenger of harmful substances. The efficiency and the competence to handle the chemicals vary with the soil and its physical, chemical and biological characteristics.

EFFECTS OF PESTICIDES

Pesticides reaching the soil in significant quantities have direct effect on soil microbiological aspects, which in turn influence plant growth. Some of the most important effects caused by pesticides are:

- Alterations hi ecological balance of the soil microflora,
- Continued application of large quantities of pesticides may cause ever lasting changes in the soil microflora,
- Adverse effect on soil fertility and crop productivity,
- Inhibition of N2 fixing soil microorganisms such as *Rhizobium, Azotobacter, Azospirillum* etc. and cellulolytic and phosphate solubilizing microorganisms,
- Suppression of nitrifying bacteria, *Nitrosomonas* and *Nitrobacter* by soil fumigants ethylene bromide, Telone, and vapam have also been reported,
- Alterations in nitrogen balance of the soil,
- Interference with ammonification in soil,
- Adverse effect on mycorrhizal symbioses in plants and nodulation in legumes, and
- Alterations in the rhizosphere microflora, both quantitatively and qualitatively.

PERSISTENCE OF PESTICIDES IN SOIL

How long an insecticide, fungicide, or herbicide persists in soil is of great importance in relation to pest management and environmental pollution. Persistence of pesticides in soil for longer period is undesirable because of the reasons:

- Accumulation of the chemicals in soil to highly toxic levels,
- May be assimilated by the plants and get accumulated in edible plant products,
- Accumulation in the edible portions of the root crops,
- To be get eroded with soil particles and may enter into the water streams, and finally leading to the soil, water and air pollutions.

The effective persistence of pesticides in soil varies from a week to several years depending upon structure and properties of the constituents in the pesticide and availability of moisture in soil. For instance, the highly toxic phosphates do not persist for more than three months while chlorinated hydrocarbon insecticides (eg. DOT, aldrin, chlordane etc) are known to persist at least for 4-5 years and some times more than 15 years.

From the agricultural point of view, longer persistence of pesticides leading to accumulation of residues in soil may result into the increased absorption of such toxic chemicals by plants to the level at which the consumption of plant products may prove deleterious/ hazardous to human beings as well as livestock's.

There is a chronic problem of agricultural chemicals, having entered in food chain at highly inadmissible levels in India, Pakistan, Bangladesh and several other developing countries in the world. For example, intensive use of DDT to control insect pests and mercurial fungicides to control diseases in agriculture had been known to persist for longer period and thereby got accumulated in the food chain leading to food contamination and health hazards. Therefore, DDT and mercurial fungicides has been, banned to use in agriculture as well as in public health department.

BIODEGRADATION OF PESTICIDES IN SOIL

Pesticides reaching to the soil are acted upon by several physical, chemical, and biological forces. However, physical and chemical forces are acting upon/ degrading the pesticides to some extent, microorganism's plays major role in the degradation of pesticides.

Many soil microorganisms have the ability to act upon pesticides and convert them into simpler non-toxic compounds. This process of degradation of pesticides and conversion into non-toxic compounds by microorganisms is known as "biodegradation".Not all pesticides reaching to the soil are biodegradable and such chemicals that show complete resistance to biodegradation are called "recalcitrant".

The chemical reactions leading to biodegradation of pesticides fall into several broad categories which are discussed in brief in the following paragraphs.

- *Detoxification:* Conversion of the pesticide molecule to a non-toxic compound. Detoxification is not synonymous with degradation. Since a single chance in the side chain of a complex molecule may render the chemical non-toxic.
- *Degradation:* The breaking down/ transformation of a complex substrate into simpler products leading finally to mineralization. Degradation is often considered to be synonymous with mineralization, *e.g.* Thirum (fungicide) is degraded by a strain of*Pseudomonas* and the degradation products are dimethlamine, proteins, sulpholipaids, etc.
- *Conjugation (complex formation or addition reaction):* In which an organism make the substrate more complex or combines the pesticide with cell metabolites. Conjugation or the formation of

addition product is accomplished by those organisms catalyzing the reaction of addition of an amino acid, organic acid or methyl crown to the substrate, for *e.g.*, in the microbial metabolism of sodium dimethly dithiocarbamate, the organism combines the fungicide with an amino acid molecule normally present in the cell and thereby inactivate the pesticides/chemical.

- *Activation:* It is the conversion of non-toxic substrate into a toxic molecule, for eg. Herbicide, 4-butyric acid (2, 4-D B) and the insecticide Phorate are transformed and activated microbiologically in soil to give metabolites that are toxic to weeds and insects.
- *Changing the spectrum of toxicity:* Some fungicides/pesticides are designed to control one particular group of organisms/ pests, but they are metabolized to yield products inhibitory to entirely dissimilar groups of organisms, for *e.g.* the fungicide PCNB fungicide is converted in soil to chlorinated benzoic acids that kill plants.

Biodegradation of pesticides/ herbicides is greatly influenced by the soil factors like moisture, temperature, PH and organic matter content, in addition to microbial population and pesticide solubility. Optimum temperature, moisture and organic matter in soil provide congenial environment for the break down or retention of any pesticide added in the soil. Most of the organic pesticides degrade within a short period (3-6 months) under tropical conditions. Metabolic activities of bacteria, fungi and actinomycetes have the significant role in the degradation of pesticides.

CRITERIA FOR BIOREMEDIATION/ BIODEGRADATION

For successful biodegradation of pesticide in soil, following aspects must be taken into consideration.

- Organisms must have necessary catabolic activity required for degradation of contaminant at fast rate to bring down the concentration of contaminant,
- The target contaminant must be bioavailability,
- Soil conditions must be congenial for microbial/plant growth and enzymatic activity and
- Cost of bioremediation must be less than other technologies of removal of contaminants.

According to Gales (1952) principal of microbial infallibility, for every naturally occurring organic compound there is a microbe/ enzyme system capable its degradation.

STRATEGIES FOR BIOREMEDIATION

For the successful biodegradation/ bioremediation of a given contaminant following strategies are needed.

- *Passive/ intrinsic Bioremediation:* It is the natural bioremediation of contaminant by tile indigenous microorganisms and the rate of degradation is very slow.
- *Biostimulation:* Practice of addition of nitrogen and phosphorus to stimulate indigenous microorganisms in soil.
- *Bioventing:* Process/way of Biostimulation by which gases stimulants like oxygen and methane are added or forced into soil to stimulate microbial activity.
- *Bioaugmentation:* It is the inoculation/introduction of microorganisms in the contaminated site/soil to facilitate biodegradation.
- *Composting:* Piles of contaminated soils are constructed and treated with aerobic thermophilic microorganisms to degrade contaminants. Periodic physical mixing and moistening of piles are done to promote microbial activity.
- *Phytoremediation:* Can be achieved directly by planting plants which hyperaccumulate heavy metals or indirectly by plants stimulating microorganisms in the rhizosphere.
- *Bioremediation:*Process of detoxification of toxic/unwanted chemicals/ contaminants in the soil and other environment by using microorganisms.
- *Mineralization:* Complete conversion of an organic contaminant to its inorganic constituent by a species or group of microorganisms.

PESTICIDE TREATMENT ON BACTERIAL SOIL

At least once a year about 85 per cent of agriculturally used soils in western Europe are treated with herbicides in order to ensure that high-quality agricultural goods will be produced at levels that cover costs. Actual registration procedures include evaluation of the impact of herbicides on the environment by testing for non-target effects on single species or microbial communities.

The impact on soil microbial communities is evaluated in view of their role in sustaining the global cycling of matter and their varied functions in supporting plant growth. In order to observe effects by quantifying changes in microbial biomass or in the general metabolic activities of microbial communities, a variety of laboratory tests have been recommended. Internationally, there is wide divergence in the extent to which the performance of such tests is obligatory for registrations. Substrate-induced-respiration (SIR), dehydrogenase activity, and nitrogen turnover are investigated in German registration procedures.

The traditional techniques used in our study can be divided into two groups. SIR and dehydrogenase activity are both strictly correlated to biomass, whereas long-term respiration and nitrogen mineralization are not. The immediate respiration of a microbial community following a glucose addition is quantified in a manner avoiding a significant contribution of cell multiplication. Testing for the activity of dehydrogenases, as determined by the accumulation of reduced triphenyl formazane, conveniently monitors the metabolic activities of microbial cells in oxidizing organic substrates. This parameter seems to indicate an impact on microbial activity most significantly. The efficiency of microbial carbon mineralization over an extended period (8 weeks in this report) can be stimulated by the mineralization of dead cells or the degradation of the pesticide itself. Carbon mineralization and nitrogen mineralization serve to complement the interpretation of the results for the biomass-related parameters.

Another approach to differentiating bacterial communities by their catabolic potentials is offered by the BIOLOG system (BIOLOG Inc., Hayward, Calif.), originally commercialized for the identification of bacterial strains. The metabolism of 95 carbon sources in microtiter plates, indicated by the reduction of a tetrazolium dye, affords a detailed view of the catabolic activities of complex bacterial communities. These features seem to be suited to monitoring of the impact of stress factors on microbial communities. In contrast to such approaches, testing for effects on a special activity of a single microbial species can be used in some cases of soil contamination. Such testing may indicate how the whole community responds to toxic interactions but might be considered to be restricted in its ecological significance.

Analysis of changes in the composition of a community can also be used to characterize its response to stress. Species abundance and diversity measurements are sensitive indicators of environmental conditions. However, the conventional microbiological investigation of community diversity has significant pitfalls. First, the time and resources needed for quantification and identification of morphologically similar microbial species preclude an extensive analysis of many samples from different treatments. The counting of colony morphotypes has not been tested as an alternative on a broader scale. Second, a significant bias invariably is introduced by the selective process of cultivation. Microbiologists are becoming increasingly aware of the fact that only a small fraction of microorganisms is analyzed by conventional cultivation methods, and frequently these species do not seem to represent the spectrum of organisms active in the habitat of choice.

A measurement more representative of the structure of communities in situ may be obtained by isolating and analyzing macromolecules of microbial origin from a habitat, as in other methods of microbial-community characterization. A community's composition and diversity are most significantly represented by the complexity of information in nucleic acid molecules. A great variety of methods of isolating DNA and RNA from different habitats, either by direct extraction or following a prior separation of microbial cells, have been published. The information in about 4,000 different bacterial chromosomes, equivalent to more than 10,000 species, seems to represent the complexity of microbial communities in healthy agricultural soils, as estimated by DNA reassociation kinetics. Wide divergence in levels of diversity among different soils can be demonstrated, but the technique does not seem to be adaptable to a measurement of induced changes, particularly with respect to its sensitivity.

A more detailed view of the composition of a community can be obtained by sampling the nucleic acid population from a habitat in order to determine the abundance of phylogenetically meaningful sequences, particularly the 16S rRNA and 16S rDNA. The origins of individual sequences are then classified by comparison to large databases and their present interpretation. The frequency of sequences within the sample represents the community composition in some detail, depending on the number of sequences analyzed. Such approaches have become an established method of molecular ecology. Unfortunately, these techniques are labour intensive and time-consuming, a major drawback for any investigation involving a large number of differently treated samples.

The sequence-dependent separation of 16S rDNA fragments in a denaturing gradient gel (denaturing gradient gel electrophoresis [DGGE]) in order to analyze the abundance of divergent sequences from particular communities was introduced by Muyzer *et al.* By superimposing a temperature gradient instead of a chemical gradient on the electric field, temperature gradient gel electrophoresis (TGGE) applies the identical principle of separating double-stranded DNA fragments.

Minor sequence variations, including point mutations, result in modified migration and distinct bands.

The technique of generating a pattern from sequence variants with well-defined phylogenetic meaning is attractive from different points of views. The pattern yields a representation of the community as a whole or of a section of it, defined by the selected primers. The bands of the pattern are not directly interpretable in phylogenetic terms. Their "fingerprint," however, represents the frequency distribution of different sequences in a sample. Variations in such fingerprints due to induced variations in community composition may be easily observed. The degrees of pattern change may be quantified to some extent to allow for ranking. Finally, band isolation and sequencing is one straightforward possibility to investigate induced changes in detail with respect to responsive community members.

We have used both the BIOLOG system and pattern generation by amplified ribosomal sequences to investigate the suitability of these approaches for the analysis of the impacts of herbicide applications on the function and structure of soil bacterial communities. The more traditional methods for analysis of such impacts serve as a well-defined background of ecotoxicological testing.

3

Beneficial Pesticides Use in Agriculture

INTRODUCTION TO PESTICIDES

Pesticides are chemicals used to eliminate or control a variety of agricultural pests that can damage crops and livestock and reduce farm productivity. The most commonly applied pesticides are insecticides (to kill insects), herbicides (to kill weeds), rodenticides (to kill rodents), and fungicides (to control fungi, mold, and mildew). The latest US Environmental Protection Agency (EPA) information on US pesticide usage, from 2007, reports that over 1 billion tons of pesticides are used in the US every year. This is 22 per cent of the estimated 5.2 billion pounds of pesticides used worldwide. Agricultural use accounted for 80 per cent of pesticide use in the US. Of the pesticide classes, herbicides are the most widely used in all US sectors.

REGULATION AND MONITORING OF PESTICIDES

Pesticides are tested and approved for use by the EPA, which establishes "tolerances," or maximum residue levels, that describe the amount of a given pesticide that can safely remain in or on a food. The US Food and Drug Administration (FDA) is then responsible for monitoring pesticide levels on fruits and vegetables, while the US Department of Agriculture (USDA) is charged with the task of surveying pesticide residues in meat, eggs, and dairy products.

There is some criticism of the EPA's methods for setting "tolerances" for pesticide exposure in humans. Short-term testing is conducted on laboratory animals exposed to only a single chemical at a time, usually in high doses. However, in real life, humans are often exposed to a number of different chemicals over long periods of time – this type of exposure can cause problems that might not be observed in a short-term study of the effects of a single chemical. The FDA has also been criticized for inadequate monitoring of pesticide levels on fruits and vegetables. Ineffective sampling of food for pesticide residue testing, problems with data analysis, and lack of authority to fine or otherwise punish growers (or importers) who use illegal pesticides are common criticisms. The Environmental Working Group compiles a yearly list of fruits and vegetables with the highest levels of pesticide residues, based on USDA and FDA testing data. The current top 12 are:

- Apples
- Celery
- Sweet Bell Peppers
- Peaches
- Strawberries
- Nectarines (imported)
- Grapes
- Spinach
- Lettuce
- Cucumbers
- Blueberries
- Potatoes

PESTICIDES AND THE ENVIRONMENT

According to Cornell entomologist David Pimentel, "[i]t has been estimated that only 0.1 per cent of applied pesticides reach the target pests, leaving the bulk of the pesticides (99.9 per cent) to impact the environment. Harmful environmental impacts of pesticide use include:

- Loss of biodiversity and elimination of key species (*e.g.*, bees)
- Water pollution
- Soil contamination

Pest resistance, resulting in the need for increased application of pesticides, or formulation of alternate pesticides.

Pesticides and Public Health

Pesticides have been linked to a number of health problems, including neurologic and endocrine (hormone) system disorders, birth defects, cancer, and other diseases. Although it is widely understood that exposure to pesticides is dangerous to humans, Centers for Disease Control and Prevention (CDC) data show that a high percentage of individuals tested had certain pesticides or the chemical breakdown of those pesticides (metabolites) in their blood and/or urine. CDC research also shows that people in the US carry levels of pesticides in their bodies that, for certain pesticides, exceed the EPA's "acceptable" levels.

For example, CDC data show that the average American child between the ages of six and eleven carries unacceptable levels of the organophosphorus pesticides, chlorpyrifos and methyl parathion, both of which are known to have neurotoxic properties. It should also be noted that the human health effects of low dose, chronic exposure to many of these pesticides is listed as "unknown" by the CDC. Children are especially susceptible to the harmful effects of pesticide residues due to their lower body mass, rapid development, and higher rates of consumption of affected products. In children, exposure to certain pesticides from residues in food can cause delayed development; disruptions to the reproductive, endocrine, and immune systems; certain types of cancer; and damage to other organs. Prenatal exposure to certain pesticides can affect cognitive development and behaviour. Several studies have found that pesticide levels in children dropped to low or undetectable levels when test subjects consumed an organic diet.

Farmworkers are also highly vulnerable to these health threats due to intensive exposure to a variety of pesticides, either from applying these chemicals or from harvesting pesticide-sprayed agricultural products.

PRODUCTION AND USAGE OF PESTICIDES IN INDIA

The production of pesticides started in India in 1952 with the establishment of a plant for the production of BHC near Calcutta, and India is now the second largest manufacturer of pesticides in Asia after China and ranks twelfth globally. There has been a steady growth in the production of technical grade pesticides in India, from 5,000 metric tons in 1958 to 102,240 metric tons in 1998. In 1996–97 the demand for pesticides in terms of value was estimated to be around ₹ 22 billion (USD 0.5 billion), which is about 2 per cent of the total world market.

The pattern of pesticide usage in India is different from that for the world in general. In India 76 per cent of the pesticide used is insecticide, as against 44 per cent globally. The use of herbicides and fungicides is correspondingly less heavy. The main use of pesticides in India is for cotton crops (45 per cent), followed by paddy and wheat.

SOURCES OF MISUSE AND/OR UNSAFE USE OF AGRICULTURAL PESTICIDES

The incidence of agricultural pesticide contamination or poisoning is exacerbated through misuse and/or unsafe use. Factors of agricultural pesticide misuse and/or unsafe use that are especially prevalent in developing countries include the absence of stringent regulations and the lack of enforcement of existing ones, the failure to follow label instructions and guidelines, and the importation of toxic agricultural pesticides that have been banned or whose use are severely restricted in developed and industrialized countries.

Others include the use of leaking equipment, the use of domestic utensils for measuring and dispensing pesticides, exposure to pesticide drifts, failure to wear personal protective equipment (PPE) such as gloves, long pants and respirators and "calendar spraying" – inflexible, scheduled application of pesticides irrespective of the level of pest infestation.

Other unsafe practices that are especially common in developing countries include storage of pesticides in family bedrooms and in unlocked cabinets that can easily be accessed by children, improper disposal of empty pesticide containers, the use of empty containers for domestic purposes – *e.g.*, storing foodstuffs and water for both humans and domestic animals. However, the misuse or unsafe use of agricultural pesticides is not limited to developing countries. Toxic agricultural pesticides that are meant for outdoor use are often used indoors even in developed and industrialized nations for the control of household pests. An extensive indoor application of methyl parathion, a restricted-use agricultural pesticide, led to the temporary relocation of more than 250 households in Jackson County, Mississippi, U. S. in 1997.

SOURCES OF HUMAN EXPOSURE TO AGRICULTURAL PESTICIDES

Human exposure to agricultural pesticides and the subsequent contamination or poisoning may be occupational, non-occupational, intentional, unintentional, or accidental. Also, exposure may be through ingestion (oral), through the skin (dermal) or through inhalation (respiratory). Occupational contamination or poisoning has been identified as the most serious problem associated with the use of agricultural pesticides, especially in developing countries. Nonetheless, the effect of accidental contamination can be very serious and even fatal. More than 400 people reportedly died in Iraq in 1972 after eating bread that had been prepared from cereals treated with a fungicide. Also, at least 37 people reportedly died of endosulfan poisoning in Republic of Benin during the 1999/2000 cotton growing season. In 1958, all members of the family of a local chief who is a prominent cocoa farmer at Okebode in southwestern Nigeria were hospitalized after eating a leaf vegetable undergrowth of a cocoa farm that was earlier sprayed with lindane. In 2004, carbofuran pesticide residues found on several batches of noodles manufactured in Nigeria may have resulted in 23 reported cases of vomiting and one death.

STRATEGIES FOR REDUCING AGRICULTURAL PESTICIDE CONTAMINATION

Several strategies that have been suggested to reduce agricultural pesticide contamination and its effects include comprehensive risk communication and education programmes, the use of appropriate personal protective clothing, the discontinuation of calendar spraying, selection of disease-resistant hybrids, and the adoption of Integrated Pesticide Management (IPM). IPM emphasizes non-chemical and cultural cultivation pest control strategies such as removal of diseased plant parts, crop rotation that may disrupt the life cycle of pests, and biological control such as the use of insect predators. While the need exists to prevent or reduce agricultural pesticide contamination and its effects world-wide, special attention needs to be paid to developing countries for several reasons. These include the observations that three-fourths of the estimated annual instances of pesticide poisoning and pesticide-related deaths occur in developing countries, and that these countries represent the fastest growing market for agricultural pesticides. Also, the adverse human health effects of pesticide poisoning are particularly high in many of these countries because of the low nutritional status, and the scarcity of heath care facilities especially in the rural areas where most of the agricultural activities occur.

DIRECT IMPACT ON HUMANS

If the credits of pesticides include enhanced economic potential in terms of increased production of food and fibre, and amelioration of vector-borne diseases, then their debits have resulted in serious health implications to man and his environment. There is now overwhelming evidence that some of these chemicals do pose a potential risk to humans and other life forms and unwanted side effects to the environment. No segment of the population is completely protected against exposure to pesticides and the potentially serious health effects, though a disproportionate burden, is shouldered by the people of developing countries and by high risk groups in each country. The world-wide deaths and chronic diseases due to pesticide poisoning number about 1 million per year.

The high risk groups exposed to pesticides include production workers, formulators, sprayers, mixers, loaders and agricultural farm workers. During manufacture and formulation, the possibility of hazards may be higher because the processes involved are not risk free. In industrial settings, workers are at increased risk since they handle various toxic chemicals including pesticides, raw materials, toxic solvents and inert carriers.

OC compounds could pollute the tissues of virtually every life form on the earth, the air, the lakes and the oceans, the fishes that live in them and the birds that feed on the fishes. The US National Academy of Sciences stated that the DDT metabolite DDE causes eggshell thinning and that the bald eagle population in the United States declined primarily because of exposure to DDT and its metabolites. Certain environmental chemicals, including pesticides termed as endocrine disruptors, are known to elicit their adverse effects by mimicking or antagonising natural hormones in the body and it has been postulated that their long-term, low-dose exposure is increasingly linked to human health effects such as immune suppression, hormone disruption, diminished intelligence, reproductive abnormalities and cancer.

A study on workers (N=356) in four units manufacturing HCH in India revealed neurological symptoms (21 per cent) which were related to the intensity of exposure. The magnitude of the toxicity risk involved in the spraying of methomyl, a carbamate insecticide, in field conditions was assessed by the National Institute of Occupational Health (NIOH). Significant changes were noticed in the ECG, the serum LDH levels, and cholinesterase (ChE) activities in the spraymen, indicating cardiotoxic effects of methomyl. Observations confined to health surveillance in male formulators engaged in production of dust and liquid formulations of various pesticides (malathion, methyl parathion, DDT and lindane) in industrial settings of the unorganised sector revealed a high occurrence of generalised symptoms (headache, nausea, vomiting, fatigue, irritation of skin and eyes) besides psychological, neurological, cardiorespiratory and gastrointestinal symptoms coupled with low plasma ChE activity.

Data on reproductive toxicity were collected from 1,106 couples when the males were associated with the spraying of pesticides (OC, OP and carbamates) in cotton fields. A study in malaria spraymen was initiated to evaluate the effects of a short-term (16 week) exposure in workers (N=216) spraying HCH in field conditions.

A study on those affected in the Seveso diaster of 1976 in Italy during the production of 2,4,5 T, a herbicide, concluded that chloracne (nearly 200 cases with a definite exposure dependence) was the only effect established with certainty as a result of dioxin formation. Early health investigations including liver function, immune function, neurologic impairment, and reproductive effects yielded inconclusive results. An excess mortality from cardiovascular and respiratory diseases was uncovered, possibly related to the psychosocial consequences of the accident in addition to the chemical contamination. An excess of diabetes cases was also found. Results of cancer incidence and mortality follow-up showed an increased occurrence of cancer of the gastrointestinal sites and of the lymphatic and haematopoietic tissue. Results cannot be viewed as conclusive, however, because of various limitations: few individual exposure data, short latency period, and small population size for certain cancer types.

A similar study in 2001 observed no increase in all-cause and all-cancer mortality. However, the results support the notion that dioxin is carcinogenic to humans and corroborate the hypotheses of its association with cardiovascular- and endocrine-related effects. During the Vietnam War, United States military forces sprayed nearly 19 million gallons of herbicide on approximately 3.6 million acres of Vietnamese and Laotian land to remove forest cover, destroy crops, and clear vegetation from the perimeters of US bases.

This effort, known as Operation Ranch Hand, lasted from 1962 to 1971. Various herbicide formulations were used, but most were mixtures of the phenoxy herbicides 2,4-dichlorophenoxyacetic acid (2,4-D) and 2,4,5-trichlorophenoxyacetic acid (2,4,5-T). Approximately 3 million Americans served in the armed forces in Vietnam during the Vietnam War. Some of them (as well as some Vietnamese combatants and civilians, and members of the armed forces of other nations) were exposed to defoliant mixtures, including Agent Orange. There was evidence on cancer risk of Vietnam veterans, workers occupationally exposed to herbicides or dioxins (since dioxins contaminated the herbicide mixtures used in Vietnam), and of the Vietnamese population.

Pesticides and Human Health

Pesticides have been linked to a wide range of human health hazards, ranging from short-term impacts such as headaches and nausea to chronic impacts like cancer, reproductive harm, and endocrine disruption.

Acute dangers - such as nerve, skin, and eye irritation and damage, headaches, dizziness, nausea, fatigue, and systemic poisoning - can sometimes be dramatic, and even occasionally fatal.

Chronic health effects may occur years after even minimal exposure to pesticides in the environment, or result from the pesticide residues which we ingest through our food and water. A July 2007 study conducted by researchers at the Public Health Institute, the California Department of Health Services, and the UC Berkeley School of Public Health found a sixfold increase in risk factor for autism spectrum disorders (ASD) for children of women who were exposed to organochlorine pesticides.

Pesticides can cause many types of cancer in humans. Some of the most prevalent forms include leukemia, non-Hodgkins lymphoma, brain, bone, breast, ovarian, prostate, testicular and liver cancers. In February 2009, the Agency for Toxic Substances and Disease Registry published a study that found that children who live in homes where their parents use pesticides are twice as likely to develop brain cancer versus those that live in residences in which no pesticides are used.

Studies by the National Cancer Institute found that American farmers, who in most respects are healthier than the population at large, had startling incidences of leukemia, Hodgkins disease, non-Hodgkins lymphoma, and many other forms of cancer.

There is also mounting evidence that exposure to pesticides disrupts the endocrine system, wreaking havoc with the complex regulation of hormones, the reproductive system, and embryonic development. Endocrine disruption can produce infertility and a variety of birth defects and developmental defects in offspring, including hormonal imbalance and incomplete sexual development, impaired brain development, behavioral disorders, and many others. Examples of known endocrine disrupting chemicals which are present in large quantities in our environment include DDT (which still persists in abundance more than 20 years after being banned in the U.S.), lindane, atrazine, carbaryl, parathion, and many others.

Multiple Chemical Sensitivity (MCS) is a medical condition characterized by the body's inability to tolerate relatively low exposure to chemicals. This condition, also referred to as Environmental Illness, is triggered by exposure to certain chemicals and/or environmental pollutants. Exposure to pesticides is a common way for individuals to develop MCS, and once the condition is present, pesticides are often a potent trigger for symptoms of the condition. The variety of these symptoms can be dizzying, including everything from cardiovascular problems to depression to muscle and joint pains. Over time, individuals suffering from MCS will begin to react adversely to substances that formerly did not affect them.

For individuals suffering from MCS, the only way to relieve their symptoms is to avoid those substances that trigger adverse reactions. For some individuals, this can mean almost complete isolation from the outside world.

Pesticides and Children

Children are particularly susceptible to the hazards associated with pesticide use. There is now considerable scientific evidence that the human brain is not fully formed until the age of 12, and childhood exposure to some of the most common pesticides on the market may greatly impact the development of the central nervous system. Children have more skin surface for their size than adults, absorb proportionally greater amounts of many substances through their lungs and intestinal tracts, and take in more air, food and water per pound than adults. Children have not developed their immune systems, nervous systems, or detoxifying mechanisms completely, leaving them less capable of fighting the introduction of toxic pesticides into their systems.

Many of the activities that children engage in - playing in the grass, putting objects into their mouth and even playing on carpet - increase their exposure to toxic pesticides. The combination of likely increased exposure to pesticides and lack of bodily development to combat the toxic effects of pesticides means that children are suffering disproportionately from their impacts.

Pesticides and the Environment

Since the publication of Rachel Carson's landmark 1962 book Silent Spring, the impacts of pesticides on the environment have been well known. Pesticides are toxic to living organisms. Some can accumulate in water systems, pollute the air, and in some cases have other dramatic environmental effects. Scientists are discovering new threats to the environment that are equally disturbing.

Pesticide use can damage agricultural land by harming beneficial insect species, soil microorganisms, and worms which naturally limit pest populations and maintain soil health;

Weakening plant root systems and immune systems; Reducing concentrations of essential plant nutrients in the soil such nitrogen and phosphorous.

The Myth of Safety: A Failed Regulatory System

Despite what government agencies and corporations tell you, pesticide products currently on the market are not safe, even when they are used legally. There are many flaws in the way that pesticides are registered and in our political process that allows corporations to influence pesticide policy to allow the continued use of their poisonous products.

Even if we know that a pesticide causes severe health and environmental impacts, including cancer and genetic damage, it may still be allowed for use. The EPA may determine that a cancer-causing chemical may be used despite its public health hazard if its "economic, social or environmental" benefits are deemed greater than its risk. According to the US EPA, more than 70 active ingredients known to cause cancer in animal tests are allowed for use. Although industry tests for a wide range of environmental and health impacts, the vast majority of pesticides currently on the market have not been fully tested.

Pesticides often contain inert ingredients in addition to the active ingredients that are designed to kill the target pest. Unfortunately, the public is not provided information about what inert ingredients are included in pesticides in most cases. At least 382 of the chemicals that the U.S. EPA lists as inert ingredients were once or are currently also registered as pesticide active ingredients. This means that the public is kept in the dark about the contents of pesticide products that may be hazardous. Among the ingredients that are listed as both inert and active ingredients are chloropicrin, which has been linked to asthma and pulmonary edema, and chlorothanonil, a probable human carcinogen.

IMPACT THROUGH FOOD COMMODITIES

For determining the extent of pesticide contamination in the food stuffs, programmes entitled 'Monitoring of Pesticide Residues in Products of Plant Origin in the European Union' started to be established in the European Union since 1996. In 1996, seven pesticides (acephate, chlopyriphos, chlopyriphos-methyl, methamidophos, iprodione, procymidone and chlorothalonil) and two groups of pesticides (benomyl group and maneb group, *i.e.* dithiocarbamates) were analysed in apples, tomatoes, lettuce, strawberries and grapes. An average of about 9 700 samples has been analysed for each pesticide or pesticide group.

For each pesticide or pesticide group, 5.2 per cent of the samples were found to contain residues and 0.31 per cent had residues higher than the respective MRL for that specific pesticide. Lettuce was the crop with the highest number of positive results, with residue levels exceeding the MRLs more frequently than in any of the other crops investigated.

The highest value found in 1996 was for a compound of the maneb group in lettuce which corresponded to a mancozeb residue of 118 mg/kg. In 1997, 13 pesticides (acephate, carbendazin, chlorothalonil, chlopyriphos, DDT, diazinon, endosulfan, methamidophos, iprodione, metalaxyl, methidathion, thiabendazole, triazophos) were assessed in five commodities (mandarins, pears, bananas,beans, and potatoes). Some 6 000 samples were analysed. Residues of chlorpyriphos exceeded MRLs most often (0.24 per cent), followed by methamidophos (0.18 per cent), and iprodione (0.13 per cent).

With regard to the commodities investigated, around 34 per cent contained pesticide residues at or below the MRL, and 1 per cent contained residues at levels above the MRL. In mandarins, pesticide residues were most frequently found at levels at or below the MRL (69 per cent), followed by bananas (51 per cent), pears (28 per cent), beans (21 per cent) and potatoes (9 per cent). MRLs were exceeded most often in beans (1.9 per cent), followed by mandarins (1.8 per cent), pears (1.3 per cent), and bananas and potatoes (0.5 per cent). Estimation of the dietary intake of pesticide residues (based on the 90th percentile) from the above-mentioned commodities, where the highest residue levels of the respective pesticides were found, shows that there is no exceeding of the ADI with all the pesticides and commodities studied (European Commission, 1999).

In 1998, four commodities (oranges, peaches, carrots, spinach) were analysed for 20 pesticides (acephate, benomyl group, chlopyriphos, chlopyriphos-methyl, deltamethrin, maneb group, diazinon, endosulfan, methamidophos, iprodione, metalaxyl, methidathion, thiabendazole, triazophos, permethrin, vinclozolin, lambdacyalothrin, pirimiphos-methyl, mercabam). With regard to all four commodities investigated in 1998 (oranges, peaches, carrots, spinach), about 32 per cent contained residues of pesticides at or below MRL, and 2 per cent above the MRL (1.8 per cent for EU-MRLs, 0.4 per cent for national MRLs). Residues at or below the MRL were found most often in oranges (67 per cent), followed by peaches (21 per cent), carrots (11 per cent) and spinach (5 per cent). MRL values were exceeded most often in spinach (7.3 per cent), followed by peaches (1.6 per cent), carrots (1.2 per cent)and oranges (0.7 per cent).

The intake of pesticide residues has not exceeded the ADI in any case. It was found to be below 10 per cent of the ADI for all pesticides. The exposure ranges from 0.35 per cent of the ADI for the benomyl group to 9.9 per cent of the ADI for the methidathion group. In 1999, four commodities (cauliflower, peppers, wheat grains, and melon) were analysed for the same 20 pesticides as in the 1998 study.

Overall, around 4700 samples were analysed. Residues of methamidophos exceeded MRLs most often (8.7 per cent), followed by the maneb group (1.1 per cent), thiabendazole (0.57 per cent), acephate (0.41 per cent) and the benomyl group (0.35 per cent). The MRL for methamidophos was exceeded most often in peppers and melons (18.7 and 3.7 per cent, respectively). The residues of the maneb group exceeded the MRL most often in cauliflower (3.9 per cent); residues of thiabendazole exceeded the MRL most often in melons (2.8 per cent of the melon samples). With regard to all the commodities investigated, around 22 per cent of samples contained residues of pesticides at or below the MRL and 8.7 per cent above the MRL. Residues at or below MRL were found most often in melons (32 per cent), followed by peppers (24 per cent), wheat grains (21 per cent) and cauliflower (17 per cent). MRL values were exceeded most often in peppers (19 per cent), followed by melons (6.1 per cent), cauliflower (3 per cent) and wheat grains (0.5 per cent). The intake of pesticide residues did not exceed the ADI in any case. It was below 1.5 per cent of the ADI for all pesticides. The exposure ranged between 0.43 per cent of the ADI for methamidophos and 1.4 per cent of the ADI for endosulfan.

The intakes for the highest residue levels in a composite sample for chlorpyriphos, deltamethrin, endosulfan and methidathion were below the ARfD for adults. They range between 1.5 per cent of the ARfD for deltamethrin and 67 per cent of the ARfD for endosulfan. In spite of food contamination, most pesticide deaths recorded in hospital surveys are the result of self-poisoning. The Global Burden of Disease Study 6 estimated that 798 000 people died from deliberate self-harm in 1990, over 75 per cent of whom were from developing countries. More recent WHO estimates showed that over 500 000 people died from self-harm in Southeast Asia and the Western Pacific during 2000 alone. Suicide is the commonest cause of death in young Chinese women and Sri Lankan men and women.

In India the first report of poisoning due to pesticides was from Kerala in 1958, where over 100 people died after consuming wheat flour contaminated with parathion. This prompted the Special Committee on Harmful Effects of Pesticides constituted by the ICAR to focus attention on the problem. In a multi-centric study to assess the pesticide residues in selected food commodities collected from different states of the country, DDT residues were found in about 82 per cent of the 2205 samples of bovine milk collected from 12 states. About 37 per cent of the samples contained DDT residues above the tolerance limit of 0.05 mg/kg (whole milk basis). The highest level of DDT residues found was 2.2 mg/kg.

The proportion of the samples with residues above the tolerance limit was highest in Maharastra (74 per cent), followed by Gujarat (70 per cent), Andhra Pradesh (57 per cent), Himachal Pradesh (56 per cent), and Punjab (51 per cent). In the remaining states, this proportion was less than 10 per cent. Data on 186 samples of 20 commercial brands of infants formulae showed the presence of residues of DDT and HCH isomers in about 70 and 94 per cent of the samples with their maximum level of 4.3 and 5.7 mg/kg (fat basis) respectively. Measurement of chemicals in the total diet provides the best estimates of human exposure and of the potential risk. The risk of consumers may then be evaluated by comparison with toxicologically acceptable intake levels. The average total DDT and BHC consumed by an adult were 19.24 mg/day and 77.15 mg/day respectively. Fatty food was the main source of these contaminants. In another study, the average daily intake of HCH and DDT by Indians was reported to be 115 and 48 mg per person respectively, which were higher than those observed in most of the developed countries.

Impact on Environment

Pesticides can contaminate soil, water, turf, and other vegetation. In addition to killing insects or weeds, pesticides can be toxic to a host of other organisms including birds, fish, beneficial insects, and non-target plants. Insecticides are generally the most acutely toxic class of pesticides, but herbicides can also pose risks to non-target organisms.

Surface Water Contamination

Pesticides can reach surface water through run-off from treated plants and soil. Contamination of water by pesticides is widespread. The results of a comprehensive set of studies done by the U.S. Geological Survey (USGS) on major river basins across the country in the early to mid- 90s yielded startling results.

More than 90 percent of water and fish samples from all streams contained one, or more often, several pesticides. Pesticides were found in all samples from major rivers with mixed agricultural and urban land use influences and 99 percent of samples of urban streams. The USGS also found that concentrations of insecticides in urban streams commonly exceeded guidelines for protection of aquatic life. Twenty-three pesticides were detected in waterways in the Puget Sound Basin, including 17 herbicides. According to USGS, more pesticides were detected in urban streams than in agricultural streams.

The herbicides 2,4-D, diuron, and prometon, and the insecticides chlorpyrifos and diazinon, all commonly used by urban homeowners and school districts, were among the 21 pesticides detected most often in surface and ground water across the nation. Trifluralin and 2,4-D were found in water samples collected in 19 out of the 20 river basins studied. The USGS also found that concentrations of insecticides in urban streams commonly exceeded guidelines for protection of aquatic life.

According to USGS, "in general more pesticides were detected in urban streams than in agricultural streams". The herbicide 2,4-D was the most commonly found pesticide, detected in 12 out of 13 streams. The insecticide diazinon, and the weed-killers dichlobenil, diuron, triclopyr, and glyphosate were detected also in Puget Sound basin streams. Both diazinon and diuron were found at levels exceeding concentrations recommended by the National Academy of Sciences for the protection of aquatic life.

Ground Water Contamination

Groundwater pollution due to pesticides is a worldwide problem. According to the USGS, at least 143 different pesticides and 21 transformation products have been found in ground water, including pesticides from every major chemical class. Over the past two decades, detections have been found in the ground water of more than 43 states. During one survey in India, 58 per cent of drinking water samples drawn from various hand pumps and wells around Bhopal were contaminated with Organo Chlorine pesticides above the EPA standards. Once ground water is polluted with toxic chemicals, it may take many years for the contamination to dissipate or be cleaned up. Cleanup may also be very costly and complex, if not impossible.

Soil Contamination

A large number of transformation products (TPs) from a wide range of pesticides have been documented. Not many of all possible pesticide TPs have been monitored in soil, showing that there is a pressing need for more studies in this field. Persistency and movement of these pesticides and their TPs are determined by some parameters, such as water solubility, soil-sorption constant (K_{oc}), the octanol/water partition coefficient (K_{ow}), and half-life in soil (DT_{50}). Pesticides and TPs could be grouped into:

- Hydrophobic, persistent, and bioaccumulable pesticides that are strongly bound to soil. Pesticides that exhibit such behaviour include the organochlorine DDT, endosulfan, endrin, heptachlor, lindane and their TPs. Most of them are now banned in agriculture but their residues are still present.
- Polar pesticides are represented mainly by herbicides but they include also carbamates, fungicides and some organophosphorus insecticide TPs.

They can be moved from soil by run-off and leaching, thereby constituting a problem for the supply of drinking water to the population. The most researched pesticide TPs in soil are undoubtedly those from herbicides. Several metabolic pathways have been suggested, involving transformation through hydrolysis, methylation, and ring cleavage that produce several toxic phenolic compounds.

The pesticides and their TPs are retained by soils to different degrees, depending on the interactions between soil and pesticide properties. The most influential soil characteristic is the organic matter content. The larger the organic matter content, the greater the adsorption of pesticides and TPs. The capacity of the soil to hold positively charged ions in an exchangeable form is important with paraquat and other pesticides that are positively charged. Strong mineral acid is required for extracting these chemicals, without any analytical improvement or study reported in recent years. Soil pH is also of some importance. Adsorption increases with decreasing soil pH for ionizable pesticides (*e.g.* 2,4-D,2,4,5-T, picloram, and atrazine).

EFFECT ON SOIL FERTILITY (BENEFICIAL SOIL MICROORGANISMS)

Heavy treatment of soil with pesticides can cause populations of beneficial soil microorganisms to decline. According to the soil scientist Dr. Elaine

Ingham, "If we lose both bacteria and fungi, then the soil degrades. Overuse of chemical fertilizers and pesticides have effects on the soil organisms that are similar to human overuse of antibiotics. Indiscriminate use of chemicals might work for a few years, but after awhile, there aren't enough beneficial soil organisms to hold onto the nutrients" (Savonen,1997). For example, plants depend on a variety of soil microorganisms to transform atmospheric nitrogen into nitrates, which plants can use. Common landscape herbicides disrupt this process: triclopyr inhibits soil bacteria that transform ammonia into nitrite (Pell *et al.*, 1998); glyphosate reduces the growth and activity of free-living nitrogen-fixing bacteria in soil (Santos and Flores, 1995) and 2,4-D reduces nitrogen fixation by the bacteria that live on the roots of bean plants (Arias and Fabra, 1993; Fabra *et al.*, 1997), reduces the growth and activity of nitrogen-fixing blue-green algae (Singh and Singh, 1989; Tözüm-Çalgan and Sivaci-Güner, 1993), and inhibits the transformation of ammonia into nitrates by soil bacteria (Frankenberger *et al.*, 1991, Martens and Bremner, 1993).

Mycorrhizal fungi grow with the roots of many plants and aid in nutrient uptake. These fungi can also be damaged by herbicides in the soil. One study found that oryzalin and trifluralin both inhibited the growth of certain species of mycorrhizal fungi (Kelley and South, 1978). Roundup has been shown to be toxic to mycorrhizal fungi in laboratory studies, and some damaging effects were seen at concentrations lower than those found in soil following typical applications (Chakravarty and Sidhu, 1987; Estok *et al.*, 1989). Triclopyr was also found to be toxic to several species of mycorrhizal fungi (Chakravarty and Sidhu,1987) and oxadiazon reduced the number of mycorrhizal fungal spores (Moorman, 1989).

THE EFFECTS OF SOIL FERTILITY

Many farmers and gardeners have observed that plant diseases have become more prevalent and soils less fertile than they used to be. Some have argued that the two phenomena are related and that diseases are more destructive because the plants have been weakened by poor mineral nutrition. A few have even maintained that there would be no serious disease problems if plants were grown in properly conditioned soil, but evidence obtained in hundreds of experiments throughout the world does not uphold this extreme viewpoint.

Soil fertility does affect the prevalence and severity of some plant diseases, but it is only one of several factors that predispose plants to infection by fungi, bacteria, viruses, and nematodes. We can make no sweeping generalizations about the effect of fertilizers on diseases because of the extreme differences in crop plants, their special nutritional requirements. the soil types upon which they are grown, and the diversity of the pathogens that attack them. Some diseases are severe on weakened, undernourished plants. Many others are most destructive when plants are growing vigourously.If wheat on a moderately fertile soil, for example, is given an extra supply of nitrogen it probably will escape seedling diseases more readily, be more subject to pythium root rot, suffer less from take-all disease, and be more subject to infection by leaf rust and powdery mildew. Phosphorus and potassium fertilizers would have an entirely different effect on those same diseases. The addition of barnyard manure to wilt-infested cotton-fields will reduce the amount of wilt in places in Arkansas where potassium and nitrogen are deficient but will increase the severity of the same disease on the delta of the Nile, where nitrogen is readily available.

Such effects of soil fertility on plant diseases must be understood in this age when fertilizer practices are being changed so rapidly. More than 18 million tons of commercial fertilizer and 25 million tons of agricultural lime were used in 1948. That is about twice as much as was required before the Second World War. Less barnyard manure is being added to the soil each year. Not enough green manure from cover crops is being plowed under to maintain the organic matter content of soils on most of our farms. Under such conditions, both the balance of the various nutrient elements and the total supply of nutrients available to the plant can be expected to fluctuate appreciably from year to year and during the growing season.

The Primary Consideration in fertilizing soil is to use such materials, in combination with suitable crop rotations and other soil management practices, as are necessary to promote the maximum productivity of the plant. Disease control is a secondary consideration beyond the fact that one must avoid certain conditions, such as an excess of nitrogen or other available nutrients, a deficiency of Potassium, or changes in soil reaction that affect diseases of a crop. Fertilizer materials that leave acid residues (such as those from ammonium sulfate, potassium sulfate, sulfur, and calcium sulfate) should be employed where neutral or alkaline soils favour diseases like potato scab.

Sodium nitrate, calcium phosphate, limestone, and similar materials that leave alkaline residues should be used where diseases such as club-root of cabbage and some wilt diseases are suppressed by alkaline soils.

Residues from organic matter are valuable for stimulating the growth of beneficial micro-organisms in the soil. They may destroy or prevent the growth of some plant parasites that are not well adjusted to living in the soil environment. When diseases become a serious problem on properly nourished, productive plants, other control measures such as spraying, crop rotation, or use of disease-escaping varieties will have to be employed. There is no sound reason for starving the plant into an unproductive state in order to escape disease. If the plant is properly nourished and capable of full development, the disease control measures, such as spraying or soil disinfestation, are fully justified. They ire a form of crop insurance. They pay the largest dividends on productive soils. Often they are not worth applying to weak and undernourished plants. Several plant diseases are influenced so seriously by soil deficiencies that much of the damage from them can be avoided by soil treatment. The outstanding examples, which are discussed here, are take-all disease of wheat, wheat root rot, Texas root rot of cotton, sugar beet seedling diseases, fusariam wilt of cotton, wildfire disease of tobacco, clubroot of cabbage, common scab of potatoes, bacterial leaf spot of peaches, and powdery mildew and rusts of cereals.

A deficiency of any plant nutrient may influence disease development. Nitrogen, phosphorus, and potassium, the primary elements, are mentioned most often. The secondary elements, calcium, sulfur, silicon, manganese, and boron also have been observed to exert appreciable influence on the prevalence of plant diseases.

Nitrogen is supplied in the form of sodium nitrate (Chilean nitrate), ammonium sulfate, urea, organic nitrogen, or anhydrous ammonia. It promotes vigourous growth and is essential for production of amino acids, growth regulants, and new protoplasm. Used to excess, it encourages rank, vegetative growth, delays maturity, and tends to cause thin cell walls. Fungi may penetrate the thin walls more readily than normal ones. Infected plants collapse more easily. Cereal plants lodge. Lesions on leaves elongate rapidly. Because nitrogen in a proper combination with other nutrients often speeds up growth of seedlings and roots, the plants escape severe damage from pathogens that develop slowly. But because nitrogen may prolong vegetative growth, leaves are exposed to infection over a longer season. The roots, water-conducting tissues, leaves, and fruits of plants that are supplied with nitrogen are more nutritious to most pathogens, which grow better in them than in nitrogen-deficient plants.

Phosphorus is applied as rock phosphate, superphosphate, ammonium phosphate, basic slag, or bonemeal. Phosphorus is essential for utilization of carbohydrates and for cell division because it combines with carbohydrate materials to form nucleic acids. Adequate supplies of phosphorus promote root growth and seed development. Applications of it are most beneficial against seedling diseases and certain root rots where vigourous development of roots permits the plants to escape destruction. It is essential for multiplication of viruses in host cells and may increase susceptibility to viruses and other disease agents if too abundant. Because of its use in building new cells, any imbalance with nitrogen may cause disease losses to increase.

Potassium (potash) is supplied to soils as potassium chloride (Kainit, muriate of potash), potassium sulfate, potassium nitrate, or wood ashes. Unlike other essential nutrients, it does not become a structural part of the plant cell. It is a mobile regulator of cell activity and promotes the reduction of nitrates and the synthesis of amino acids from carbohydrate and inorganic nitrogen. Potassium promotes the development of thicker outer walls in the epidermal cells and firm tissues, which are less subject to collapse. A deficiency of potassium enforces the accumulation of carbohydrates and inorganic nitrogen in the plant. Eventually it retards photosynthesis and production of new tissues. More plant diseases have been retarded by use of potash fertilizers than any other substance, perhaps because potassium is so essential for catalyzing cell activities. It is unavailable in many soils of light texture because its salts are so water-soluble they readily leach from the soil.

Calcium is added to the soil as ground limestone, hydrated lime, gypsum, or calcium phosphate. It is essential for normal growth since it regulates chromosome development in cell division and is assimilated into the middle lamellae of new cell walls. It is therefore important for cell division and cell development. It also may neutralize acid byproducts of cell metabolism that could become harmful if not precipitated in an insoluble condition. Calcium also influences plant diseases indirectly by its effect on soil acidity, by neutralizing toxins produced by wilt-inducing fungi, and by affecting cell division in those diseases where abnormal growth of tissues is important. Its balance with potassium becomes a primary consideration in gall development because both materials contribute to the growth and division of cells.

Silicon affects the availability of potassium. It also may be combined with other materials to give the cell walls greater structural strength. The primary effect on plant diseases apparently is in the prevention of infection by powdery mildew, a disease in which the fungus develops externally and usually penetrates the host cell through the outer wall. Sulfur is oxidized to sulfates and thereby promotes soil acidity, which discourages growth and survival of some bacteria and fungi.

Soil reaction (the hydrogen-ion concentration, expressed in terms of pH units) influences the growth and persistence of some fungi and bacteria. The parasitic organisms that depend on delicate bodies, such as thin-walled swarmspores, may have difficulty in multiplying and infecting roots in soils at unfavourable hydrogen-ion concentrations. Hydrogen-ion concentration may change the ability of pine tree roots to resist invasion, but that, if true, is an exceptional case. Soil reaction also may affect the availability of essential nutrients in the soil and the biological balance between plant parasites and saprophytic soil-inhabiting fungi and bacteria.

Although organic matter contributes essential nutrients to the crop, it probably exerts more influence through the physical and biological changes chat it brings about in the soil. The carbohydrates and proteins in animal and plant byproducts provide nourishment to soil organisms that fix nitrogen from the air, tie up available nitrates in the soil, and frequently suppress plant parasites by antibiotic activity.

The humus from lignified plant tissues and other residual products promotes the massing together of soil particles and thereby improves aeration and water-holding capacity of soils. Organic matter may promote the growth of plant parasites or even facilitate their dispersal. Some of the decomposition products may increase the susceptibility of roots to invasion. Young Seedlings of most crops are attacked by fungi that live in the soil or are carried on the seed.

CONTAMINATION OF AIR, SOIL, AND NON-TARGET VEGETATION

Pesticide sprays can directly hit non-target vegetation, or can drift or volatilize from the treated area and contaminate air, soil, and non-target plants. Some pesticide drift occurs during every application, even from ground equipment (Glotfelty and Schomburg, 1989). Drift can account for a loss of 2 to 25 per cent of the chemical being applied, which can spread over a distance of a few yards to several hundred miles. As much as 80–90 per cent of an applied pesticide can be volatilised within a few days of application (Majewski,1995). Despite the fact that only limited research has been done on the topic, studies consistently find pesticide residues in air. According to the USGS, pesticides have been detected in the atmosphere in all sampled areas of the USA (Savonen, 1997).

Nearly every pesticide investigated has been detected in rain, air, fog, or snow across the nation at different times of the year (U.S. Geological Survey, 1999). Many pesticides have been detected in air at more than half the sites sampled nationwide. Herbicides are designed to kill plants, so it is not surprising that they can injure or kill desirable species if they are applied directly to such plants, or if they drift or volatilise onto them. Many ester-formulation herbicides have been shown to volatilise off treated plants with vapors sufficient to cause severe damage to other plants (Straathoff, 1986). In addition to killing non-target plants outright, pesticide exposure can cause sublethal effects on plants.

Phenoxy herbicides, including 2,4-D, can injure nearby trees and shrubs if they drift or volatilise onto leaves (Dreistadt *et al.*, 1994). Exposure to the herbicide glyphosate can severely reduce seed quality (Locke *et al.*, 1995). It can also increase the susceptibility of certain plants to disease (Brammall and Higgins, 1998).

This poses a special threat to endangered plant species. The U.S. Fish and Wildlife Service has recognized 74 endangered plants that may be threatened by glyphosate alone (U.S. EPA Office of Pesticides and Toxic Substances, 1986). Exposure to the herbicide clopyralid can reduce yields in potato plants (Lucas and Lobb, 1987). EPA calculated that volatilisation of only 1 per cent of applied clopyralid is enough to damage non-target plants (US EPA, 1990). Some insecticides and fungicides can also damage plants (Dreistadt *et al.*, 1994). Pesticide damage to plants is commonly reported to state agencies in the Northwest. (Oregon Dept. of Agriculture, 1999; Washington Dept. of Health, 1999).

Plants can also suffer indirect consequences of pesticide applications when harm is done to soil microorganisms and beneficial insects. Pesticides including those of new the generation, *e.g.*, dacthal, chlorothalonil, chlorpyrifos, metolachlor, terbufos and trifluralin have been detected in Arctic environmental samples (air, fog, water, snow) (Rice and Cherniak, 1997), and (Garbarino *et al.*, 2002).

Other studies have identified the ability of some of these compounds to undergo short-range atmospheric transport (Muir *et al.*, 2004) to ecologically sensitive regions such as the Chesapeake Bay and the Sierra Nevada mountains (LeNoir *et al.*, 1999; McConnell *et al.*, 1997; Harman-Fetcho *et al.*,2000, Thurman and Cromwell, 2000). One long-term study that investigated pesticides in the atmosphere of British Columbia (BC), dating from 1996 (Belzer *et al.*, 1998) showed that 57 chemicals were investigated at two sampling sites (Agassiz and Abbotsford) in the Fraser Valley, from February 1996 until March 1997.

Atrazine, malathion, and diazinon, highly toxic chemicals identified as high-priority pesticides by Verrin *et al.* (2004), were detected as early as the end of February (72 pg/m^3) until mid-October (253 pg/m^3), with a peak concentration in mid-June of 42.7 ngm”^{3}.

Dichlorvos is a decomposition product of another pesticide, Naled (Dibrom) (Hall *et al.*, 1997). Captan and 2,4-D showed the highest concentrations and deposition rates at these two sites, followed by dichlorvos and diazinon (Dosman and Cockcraft, 1989). Air concentrations of currently used pesticides in Alberta were investigated in 1999 at four sampling sites that were chosen according to geography and pesticide sales data (Kumar, 2001).

Triallate and trifluralin were the two mostly detected pesticides at the four sites. Insecticides (malathion, chlorpyrifos, diazinon and endosulfan) were detected intermittently with concentrations in the range 20–780 pg/m^3. South of Regina, Saskatchewan, in 1989 and 1990, 2,4-D reached 3.9 and 3.6 ng/m^3 at the end of June (Waite *et al.*, 2002a).

Triallate, dicamba, bromoxynil concentrations were also higher in 1989 (peak concentration of 4.2 ng/m^3 in mid-June) compared with 1990 (600–700 pg/m^3 in mid-June). In a more recent study, Waite *et al.* (2005) studied spatial variations of selected herbicides on a threesite, 500km transect that included two agricultural sites—Bratt's Lake, located 35 km southwest of Regina and Hafford to the North—and a background site at Waskesiu.

Some acid herbicides were also investigated in South Tobacco Creek, Manitoba during 1993–1996. Once again, maximum concentrations occurred during periods of local use (Rawn *et al.*, 1999a).

A neutral herbicide, atrazine, was also investigated in 1995 (Rawn *et al.*, 1998). It was first detected in mid-April, peaked mid- June at about 300 pg/m^3, and was detected until the end of October. The insecticide dacthal was identified throughout the sampling periods in 1994, 1995 and 1996 (Rawn and Muir, 1999) even though it was not used in this area (<20–300 pg/m^3).

Non-target Organisms

Pesticides are found as common contaminants in soil, air, water and on non-target organisms in our urban landscapes. Once there, they can harm plants and animals ranging from beneficial soil microorganisms and insects, non-target plants, fish, birds, and other wildlife. Chlorpyrifos, a common contaminant of urban streams (U.S. Geological Survey, 1999), is highly toxic to fish, and has caused fish, kills in waterways near treated fields or buildings (US EPA, 2000). Herbicides can also be toxic to fish. According to the EPA, studies show that trifluralin, an active ingredient in the weed-killer Snapshot, "is highly to very highly toxic to both cold and warm water fish" (U.S. EPA, 1996). In a series of different tests it was also shown to cause vertebral deformities in fish (Koyama, 1996).

The weed-killers Ronstar and Roundup are also acutely toxic to fish (Folmar *et al.*, 1979; Shafiei and Costa, 1990). The toxicity of Roundup is likely due to the high toxicity of one of the inert ingredients of the product (Folmar *et al.*, 1979). In addition to direct acute toxicity, some herbicides may produce sublethal effects on fish that lessen their chances for survival and threaten the population as a whole. Glyphosate or glyphosate-containing products can cause sublethal effects such as erratic swimming and labored breathing, which increase the fish's chance of being eaten (Liong *et al.*, 1988). 2,4-D herbicides caused physiological stress responses in sockeye salmon (McBride *et al.*, 1981) and reduced the food-gathering abilities of rainbow trout (Little, 1990).

Several cases of pesticide poisoning of dolphins have been reported worldwide. Because of their high trophic level in the food chain and relatively low activities of drug-metabolising enzymes, aquatic mammals such as dolphins accumulate increased concentrations of persistent organic pollutants (Tanabe *et al.*, 1988) and are thereby vulnerable to toxic effects from contaminant exposures. Dolphins inhabiting riverine and estuarine ecosystems are particularly vulnerable to the activities of humans because of the restricted confines of their habitat, which is in close proximity to point sources of pollution.

River dolphins are among the world's most seriously endangered species. Populations of river dolphins have been dwindling and face the threat of extinction; the Yangtze river dolphin (Lipotes vexillifer) in China and the Indus river dolphin (Platanista minor) in Pakistan are already close to extinction (Renjun, 1990; Perrin *et al.*, 1989; Reeves *et al.*, 1991; Reeves and Chaudhry,1998). In addition to habitat degradation (such as construction of dams) (Reeves and Leatherwood,1994), boat traffic, fishing, incidental and intentional killings, and chemical pollution have been threats to the health of river dolphins (Kannan *et al.*, 1993b, 1994, 1997; Senthilkumar *et al.*, 1999). Earlier studies reported concentrations of heavy metals (Kannan *et al.*, 1993), organochlorine pesticides and polychlorinated biphenyls (PCBs) (Kannan *et al.*, 1994), and butyltin compounds (Kannan *et al.*, 1997) in Ganges river dolphins and their prey. The continuing use of organochlorine pesticides and PCBs in India is of concern (Kannan *et al.*, 1992; Kannan *et al.*, 1997a; Kannan *et al.*, 1997b; Tanabe *et al.*,1998). The Ganges river basin is densely populated and heavily polluted by fertilizers, pesticides, and industrial and domestic effluents (Mohan, 1989). In addition to fish, other marine or freshwater animals are endangered by pesticide contamination. Exposure to great concentrations of persistent, bioaccumulative, and toxic contaminants such as DDT (1,1,1-trichloro-2,2-bis[*p*-chlorophenyl]ethane) and PCBs has been shown to elicit adverse effects on reproductive and immunological functions in captive or wild aquatic mammals (Helle *et al.*, 1976; Reijnders, 1986; Ross *et al.*, 1995; Martineau *et al.*,1987; Kannan *et al.*, 1993; Colborn and Smolen, 1996). Aquatic mammals inhabiting freshwater systems, such as otters and mink, have been reported to be sensitive to chemical contamination (Leonards *et al.*, 1995; Leonards *et al.*, 1997). 2,4-D or 2,4-D containing products have been shown to be harmful to shellfish (Cheney *et al.*, 1997) and other aquatic species (U.S. EPA, 1989; Sanders, 1989) The weed-killer trifluralin is moderately to highly toxic to aquatic invertebrates, and highly toxic to estuarine and marine organisms like shrimp and mussels (U.S. EPA, 1996).

Since herbicides are designed to kill plants, it makes sense that herbicide contamination of water could have devastating effects on aquatic plants. In one study, oxadiazon was found to severely reduce algae growth (Ambrosi *et al.*, 1978). Algae is a staple organism in the food chain of aquatic ecosystems. Studies looking at the impacts of the herbicides atrazine and alachlor on algae and diatoms in streams showed that even at fairly low levels, the chemicals damaged cells, blocked photosynthesis, and stunted growth in varying ways (U.S. Water News Online, 2000).

The herbicide oxadiazon is also toxic to bees, which are pollinators (Washington State Department of Transportation, 1993). Herbicides may hurt insects or spiders also indirectly when they destroy the foliage that these animals need for food and shelter. For example spider and carabid beetle populations declined when 2,4-D applications destroyed their natural habitat (Asteraki *et al.*,1992). Non-target birds may also be killed if they ingest poisoned grains set out as bait for pigeons and rodents (US EPA, 1998). Avitrol, a commonly used pigeon bait, poses a large potential for ingestion by non- target grain feeding birds. It can be lethal to small seed-eating birds (Extoxnet, 1996). Brodifacoum, a common rodenticide, is highly toxic to birds. It also poses a secondary poisoning hazard to birds that may feed on poisoned rodents (US EPA, 1998). Herbicides can also be toxic to birds. Although trifluralin was considered "practically non-toxic to birds" in studies of acute toxicity, birds exposed multiple times to the herbicide experienced diminished reproductive success in the form of cracked eggs (U.S. EPA, 1996). Exposure of eggs to 2,4-D reduced successful hatching of chicken eggs (Duffard *et al.*, 1981) and caused feminisation or sterility in pheasant chicks (Lutz *et al.*, 1972). Herbicides can also adversely affect birds by destroying their habitat.

Glyphosate treatment in clear cuts caused dramatic decreases in the populations of birds that lived there (MacKinnon *et al.*, 1993) Effects of some organochlorines (OCs) on fish-eating water birds and marine mammals have been documented in North America and Europe (Barron *et al.*, 1995; Cooke, 1979; Kubiak *et al.*, 1989). Despite the continuing usage, little is known about the impacts of OCs in bird populations in developing countries. Among the countries that continue to use OCs, India has been one of the major producers and consumers in recent years. As a consequence, wild birds in India are exposed to great amounts of OC pesticides (Tanabe *et al.*, 1998). Use of OCs in tropical countries may not only result in exposure of resident birds but also of migratory birds when they visit tropical regions in winter. The Indian sub-continent is a host to a multitude of birds from western Asia, Europe and Arctic Russia in winter(Woodcock, 1980). Hundreds of species of waterfowl, including wading birds such as plovers, terns and sandpipers, migrate each winter to India covering long distances (Grewal, 1990). While concentrations of OC pesticides in wholebody homogenates of birds have been reported elsewhere (Tanabe *et al.*, 1998), concentrations of OCs in prey items and in eggs of Indian birds have not been reported.

A few studies related to the decline in the populations of bats in various parts of the world to OC exposure were also being conducted (Altenbach *et al.*, 1979; Clark, 1976; Clark, 1983; Clark, 1981; Geluso *et al.*, 1976; Jefferies, 1976; Thies and Mc Bee, 1994). The world population of bats was estimated to be 8.7 million during 1936 and it declined to approximately 200,000 in 1973 (Geluso *et al.*, 1976) It has recovered slightly to an estimated number of 700,000 in 1991 (Geluso *et al.*, 1976; Thies and Mc Bee, 1994). High tissue concentrations of *p,p'*-dichlorodiphenyldichloroethene (*p,p'*–DDE) have been found in bats in Carlsbad Caverns in Mexico and in New Mexico in the USA (Geluso *et al.*, 1976; Thies and Mc Bee, 1994).

Occurrence of stillbirths in little brown bats exposed to high concentrations of PCBs,*p,p'*–DDE, and/or oxychlordane was documented (Clark, 1976; Jefferies, 1976). These observations indicate that bats can accumulate high concentrations of OCs and may be affected by their potential toxic effects. The flying fox or the new world fruit bat, short-nosed fruit bat and Indian pipistrelle bat are resident species and are very common in South India. Their habitat is mainly agricultural areas, rock caves, and abandoned houses in domesticated areas. Insects constitute an important diet for many bats, allowing the passage of OCs in their body (Mc Bee *et al.*, 1992). Several studies found OC pesticides and PCBs in livers and eggs of birds in developed countries (Becker, 1989; Bernardz *et al.*, 1990; Cade *et al.*,1989; Castillo *et al.*, 1994; Mora, 1996; Mora, 1997). Similarly, several studies reported OCs in a variety of biota including humans and wildlife from India (Senthilkumar *et al.*, 2000). However, no study has used whole body homogenates of birds, which is important to evaluate biomagnification features and body burdens of OCs (Mc Bee *et al.*, 1992). Earlier studies used specific body tissues to estimate biomagnification of OCs. However theoretically, estimation of biomagnification factors requires whole body concentrations rather than specific tissue concentrations.

THE CONCEPT OF EFTECTIVE MICROORGANISMS: THEIR ROLE AND APPLICATION

The concept of effective microorganisms (EM) was developed by Professor Teruo Higa, University of the Ryukyus, Okinawa, Japan (Higa, 1991; Higa and Wididana, 1991a). EM consists of mixed cultures of beneficial an naturally-occurring microorganisms that can be applied as inoculants to increase the microbial diversity of soils and plant. Research has shown that the inoculation of EM cultures to the soil/plant ecosystem can improve soil quality, soil health, and the growth, yield, and quality of crops. EM contains selected species of microorganisms including predominant populations of lactic acid bacteria and yeasts and smaller numbers of photosynthetic bacteria, actinomycetes and other types of organisms. All of these are mutually compatible with one another and can coexist in liquid culture.

EM is not a substitute for other management practices. It is, however, an added dimension for optimizing our best soil and crop management practices such as crop rotations, use of organic amendments, conservation tillage, crop residue recycling, and biocontrol of pests. If used properly, EM can significantly enhance the beneficial effects of these practices (Higa and Wididana, 1991b).

Throughout the discussion which follows, we will use the term "beneficial microorganisms" In a general way to designate a large group of often unknown or ill-defined microorganisms that interact favourably in soils and with plants to render beneficial effects which are sometimes difficult to predict. We use the term "effective microorganisms" or EM to denote specific mixed cultures of known, beneficial microorganisms that are being used effectively as microbial inoculants.

UTILIZATION OF BENEFICIAL MICROORGANISMS IN AGRICULTURE

Conceptual design is important in developing new technologies for utilizing beneficial and effective microorganisms for a more sustainable agriculture and environment. The basis of a conceptual design is imply to first conceive an ideal or model and then to devise a strategy and method for achieving the reality. However it is necessary to carefully coordinate the materials, the environment, and the technologies constituting the method. Moreover one should adopt a philosophical attitude in applying microbial technologies to agricultural production and conservation systems. There are many opinions on what an ideal agricultural system is. Many would agree that such an idealized system should produce food on a long-term sustainable basis. Many would also insist that it should maintain and improve human health, be economically and spiritually beneficial to both producers and consumers, actively preserve and protect the environment, be self-contained and regenerative, and produce enough food for an increasing world population (Higa, 1991).

Efficient Utilization and Recycling of Energy

Agricultural production begins with the process of photosynthesis by green plants which requires solar energy, water, and carbon dioxide. It occurs through the plants ability to utilize solar energy in "fixing" atmospheric carbon into carbohydrates. The energy obtained is used for further biosynthesis in the plant, including essential amino acids and proteins. The materials used for agricultural production are abundantly available with little initial cost. However, when it is observed as an economic activity, the fixation of carbon by photosynthesis has an extremely low efficiency mainly because of the low utilization rate of solar energy by green plants. Therefore, an integrated approach is needed to increase the level of solar energy utilization by plants so that greater amounts of atmospheric carbon can be converted into useful substrates (Higa and Wididana, 1991a).

Although the potential utilization rate of solar energy by plants has been estimated theoretically at between 10 and 20 per cent, the actual utilization rate is less than 1 per cent. Even the utilization rate of C4 plants, such as sugar cane whose photosynthetic efficiency is very high, barely exceeds 6 or 7 per cent during the maximum growth period. The utilization rate is normally less than 3 per cent even for optimum crop yields.

Past studies have shown that photosynthetic efficiency of the chloroplasts of host crop plants cannot be increased much further; this means that their biomass production has reached a maximum level. Therefore, the best opportunity for increasing biomass production is to somehow utilize the visible light, which chloroplasts cannot presently use, and the infrared radiation; together, these comprise about 80 per cent of the total solar energy. Also, we must explore ways of recycling organic energy contained in plant and animal residues through direct utilization of organic molecules by plants (Higa and Wididana, 1991a). Thus, it is difficult to exceed the existing limits of crop production unless the efficiency of utilizing solar energy is increased, and the energy contained in existing organic molecules (amino acids, peptides and carbohydrates) is utilized either directly or indirectly by the plant. This approach could help to solve the problems of environmental pollution and degradation caused by the misuse and excessive application of chemical fertilizers and pesticides to soils. Therefore, new technologies that can enhance the economic-viability of farming systems with little or no use of chemical fertilizers and pesticides are urgently needed and should be a high priority of agricultural research both now and in the immediate future (National Academy of Sciences, 1989; 1993).

Preservation of Natural Resources and the Environment

The excessive erosion of topsoil from farmland caused by intensive tillage and row-crop production has caused extensive soil degradation and also contributed to the pollution of both surface and groundwater. Organic wastes from animal production, agricultural and marine processing industries, and municipal wastes (e.i., sewage and garbage), have become major sources of environmental pollution in both developed and developing countries. Furthermore, the production of methane from paddy fields and ruminant animals and of carbon dioxide from the burning of fossil fuels, land clearing and organic matter decomposition have been linked to global warming as "greenhouse gases" (Parr and Hornick, 1992b).

Chemical-based, conventional systems of agricultural production have created many sources of pollution that, either directly or indirectly, can contribute to degradation of the environment and destruction of our natural resource base. This situation would change significantly if these pollutants could be utilized in agricultural production as sources of energy.

Therefore, it is necessary that future agricultural technologies be compatible with the global ecosystem and with solutions to such problems in areas different from those of conventional agricultural technologies. An area that appears to hold the greatest promise for technological advances in crop production, crop protection, and natural resource conservation is that of beneficial and effective microorganisms applied as soil, plant and environmental inoculants (Higa, 1995).

Beneficial and Effective Microorganisms for a Sustainable Agriculture

Agriculture in a broad sense, is not an enterprise which leaves everything to nature without intervention. Rather it is a human activity in which the farmer attempts to integrate certain agroecological factors and production inputs for optimum crop and livestock production. Thus, it is reasonable to assume that farmers should be interested in ways and means of controlling beneficial soil microorganisms as an important component of the agricultural environment. Nevertheless, this idea has often been rejected by naturalists and proponents of nature farming and organic agriculture. They argue that beneficial soil microorganisms will increase naturally when organic amendments are applied to soils as carbon, energy and nutrient sources. This indeed may be true where an abundance of organic materials are readily available for recycling which often occurs in small-scale farming. However, in most cases, soil microorganisms, beneficial or harmful, have often been controlled advantageously when crops in various agroecological zones are grown and cultivated in proper sequence (*i.e.*, crop rotations) and without the use of pesticides. This would explain why scientists have long been interested in the use of beneficial microorganisms as soil and plant inoculants to shift the microbiological equilibrium in a way that enhances soil quality and the yield and quality of crops (Higa and Wididana, 1991b; Higa, 1994:1995).

Most would agree that a basic rule of agriculture is to ensure that specific crops are grown according to their agroclimatic and agroecological requirements. However, in many cases the agricultural economy is based on market forces that demand a stable supply of food, and thus, it becomes necessary to use farmland to its full productive potential throughout the year.

The purpose of crop breeding is to improve crop production, crop protection, and crop quality. Improved crop cultivars along with improved cultural and management practices have made it possible to grow a wide variety of agricultural and horticultural crops in areas where it once would not have been culturally or economically feasible. The cultivation of these crops in such diverse environments has contributed significantly to a stable food supply in many countries. However, it is somewhat ironic that new crop cultures are almost never selected with consideration of their nutritional quality or bioavailability after ingestion (Hornick, 1992).

As will be discussed later, crop growth and development are closely related to the nature of the soil microflora, especially those in close proximity to plant roots, *i.e.*, the rhizosphere. Thus, it will be difficult to overcome the limitations of conventional agricultural technologies without controlling soil microorganisms. This particular tenet is further reinforced because the evolution of most forms of life on earth and their environments are sustained by microorganisms. Most biological activities are influenced by the state of these invisible, minuscule units of life. Therefore, to significantly increase food production, it is essential to develop crop cultivars with improved genetic capabilities (*i.e.*, greater yield potential, disease resistance, and nutritional quality) and with a higher level of environmental competitiveness, particularly under stress conditions (*i.e.*, low rainfall, high temperatures, nutrient deficiencies, and agressive weed growth).

To enhance the concept of controlling and utilizing beneficial microorganisms for crop production and protection, one must harmoniously integrate the essential components for plant growth and yield including light (intensity, photoperiodicity and quality), carbon dioxide, water, nutrients (organic-inorganic) soil type, and the soil microflora. Because of these vital interrelationships, it is possible to envision a new technology and a more energy-efficient system of biological production. Low agricultural production efficiency is closely related to a poor coordination of energy conversion which, in turn, is influenced by crop physiological factors, the environment, and other biological factors including soil microorganisms. The soil and rhizosphere microflora can accelerate the growth of plants and enhance their resistance to disease and harmful insects by producing bioactive substances. These microorganisms maintain the growth environment of plants, and may have secondary effects on crop quality. A wide range of results are possible depending on their predominance and activities at any one time. Nevertheless, there is a growing consensus that it is possible to attain maximum economic crop yields of high quality, at higher net returns, without the application of chemical fertilizers and pesticides. Until recently, this was not thought to be a very likely possibility using conventional agricultural methods. However, it is important to recognize that the best soil and crop management practices to achieve a more sustainable agriculture will also enhance the growth, numbers and activities of beneficial soil microorganisms that, in turn, can improve the growth, yield and quality of crops (National Academy of Sciences, 1989; Hornick, 1992; Parr *et al.*, 1992).

CONTROLLING THE SOIL MICROFLORA: PRINCIPLES AND STRATEGIES

The misuse and excessive use of chemical fertilizers and pesticides have often adversely affected the environment and created many a) food safety and quality and b) human and animal health problems. Consequently, there has been a growing interest in nature farming and organic agriculture by consumers and environmentalists as possible alternatives to chemical-based, conventional agriculture.

Agricultural systems which conform to the principles of natural ecosystems are now receiving a great deal of attention in both developed and developing countries. A number of books and journals have recently been published which deal with many aspects of natural farming systems. New concepts such as alternative agriculture, sustainable agriculture, soil quality, integrated pest management, integrated nutrient management and even beneficial microorganisms are being explored by the agricultural research establishment (National Academy of Sciences, 1989; Reganold *et al.*, 1990; Parr *et al.*, 1992). Although these concepts and associated methodologies hold considerable promise, they also have limitations. For example, the main limitation in using microbial inoculants is the problem of reproducibility and lack of consistent results.

Unfortunately certain microbial cultures have been promoted by their suppliers as being effective for controlling a wide range of soil-borne plant diseases when in fact they were effective only on specific pathogens under very specific conditions. Some suppliers have suggested that their particular microbial inoculant is akin to a pesticide that would suppress the general soil microbial population while increasing the population of a specific beneficial microorganism. Nevertheless, most of the claims for these single-culture microbial inoculants are greatly exaggerated and have not proven to be effective under field conditions. One might speculate that if all of the microbial cultures and inoculants that are available as marketed products were used some degree of success might be achieved because of the increased diversity of the soil microflora and stability that is associated with mixed cultures. While this, of course, is a hypothetical example, the fact remains that there is a greater likelihood of controlling the soil microflora by introducing mixed, compatible cultures rather than single pure cultures (Higa, 1991).

Even so, the use of mixed cultures in this approach has been criticized because it is difficult to demonstrate conclusively which microorganisms are responsible for the observed effects, how the introduced microorganisms interact with the indigenous species, and how these new associations affect the soil/plant environment. Thus, the use of mixed cultures of beneficial microorganisms as soil inoculants to enhance the growth, health, yield, and quality of crops has not gained widespread acceptance by the agricultural research establishment because conclusive scientific proof is often lacking.

The use of mixed cultures of beneficial microorganisms as soil inoculants is based on the principles of natural ecosystems which are sustained by their constituents; that is, by the quality and quantity of their inhabitants and specific ecological parameters, *i.e.*, the greater the diversity and number of the inhabitants, the higher the order of their interaction and the more stable the ecosystem. The mixed culture approach is simply an effort to apply these principles to natural systems such as agricultural soils, and to shift the microbiological equilibrium in favour of increased plant growth, production and protection (Higa, 1991; 1994;Parr *et al.*, 1994).

It is important to recognize that soils can vary tremendously as to their types and numbers of microorganisms. These can be both beneficial and harmful to plants and often the predominance of either one depends on the cultural and management practices that are applied. It should also be emphasized that most fertile and productive soils have a high content of organic matter and, generally, have large, populations of highly diverse microorganisms (*i.e.*, both species and genetic diversity). Such soils will also usually have a wide ratio of beneficial to harmful microorganisms (Higa and Wididana, 1991b).

Controlling the Soil Microflora

The idea of controlling and manipulating the soil microflora through the use of inoculants organic amendments and cultural and management practices to create a more favourable soil microbiological environment for optimum crop production and protection is not new. For almost a century, microbiologists have known that organic wastes and residues, including animal manures, crop residues, green manures, municipal wastes (both raw and composted), contain their own indigenous populations of microorganisms often with broad physiological capabilities.

It is also known that when such organic wastes and residues are applied to soils many of these introduced microorganisms can function as biocontrol agents by controlling or suppressing soil-borne plant pathogens through their competitive and antagonistic activities. While this has been the theoretical basis for controlling the soil microflora, in actual practice the results have been unpredictable and inconsistent, and the role of specific microorganisms has not been well-defined.

For, many years microbiologists have tried to culture beneficial microorganisms for use as soil inoculants to overcome the harmful effects of phytopathogenic organisms, including bacteria, fungi and nematodes. Such attempts have usually involved single applications of pure cultures of microorganisms which have been largely unsuccessful for several reasons. First, it is necessary to thoroughly understand the individual growth and survival characteristics of each particular beneficial microorganism, including their nutritional and environmental requirements. Second, we must understand their ecological relationships and interactions with other microorganisms, including their ability to coexist in mixed cultures and after application to soils (Higa, 1991; 1994).

There are other problems and constraints that have been major obstacles to controlling the microflora of agricultural soils. First and foremost is the large number of types of microorganisms that are present at any one time, their wide range of physiological capabilities, and the dramatic fluctuations in their populations that can result from manâ™s cultural and management practices applied to a particular farming system. The diversity of the total soil microflora depends on the nature of the soil environment and those factors which affect the growth and activity of each individual organism including temperature, light, aeration, nutrients, organic matter, pH and water. While there are many microorganisms that respond positively to these factors, or a combination thereof, there are many that do not. Microbiologists have actually studied relatively few of the microorganisms that exist in most agricultural soil, mainly because we don't know how to culture them; *i.e.*, we know very little about their growth, nutritional, and ecological requirements.

The "diversity" and "population" factors associated with the soil microflora have discouraged scientists from conducting research to develop control strategies. Many believe that, even when beneficial microorganisms are cultured and inoculated into soils, their number is relatively small compared with the indigenous soil inhabitants, and they would likely be rapidly overwhelmed by the established soil microflora. Consequently, many would argue that even if the application of beneficial microorganisms is successful under limited conditions (*e.g.*, in the laboratory) it would be virtually impossible to achieve the same success under actual field conditions. Such thinking still exists today, and serves as a principle constraint to the concept of controlling the soil microflora (Higa, 1994).

It is noteworthy that most of the microorganisms encountered in any particular soil are harmless to plants with only a relatively few that function as plant pathogens or potential pathogens. Harmful microorganisms become dominant if conditions develop that are favourable to their growth, activity and reproduction. Under such conditions, soil-borne pathogens (*e.g.*, fungal pathogens) can rapidly increase their populations with devastating effects on the crop. If these conditions change, the pathogen population declines just as rapidly to its original state. Conventional farming systems that tend towards the consecutive planting of the same crop (*i.e.*, monoculture) necessitate the heavy use of chemical fertilizers and pesticides. This, in turn, generally increases the probability that harmful, disease-producing, plant pathogenic microorganisms will become more dominant in agricultural soils (Higa, 1991; 1994; Parr and Hornick, 1994).

Chemical-based conventional farming methods are not unlike symptomatic therapy. Examples of this are applying fertilizers when crops show symptoms of nutrient-deficiencies, and applying pesticides whenever crops are attacked by insects and diseases. In efforts to control the soil microflora some scientists feel that the introduction of beneficial microorganisms should follow a symptomatic approach. However, we do not agree. The actual soil conditions that prevail at any point in time may be most unfavourable to the growth and establishment of laboratory-cultured, beneficial microorganisms. To facilitate their establishment, it may require that the farmer make certain changes in his cultural and management practices to induce conditions that will (a) allow the growth and survival of the inoculated microorganisms and (b) suppress the growth and activity of the indigenous plant pathogenic microorganisms (Higa, 1994; Parr *et al.*, 1994).

An example of the importance of controlling the soil microflora and how certain cultural and management practices can facilitate such control is useful here. Vegetable cultivars are often selected on their ability to grow and produce over a wide range of temperatures. Under cool, temperate conditions there are generally few pest and disease problems. However, with the onset of hot weather, there is a concomitant increase in the incidence of diseases and insects making it rather difficult to obtain acceptable yields without applying pesticides. With higher temperatures, the total soil microbial population increases as does certain plant pathogens such as Fusarium, which is one of the main putrefactive, fungal pathogens in soil.

The incidence and destructive activity of this pathogen can be greatly minimized by adopting reduced tillage methods and by shading techniques to keep the soil cool during hot weather. Another approach is to inoculate the soil with beneficial, antagonistic, antibiotic-producing microorganisms such as actinomycetes and certain fungi (Higa and Wididana, 1991a; 1991b).

APPLICATION OF BENEFICIAL AND EFFECTIVE MICROORGANISMS

Many microbiologists believe that the total number of soil microorganisms can be increased by applying organic amendments to the soil. This is generally true because most soil microorganisms are heterotrophic, *i.e.*, they require complex organic molecules of carbon and nitrogen for metabolism and biosynthesis. Whether the regular addition of organic wastes and residues will greatly increase the number of beneficial soil microorganisms in a short period of time is questionable. However, we do know that heavy applications of organic materials, such as seaweed, fish meal, and chitin from crushed crab shells, not only helps to balance the micronutrient content of a soil but also increases the population of beneficial antibiotic-producing actinomycetes. This changes the soil to a disease-suppressive condition within a relatively short period.

The probability that a particular beneficial microorganism will become predominant, even with organic farming or nature farming methods, will depend on the ecosystem and environmental conditions. It can take several hundred years for various species of higher and lower plants to interact and develop into a definable and stable ecosystem. Even if the population of a specific microorganism is increased through cultural and management practices, whether it will be beneficial to plants is another question. Thus, the likelihood of a beneficial, plant-associated microorganism becoming predominant under conservation-based farming systems is virtually impossible to predict. Moreover, it is very unlikely that the population of useful anaerobic microorganisms, which usually comprise only a small part of the soil microflora, would increase significantly even under natural farming conditions.

This information then emphasizes the need to develop methods for isolating and selecting different microorganisms for their beneficial effects on soils and plants. The ultimate goal is to select microorganisms that are physiologically and ecologically compatible with each other and that can be introduced as mixed cultures into soil where their beneficial effects can be realized (Higa, 1991; 1994; 1995).

Application of Beneficial and Effective Microorganisms

Microorganisms are utilized in agriculture for various purposes; as important components of organic amendments and composts, as legume inoculants for biological nitrogen fixation as a means of suppressing insects and plant diseases to Improve crop quality and yields, and for reduction of labour. All of these are closely related to each other. An important consideration in the application of beneficial microorganisms to soils is the enhancement of their synergistic effects. This is difficult to accomplish if these microorganisms are applied to achieve symptomatic therapy, as in the case of chemical fertilizers and pesticides (Higa, 1991; 1994).

If cultures of beneficial microorganisms are to be effective after inoculation into soil, it is important that their initial populations be at a certain critical threshold level. This helps to ensure that the amount of bioactive substances produced by them will be sufficient to achieve the desired positive effects on crop production and/or crop protection. If these conditions are not met, the introduced microorganisms, no matter how useful they are, will have little if any effect. At present, there are no chemical tests that can predict the probability of a particular soil-inoculated microorganism to achieve a desired result. The most reliable approach is to inoculate the beneficial microorganism into soil as part of a mixed culture, and at a sufficiently high inoculum density to maximize the probability of its adaptation to environmental and ecological conditions (Higa and Wididana, 1991b; Parr *et al.*, 1994).

The application of beneficial microorganisms to soil can help to define the structure and establishment of natural ecosystems. The greater the diversity of the cultivated plants that are grown and the more chemically complex the biomass, the greater the diversity of the soil microflora as to their types, numbers and activities. The application of a wide range of different organic amendments to soils can also help to ensure a greater microbial diversity. For example, combinations of various crop residues, animal manures, green manures, and municipal wastes applied periodically to soil will provide a higher level of microbial diversity than when only one of these materials is applied. The reason for this is that each of these organic materials has its own unique indigenous microflora which can greatly affect the resident soil microflora after they are applied, at least for a limited period.

BENEFITS OF PESTICIDES

The primary benefits are the consequences of the pesticides' effects – the direct gains expected from their use. For example the effect of killing caterpillars feeding on the crop brings the primary benefit of higher yields and better quality of cabbage. The three main effects result in 26 primary benefits ranging from protection of recreational turf to saved human lives. The secondary benefits are the less immediate or less obvious benefits that result from the primary benefits. They may be subtle, less intuitively obvious, or of longer term. It follows that for secondary benefits it is therefore more difficult to establish cause and effect, but nevertheless they can be powerful justifications for pesticide use. For example the higher cabbage yield might bring additional revenue that could be put towards children's education or medical care, leading to a healthier, better educated population. There are various secondary benefits identified, ranging from fitter people to conserved biodiversity.

Improving productivity

Tremendous benefits have been derived from the use of pesticides in forestry, public health and the domestic sphere – and, of course, in agriculture, a sector upon which the Indian economy is largely dependent. Food grain production, which stood at a mere 50 million tons in 1948–49, had increased almost fourfold to 198 million tons by the end of 1996–97 from an estimated 169 million hectares of permanently cropped land. This result has been achieved by the use of high-yield varieties of seeds, advanced irrigation technologies and agricultural chemicals (Employment Information: Indian Labour Statistics, 1994). Similarly outputs and productivity have increased dramatically in most countries, for example wheat yields in the United Kingdom, corn yields in the USA.

Increases in productivity have been due to several factors including use of fertilizer, better varieties and use of machinery. Pesticides have been an integral part of the process by reducing losses from the weeds, diseases and insect pests that can markedly reduce the amount of harvestable produce. Warren (1998) also drew attention to the spectacular increases in crop yields in the United States in the twentieth century.

Webster *et al.* (1999) stated that "considerable economic losses" would be suffered without pesticide use and quantified the significant increases in yield and economic margin that result from pesticide use. Moreover, in the environment most pesticides undergo photochemical transformation to produce metabolites which are relatively non-toxic to both human beings and the environment (Kole *et al.*, 1999).

Protection of Crop Losses/yield Reduction

In medium land, rice even under puddle conditions during the critical period warranted an effective and economic weed control practice to prevent reduction in rice yield due to weeds that ranged from 28 to 48 per cent, based on comparisons that included control (weedy) plots (Behera and Singh, 1999). Weeds reduce yield of dry land crops (Behera and Singh, 1999) by 37–79 per cent. Severe infestation of weeds, particularly in the early stage of crop establishment, ultimately accounts for a yield reduction of 40 per cent. Herbicides provided both an economic and labour benefit.

Vector Disease Control

Vector-borne diseases are most effectively tackled by killing the vectors. Insecticides are often the only practical way to control the insects that spread deadly diseases such as malaria, resulting in an estimated 5000 deaths each day (Ross, 2005). In 2004, Bhatia wrote that malaria is one of the leading causes of morbidity and mortality in the developing world and a major public health problem in India. Disease control strategies are crucially important also for livestock.

Quality of Food

In countries of the first world, it has been observed that a diet containing fresh fruit and vegetables far outweigh potential risks from eating very low residues of pesticides in crops (Brown, 2004). Increasing evidence (Dietary Guidelines, 2005) shows that eating fruit and vegetables regularly reduces the risk of many cancers, high blood pressure, heart disease, diabetes, stroke, and other chronic diseases.

Lewis *et al.* (2005) discussed the nutritional properties of apples and blueberries in the US diet and concluded that their high concentrations of antioxidants act as protectants against cancer and heart disease. Lewis attributed doubling in wild blueberry production and subsequent increases in consumption chiefly to herbicide use that improved weed control.

Other areas – transport, Sport Complex, Building

The transport sector makes extensive use of pesticides, particularly herbicides. Herbicides and insecticides are used to maintain the turf on sports pitches, cricket grounds and golf courses. Insecticides protect buildings and other wooden structures from damage by termites and woodboring insects.

4

Beneficial Algae and its Importance

INTRODUCTION

Algae (singular alga) encompass several different groups of living organisms that capture light energy through photosynthesis, converting inorganic substances into simple sugars using the captured energy. Algae have been traditionally regarded as simple plants, and indeed some are closely related to the higher plants. Others appear to represent different protist groups, alongside other organisms that are traditionally considered more animal-like (that is, protozoa). Thus algae do not represent a single evolutionary direction or line, but a level of organization that may have developed several times in the early history of life on earth.

Algae range from single-celled organisms to multi-cellular organisms, some with fairly complex differentiated form and (if marine) called seaweeds. All lack leaves, roots, flowers, and other organ structures that characterize higher plants. They are distinguished from other protozoa in that they are photoautotrophic, although this is not a hard and fast distinction as some groups contain members that are mixotrophic, deriving energy both from photosynthesis and uptake of organic carbon either by osmotrophy, myzotrophy, or phagotrophy. Some unicellular species rely entirely on external energy sources and have reduced or lost their photosynthetic apparatus.

All algae have photosynthetic machinery ultimately derived from the cyanobacteria, and so produce oxygen as a by-product of photosynthesis, unlike non-cyanobacterial photosynthetic bacteria. It is estimated that algae produce about 73 to 87 percent of the net global production of oxygen--which is available to humans and other terrestrial animals for respiration.Algae are usually found in damp places or bodies of water and thus are common in terrestrial as well as aquatic environments.

However, terrestrial algae are usually rather inconspicuous and far more common in moist, tropical regions than dry ones, because algae lack vascular tissues and other adaptions to live on land. Algae can endure dryness and other conditions in symbiosis with a fungus as lichen.

The various sorts of algae play significant roles in aquatic ecology. Microscopic forms that live suspended in the water column - called phytoplankton - provide the food base for most marine food chains. In very high densities (so-called algal blooms) these algae may discolor the water and outcompete or poison other life forms. Seaweeds grow mostly in shallow marine waters. Some are used as human food or harvested for useful substances such as agar or fertilizer. The study of algae is called phycology or algology.

RELATIONSHIPS AMONG ALGAL GROUPS

PROKARYOTIC ALGAE

Traditionally the cyanobacteria have been included among the algae, referred to as the cyanophytes or Blue-green Algae, (the term "algae" refers to any aquatic organisms capable of photosynthesis), though some recent treatises on algae specifically exclude them. Cyanobacteria are some of the oldest organisms to appear in the fossil record, dating back about 3.8 billion years (Precambrian). Ancient cyanobacteria likely produced much of the oxygen in the Earth's atmosphere.

Cyanobacteria can be unicellular, colonial, or filamentous. They have a prokaryotic cell structure typical of bacteria and conduct photosynthesis directly within the cytoplasm, rather than in specialized organelles. Some filamentous blue-green algae have specialized cells, termed heterocysts, in which nitrogen fixation occurs.

EUKARYOTIC ALGAE

All other algae are eukaryotes and conduct photosynthesis within membrane-bound structures (organelles) called chloroplasts. Chloroplasts contain DNA and are similar in structure to cyanobacteria, presumably representing reduced cyanobacterial endosymbionts. The exact nature of the chloroplasts is different among the different lines of algae, reflecting different endosymbiotic events.

There are three groups (Archaeplastida) that have primary chloroplasts:

- Green algae, together with higher plants
- Red algae
- Glaucophytes

In these groups, the chloroplast is surrounded by two membranes and probably developed through a single endosymbiosis. The chloroplasts of red algae have a more or less typical cyanobacterial pigmentation, while those of the green alga have chloroplasts with chlorophyll a and b, the latter found in some cyanobacteria and not most. Higher plants are pigmented similarly to green algae and probably developed from them.

Two other groups of algae have green chloroplasts containing chlorophyll b:

- Euglenids and
- Chlorarachniophytes.

These are surrounded by three and four membranes, respectively, and were probably retained from an ingested green alga. Those of the chlorarchniophytes contain a small nucleomorph, which is the remnant of the alga's nucleus. It has been suggested that the euglenid chloroplasts only have three membranes because they were acquired through myzocytosis rather than phagocytosis.

The remaining algae all have chloroplasts containing chlorophylls a and c. The latter chlorophyll type is not known from any prokaryotes or primary chloroplasts, but genetic similarities with the red algae suggest a relationship there. These groups include:

Heterokonts (*e.g.*, golden algae, diatoms, brown algae)
Haptophytes (*e.g.*, coccolithophores)
Cryptomonads
Dinoflagellates

In the first three of these groups (Chromista), the chloroplast has four membranes, retaining a nucleomorph in cryptomonads, and they likely share a common pigmented ancestor. The typical dinoflagellate chloroplast has three membranes, but there is considerable diversity in chloroplasts among the group, as some members have acquired theirs from different sources. The Apicomplexa, a group of closely related parasites, also have plastids though not actual chloroplasts, which appear to have a common origin with those of the dinoflagellates.

Forms of algae

Most of the simpler algae are unicellular flagellates or amoeboids, but colonial and non-motile forms have developed independently among several of the groups. Some of the more common organizational levels, more than one of which may occur in the life cycle of a species, are:

- Colonial - small, regular groups of motile cells
- Capsoid - individual non-motile cells embedded in mucilage
- Coccoid - individual non-motile cells with cell walls
- Palmelloid - non-motile cells embedded in mucilage
- Filamentous - a string of non-motile cells connected together, sometimes branching
- Parenchymatous - cells forming a thallus with partial differentiation of tissues

In three lines even higher levels of organization have been reached, leading to organisms with full tissue differentiation. These are the brown algae - some of which may reach 70 m in length (kelps) - the red algae, and the green algae. The most complex forms are found among the green algae in a lineage that eventually led to the higher land plants. The point where these non-algal plants begin and algae stop is usually taken to be the presence of reproductive organs with protective cell layers, a characteristic not found in the other alga groups.

ALGAE AND SYMBIOSES

Some species of algae form symbiotic relationships with other organisms. In these symbioses, the algae supply photosynthates (organic substances) to the host organism providing protection to the algal cells. The host organism derives some or all of its energy requirements from the alga. Examples include:

- lichens - a fungus is the host, usually with a green alga or a cyanobacterium as its symbiont. Both fungal and algal species found in lichens are capable of living independently, although habitat requirements may be greatly different from those of the lichen pair.
- Corals - algae known as zooxanthellae are symbionts with corals. Notable amongst these is the dinoflagellate Symbiodinium, found in many hard corals. The loss of Symbiodinium, or other zooxanthellae, from the host is known as coral bleaching.
- Sponges - green algae live close to the surface of some sponges, for example, breadcrumb sponge (Halichondria panicea). The alga is thus protected from predators; the sponge is provided with oxygen and sugars which can account for 50 to 80% of sponge growth in some species

USES OF ALGAE

Algae are used by man in a great many ways. Because many species are aquatic and microscopic, they are cultured in clear tanks or ponds and either harvested or used to treat effluents pumped through the ponds. Algaculture on a large scale is an important type of aquaculture in some places.

ENERGY SOURCE

Algae can be used to produce biodiesel and by some estimates can produce vastly superior amounts of oil, compared to terrestrial crops grown for the same purpose. Because algae grown to produce biodiesel does not need to meet the requirements of a food crop, it is much cheaper to produce. Also it does not need fresh water or fertilizer (both of which are quite expensive).Algae can be grown to produce hydrogen.

In 1939 a German researcher named Hans Gaffron, while working at the University of Chicago, observed that the algae he was studying, Chlamydomonas reinhardtii (a green-algae), would sometimes switch from the production of oxygen to the production of hydrogen Gaffron never discovered the cause for this change and for many years other scientists failed in their attempts at its discovery. In the late 1990s professor Anastasios Melis a researcher at the University of California at Berkeley discovered that if the algae culture medium is deprived of sulfur it will switch from the production of oxygen (normal photosynthesis), to the production of hydrogen.

He found that the enzyme responsible for this reaction is hydrogenase, but that the hydrogenase lost this function in the presence of oxygen. Melis found that depleting the amount of sulfur available to the algae interrupted its internal oxygen flow, allowing the hydrogenase an environment in which it can react, causing the algae to produce hydrogen. Chlamydomonas moeweesi is also a good strain for the production of hydrogen. Algae can be grown to produce biomass, which can be burned to produce heat and electricity.

POLLUTION CONTROL

Algae are used in wastewater treatment facilities, reducing the need for more dangerous chemicals. Algae can be used to capture fertilizers in runoff from farms. If this algae is then harvested, it itself can be used as fertilizer. Algae bioreactors are used by some powerplants to reduce CO_2 emissions. The CO_2 can be pumped into a pond, or some kind of tank, on which the algae feed. Alternatively, the bioreactor can be installed directly on top of a smokestack. This techology has been pioneered by Massachusetts-based GreenFuelTechnologies

NUTRITIONAL VALUE OF ALGAE

Algae is commercially cultivated as a nutritional supplement. One of the most popular microalgal species is Spirulina (Arthrospira platensis), which is a Cyanobacteria (known as blue-green algae), and has been hailed by some as a superfood. Other algal species cultivated for their nutritional value include; Chlorella (a green algae), and Dunaliella (Dunaliella salina), which is high in beta-carotene and is used in vitamin C supplements.

Algae is sometimes also used as a food, as in the Chinese "vegetable" known as fat choy (which is actually a cyanobacterium). The oil from some algae have high levels of unsaturated fatty acids. Arachidonic acid(a polyunsaturated fatty acid), is very high in parietochloris incisa, (a green algae) where it reaches up to 47% of the triglyceride pool

The natural pigments produced by algae can be used as an alternative to chemical dyes and colouring agentsMany of the paper products used today are not recyclable because of the chemical inks that they use, paper recyclers have found that inks made from algae are much easier to break down. There is also much interest in the food industry into replacing the colouring agents that are currently used with colouring derived from algal pigments.

KINDS OF ALGAE

GREEN ALGAE

The Green Algae (singular: Green Alga) are the large group of algae from which the embryophytes (higher plants) emerged. As such they form a paraphyletic group, variously included among the Plantae or with the Protista. The green algae include unicellular and colonial flagellates, usually but not always with two flagella per cell, as well as various colonial, coccoid, and filamentous forms. In the Charales, the closest relatives of higher plants, full differentiation of tissues occurs.

Almost all forms have chloroplasts. These contain chlorophylls a and b, giving them a bright green colour (as well as the accessory pigments beta carotene and xanthophylls), and have stacked thylakoids. They are bound by a double membrane, so presumably were acquired by direct endosymbiosis of cyanobacteria. A number of cyanobacteria show similar pigmentation, but this appears to have arisen more than once, and the chloroplasts of green algae are no longer considered closely related to such forms. Instead, the green algae probably share a common origin with the red algae; the two are grouped as the Archaeplastida or Plantae sensu lato.

All green algae have mitochondria with flat cristae. When present flagella are typically anchored by a cross-shaped system of microtubules, but these are absent among the higher plants and charophytes. They usually have cell walls containing cellulose, and undergo open mitosis without centrioles. Sexual reproduction varies from fusion of identical cells (isogamy) to fertilization of a large non-motile cell by a smaller motile one (oogamy). However, these traits show some variation, most notably among the basal green algae, called prasinophytes.

The remaining forms are usually classified as follows:

- Chlorophyta
- Chlorophyceae
- Ulvophyceae
- Trebouxiophyceae
- Chlorokybales
- Klebsormidiales
- Zygnematales
- Desmidiales
- Coleochaetales
- Charales

The orders outside the Chlorophyta are often grouped as the division Charophyta, which is paraphyletic to higher plants, together comprising the Streptophyta. Sometimes the Charophyta is restricted to the Charales, and a division Gamophyta is introduced for the Zygnematales and Desmidiales. In older systems the Chlorophyta may be taken to include all the green algae, but taken as above they appear to form a monophyletic group.

A few other organisms rely on green algae to conduct photosynthesis for them. The chloroplasts in euglenids and chlorarachniophytes were presumably acquired from ingested green algae, and in the latter retain a vestigial nucleus (nucleomorph). Some species of green algae, particularly of genera Trebouxia or Pseudotrebouxia (Trebouxiophyceae), can be found in symbiotic associations with fungi to form lichens. In general the fungal species that partner in lichens cannot live on their own, while the algal species is often found living in nature without the fungus. The first land plants probably evolved from green algae.

RED TIDE

Red tide is estuarine or marine algal bloom. Red tide is caused by species of dinoflagellates, often present in sufficient numbers (thousands or millions of cells per millilitre) to turn the water red or brown. The species responsible for red tides on the gulf coast of Florida is a dinoflagellate called Karenia brevis It produces brevetoxins which produce respiratory irritation in humans.Coastal water pollution produced by humans and systematic increase in sea water temperature appear to be causal factors in red tides. On the Pacific Coast of the U.S. there have been apparent increases in the occurrence of red tides since about 1991.

These increases are correlated with a marine temperature rise of about one degree Celsius, and also with increased nutrient loading into ocean waters. Additionally, red tide can assume its dormant form if the water cools to a sufficient degree. When it does this it forms cysts which drop to the ocean floor and rebloom if the right conditions occur. As another example, non- red tide algal blooms in lakes and rivers may be caused by greatly increased amounts of phosphorus or nitrogen entering an aquatic ecosystem from either sewage systems or surface runoff of agricultural fertilizers. Some red tides on the Pacific coast have also been linked to occurrences of El Nino events. Red tides also occur in places where there are no obvious associated human activities.

Some red tides produce large quantities of toxins, such as saxitoxin, which disrupt the proper function of ion channels in neurons. Domoic acid, causative of neurological damage in certain marine mammals, is one toxin associated with red tides. The red tide toxins can induce neurological damage and death in marine mammals which feed on affected filter feeders, these include whales, seals, and otters. Red tide can also increase fish mortality; they are also accumulated in the digestive tracts of filter feeders. This bioaccumulation of toxins causes bivalves - like oysters and clams - collected in areas affected by algal blooms to be potentially dangerous for human consumption. Initial signs of poisoning from this toxin is tingling in the lips followed by a reduction of motor abilities and difficulty breathing and can be fatal if consumed in sufficient amounts. If these symptoms occur after eating shellfish, seek immediate medical treatment. Standard medical treatment is to give victims oxygen, or to hook them up to a breather. There exists no antidote, and the idea is to keep the person alive until the toxin has passed from the system.

There is more than one type of red tide, something that frequently goes unnoticed. For example, the red tide that effects the Gulf Region produces, in addition to the toxins that accumulate within filter feeders, an airborne vector that can cause difficulty breathing, especially for those with pre-existing conditions whereas that which occurs in the Northeast US doesn't have an airborne vector. It should also be noted that red tide is far from being entirely understood. Gymnocin A is a cytotoxic polyether and a typical representative of toxins associated with red tide

RED ALGAE

The red algae are a large group of mostly multicellular, marine algae, including many notable seaweeds. Most of the coralline algae, which secrete calcium carbonate and play a major role in building coral reefs, belong here. Red algae such as dulse and nori are a traditional part of European and Asian cuisine and are used to make certain other products like agar and food additives.

Biology

Many red algae have multicellular stages but these lack differentiated tissues and organs. Unlike most other algae, no cells with a flagellum are found in any member of the group. Unicellular forms typically live attached to surfaces rather than floating among the plankton, and both the larger female and smaller male gametes are non-motile, so that most have a low chance of fertilization. They have cell walls are made out of cellulose and thick gelatinous polysaccharides, which are the basis for most of the industrial products made from red algae. The chloroplasts of red algae are bound by a double membrane, like those of green plants; both groups (Archaeplastida) probably share a common origin. Their plastids formed by direct endosymbiosis of a cyanobacteria, and in red algae are pigmented with chlorophyll a and various proteins called phycobilins, which are responsible for their reddish colour. Other algae that lack chlorophyll b appear to have acquired their chloroplasts from red algae, although their pigmentations are somewhat different.

Taxonomy

The diverse eukaryotes that constitute the red algae have been the focus of numerous recent molecular surveys and remain a rich source of undescribed and little known species for the traditional taxonomist. Molecular studies place the red algae in the Archaeplastida (Plantae sensu lato); however, supraordinal classification has been largely confined to debate on subclass vs. class level status for the two recognized subgroups, one of which is widely acknowledged as paraphyletic. This narrow focus has generally masked the extent to which red algal classification needs modification.The most current classification system is below (Saunders et. al., 2004). Note that while this is a valid published taxonomy it does not necessarily have to be used as the taxonomy of the algae is still in a state of flux.

- Kingdom Protist
- Subkingdom Rhodoplantae
- Phylum Cyanidiophyta
- Class Cyanidiophyceae
- Phylum Rhodophyta
- Subphylum Rhodellophytina
- Class Rhodellophyceae
- Subphylum Metarhodophytina
- Class Compsopogonophyceae
- Subphylum Eurhodophytina
- Class Bangiophyceae
- Class Florideophyceae
- Subclass Hildenbrandiophycidae
- Subclass Nemaliophycidae
- Subclass Ahnfeltiophycidae
- Subclass Rhodymeniophycidae

BROWN ALGAE

The brown algae or phaeophytes are a large group of multicellular algae, including many notable seaweeds. They play an important role in marine environments. For instance Macrocystis kelp, which may reach 60 metres in length, form prominent underwater forests. Another notable example is Sargassum, which creates unique habitats in the Sargasso Sea (hence the name Sargassum). Many brown algae, such as Fucus, of which bladder wrack is one species, are found along the seashore and some are used as food. Brown algae belong to a very large group called the heterokonts, most of which are coloured flagellates. Most contain the pigment fucoxanthin, which is responsible for the distinctive greenish-brown colour that gives brown algae their name. Brown algae are unique among heterokonts in developing into multicellular forms with differentiated tissues, but they reproduce by means of flagellate spores, which closely resemble other heterokont cells. Genetic studies show their closest relatives are the yellow-green algae.

CYANOBACTERIA

Cyanobacteria are a phylum of Bacteria that obtain their energy through photosynthesis. They are often referred to as blue-green algae, even though it is now known that they are not algae. Nonetheless, the description is still sometimes used to reflect their appearance and ecological role. Fossil traces of cyanobacteria have been found from around 3.8 billion years ago As soon as these blue-green bacteria evolved, they became the dominant metabolism for producing fixed carbon in the form of sugars from carbon dioxide. Cyanobacteria are now one of the largest and most important groups of bacteria on earth.

FORMS

Cyanobacteria include unicellular, colonial, and filamentous forms. Some filamentous cyanophytes form differentiated cells, called heterocysts, that are specialized for nitrogen fixation, and resting or spore cells called akinetes. Each individual cell typically has a thick, gelatinous cell wall, which stains gram-negative. The cyanophytes lack flagella, but may move about by gliding along surfaces. Most are found in fresh water, while others are marine, occur in damp soil, or even temporarily moistened rocks in deserts. A few are endosymbionts in lichens, plants, various protists, or sponges and provide energy for the host. Some live in the fur of sloths, providing a form of camouflage.

PHOTOSYNTHESIS

Photosynthesis in cyanobacteria generally uses water as an electron donor and produces oxygen as a by-product, though some may also use hydrogen sulfide as occurs among other photosynthetic bacteria. Carbon dioxide is reduced to form carbohydrates via the Calvin cycle. In most forms the photosynthetic machinery is embedded into folds of the cell membrane, called thylakoids. The large amounts of oxygen in the atmosphere are considered to have been first created by the activities of ancient cyanobacteria. Due to their ability to fix nitrogen in aerobic conditions they are often found as symbionts with a number of other groups of organisms such as fungi (lichens), corals, pteridophytes (Azolla), angiosperms (Gunnera) etc.

Cyanobacteria are the only group of organisms that are able to reduce nitrogen and carbon in aerobic conditions, a fact that may be responsible for their evolutionary and ecological success. The water-oxidizing photosynthesis is accomplished by coupling the activity of photosystem (PS) II and I. They are also able to use in anaerobic conditions only PS I-cyclic photophosphorylation-with electron donors other than water (hydrogen sulfide, thiosulphate, or even molecular hydrogen) just like purple photosynthetic bacteria.

Furthermore, they share an archaebacterial property-the ability to reduce elemental sulfur by anaerobic respiration in the dark. Perhaps the most intriguing thing about these organisms is that their photosynthetic electron transport shares the same compartment as the components of respiratory electron transport. Actually, their plasma membrane contains only components of the respiratory chain, while the thylakoid membrane hosts both respiratory and photosynthetic electron transport.Attached to thylakoid membrane, phycobilisomes act as light harvesting antennae for photosystem II.

The phycobilisome components (phycobilin) are responsible for the blue-green pigmentation of most cyanobacteria. The variations to this theme is mainly due to carotenoids and phycoerythrins which give the cells the red-brownish coloration. In some cyanobacteria, the colour of light influences the composition of phycobilisomes. In green light, the cells accumulates more phycoerythrin, whereas in red light they produce more phycocyanin. Thus the bacteria appears green in red light and red in green light. This process is known as complementary chromatic adaptation and is a way for the cells maximize the use of available light for photosynthesis.

A few genera, however, lack phycobilins and have chlorophyll b as well as chlorophyll a, giving them a bright green colour. These were originally grouped together as the prochlorophytes or chloroxybacteria, but appear to have developed in several different lines of cyanobacteria.

RELATIONSHIP TO CHLOROPLASTS

Chloroplasts found in eukaryotes (algae and higher plants) most likely represent reduced endosymbiotic cyanobacteria. This endosymbiotic theory is supported by various structural and genetic similarities.

Primary chloroplasts are found among the green plants, where they contain chlorophyll b, and among the red algae and glaucophytes, where they contain phycobilins. It now appears that these chloroplasts probably had a single origin. Other algae likely took their chloroplasts from these forms by secondary endosymbiosis or ingestion.

CLASSIFICATION

The cyanobacteria were traditionally classified by morphology into five sectionsThe first three - Chroococcales, Pleurocapsales, and Oscillatoriales - are not supported by phylogenetic studies. However, the latter two - Nostocales and Stigonematales - are monophyletic, and make up the heterocystous cyanobacteria. Most taxa included in the phylum or division Cyanobacteria have not been validly published under the Bacteriological Code. Except: The classes Chroobacteria, Hormogoneae and Gloeobacteria. The orders Chroococcales, Gloeobacterales, Nostocales, Oscillatoriales, Pleurocapsales and Stigonematales. The families Prochloraceae and Prochlorotrichaceae. The genera Halospirulina, Planktothricoides, Prochlorococcus, Prochloron, Prochlorothrix.

BIOTECHNOLOGY AND APPLICATIONS OF ALGAE

Certain cyanobacteria produce cyanotoxins like Anatoxin-a, Anatoxin-as, Aplysiatoxin, Cylindrospermopsin, Domoic acid, Microcystin LR, Nodularin R (from Nodularia), or Saxitoxin. Sometimes a mass-reproduction of cyanobacteria results in algal blooms. The unicellular cyanobacterium Synechocystis sp. PCC 6803 was the first photosynthetic organism whose genome was completely sequenced It continues to be an important model organism.

At least one secondary metabolite, cyanovirin, has shown to possess anti-HIV activity. Some cyanobacteria are sold as food, notably Aphanizomenon flos-aquae (E3live) and Arthrospira platensis (Spirulina). It has been suggested that they could be a much more substantial part of human food supplies, as a kind of "superfood".

DIATOM

Diatoms (Gr. dia 'through'; tomos 'cutting', *i.e.*, "cut in half") are a major group of eukaryotic algae, and are one of the most common types of phytoplankton. Most diatoms are unicellular, although some form chains or simple colonies. A characteristic feature of diatom cells is that they are encased within a unique cell wall made of silica. These walls show a wide diversity in form, some quite beautiful and ornate, but usually consist of two symmetrical sides with a split between them, hence the group name.

GENERAL BIOLOGY

There are more than 200 genera of living diatoms, and it is estimated that there are approximately 100 000 extant species (Round & Crawford, 1990). Diatoms are a widespread group and can be found in the oceans, in freshwater, in soils and on damp surfaces. Most live pelagically in open water, although some live as surface films at the water-sediment interface (benthic), or even under damp atmospheric conditions. They are especially important in oceans, where they are estimated to contribute up to 45% of the total oceanic primary production

Diatoms belong to a large group called the heterokonts, including both autotrophs (*e.g.* golden algae, kelp) and heterotrophs (*e.g.* water moulds). Their chloroplasts are typical of heterokonts, with four membranes and containing pigments such as fucoxanthin. Individuals usually lack flagella, but they are present in gametes and have the usual heterokont structure, except they lack the hairs (mastigonemes) characteristic in other groups.

Most diatom species are non-motile but some are capable of an oozing motion. As their relatively dense cell walls cause them to readily sink, planktonic forms in open water usually rely on turbulent mixing of the upper layers by the wind to keep them suspended in sunlit surface waters. Some species actively regulate their buoyancy to counter sinking. Diatoms cells are contained within a unique silicate (silicic acid) cell wall comprised of two separate valves (or shells). The biogenic silica that the cell wall is composed of is synthesised intracellularly by the polymerisation of silicic acid monomers. This material is then extruded to the cell exterior and added to the wall. Diatom cell walls are also called frustules or tests, and their two valves typically overlap one other like the two halves of a petri dish. In most species, when a diatom divides to produce two daughter cells, each cell keeps one of the two valves and grows a smaller valve within it. As a result, after each division cycle the average size of diatom cells in the population gets smaller. Once such cells reach a certain minimum size, rather than simply divide vegetatively, they reverse this decline by forming an auxospore. This expands in size to give rise to a much larger cell, which then returns to size-diminishing divisions. Auxospore production is almost always linked to meiosis and sexual reproduction.

Decomposition and decay of diatoms leads to organic and inorganic (in the form of silicates) sediment, the inorganic component of which can lead to a method of analyzing past marine environments by corings of ocean floors or bay muds, since the inorganic matter is embedded in deposition of clays and silts and forms a permanent geological record of such marine strata.

Classification

The classification of heterokonts is still unsettled, and they may be treated as a division (or phylum), kingdom, or something in-between. Accordingly, groups like the diatoms may be ranked anywhere from class (usually called Bacillariophyceae) to division (usually called Bacillariophyta), with corresponding changes in the ranks of their subgroups.

Diatoms are traditionally divided into two orders: centric diatoms (Centrales), which are radially symmetric, and pennate diatoms (Pennales), which are bilaterally symmetric. The former are paraphyletic to the latter. A more recent classification is that of Round & Crawford (1990), who divide the diatoms into three classes: centric diatoms (Coscinodiscophyceae), pennate diatoms without a raphe (Fragilariophyceae), and pennate diatoms with a raphe (Bacillariophyceae). It is probable there will be further revisions as our understanding of their relationships increases.

Ecology

Planktonic forms in freshwater and marine environments typically exhibit a "bloom and bust" lifestyle. When conditions in the upper mixed layer (nutrients and light) are favourable (*e.g.* at the start of spring) their competitive edge allows them to quickly dominate phytoplankton communities ("bloom"). As such they are often classed as opportunistic r-strategists (*i.e.* those organisms whose ecology is defined by a high growth rate, r).

When conditions turn unfavourable, usually upon depletion of nutrients, diatom cells typically increase in sinking rate and exit the upper mixed layer ("bust"). This sinking is induced by either a loss of buoyancy control, the synthesis of mucilage that sticks diatoms cells together, or the production of heavy resting spores. Sinking out of the upper mixed layer removes diatoms from conditions inimical to growth, including grazer populations and higher temperatures (which would otherwise increase cell metabolism). Cells reaching deeper water or the shallow seafloor can then rest until conditions become more favourable again. In the open ocean, many sinking cells are lost to the deep, but refuge populations can persist near the thermocline.

Ultimately, diatom cells in these resting populations re-enter the upper mixed layer when vertical mixing entrains them. In most circumstances, this mixing also replenishes nutrients in the upper mixed layer. In the open ocean (away from areas of continuous upwellingthis cycle of bloom, bust, then return to pre-bloom conditions typically occurs over an annual cycle, with diatoms only being prevalent during the spring and early summer.

In some locations, however, an autumn bloom may occur, caused by the breakdown of summer stratification and the entrainment of nutrients while light levels are still sufficient for growth. Since vertical mixing is increasing, and light levels are falling as winter approaches, these blooms are smaller and shorter-lived than their spring equivalents.

In the open ocean, the condition that typically causes diatom (spring) blooms to end is a lack of silicon. Unlike other nutrients, this is only a major requirement of diatoms so it is not regenerated in the plankton ecosystem as efficiently as, for instance, nitrogen or phosphorus nutrients. This can be seen in maps of surface nutrient concentrations - as nutrients decline along gradients, silicon is usually the first to be exhausted (followed normally by nitrogen then phosphorus).

Because of this boom-and-bust lifestyle, diatoms are believed to play a disproportionately important role in the export of carbon from oceanic surface waters. Significantly, they also play a key role in the regulation of the biogeochemical cycle of silicon in the modern ocean The use of silicon by diatoms is believed by many researchers to be the key to their ecological success.

In a now classic study, Egge & Aksnes (1992) found that diatom dominance of mesocosm communities was directly related to the availability of silicate. Above concentrations of silicon of 2 mmol m-3, diatoms typically comprised more than 70% of the phytoplankton community. Raven (1983) noted that, relative to organic cell walls, silica frustules require less energy to synthesize (approximately 8%), potentially a significant saving on the overall cell energy budget.

Other researchers (Milligan & Morel, 2002) have suggested that the biogenic silica in diatom cell walls acts as an effective pH buffer, facilitating the conversion of bicarbonate to dissolved CO_2 (which is more readily assimilated). Notwithstanding the possible advantages conferred by silicon, diatoms typically have higher growth rates than other algae of a corresponding size

EVOLUTIONARY HISTORY

Heterokont chloroplasts appear to be derived from those of red algae, rather than directly from prokaryotes as occurs in plants. This suggests they had a more recent origin than many other algae. However, fossil evidence is scant, and it is really only with the evolution of the diatoms themselves that the heterokonts make a serious impression on the fossil record.

The earliest known fossil diatoms date from the early Jurassic although recent genetic and sedimentary evidence suggests an earlier origin. Medlin *et al*. (1997) suggest that their origin may be related to the end-Permian mass extinction after which many marine niches were opened. The gap between this event and the time that fossil diatoms first appear may indicate a period when diatoms were unsilicified and their evolution was cryptic Since the advent of silicification, diatoms have made a significant impression on the fossil record, with major deposits of fossil diatoms found as far back as the early Cretaceous, and some rocks (diatomaceous earth, diatomite, kieselguhr) being composed almost entirely of them.

Although the diatoms may have existed since the Triassic, the timing of their ascendancy and "take-over" of the silicon cycle is more recent. Prior to the Phanerozoic it is believed that microbial or inorganic processes weakly regulated the ocean's silicon cycle Subsequently, the cycle appears dominated (and more strongly regulated) by the radiolarians and siliceous sponges, the former as zooplankton, the latter as sedentary filter feeders primarily on the continental shelves Within the last 100 My, it is thought that the silicon cycle has come under even tighter control, and that this derives from the ecological ascendancy of the diatoms. However, the precise timing of the "take-over" is unclear, and different authors have conflicting interpretations of the fossil record. Some evidence, such as the eviction of siliceous sponges from the shelves suggests that this takeover began in the Cretaceous while evidence from radiolarians suggests "take-over" did not begin until the Cenozoic (65 Ma to present). Nevertheless, regardless of the details of the "take-over" timing, it is clear that this most recent revolution has installed much tighter biological control over the biogeochemical cycle of silicon.

COLLECTION

Living diatoms are often found clinging in great numbers to filamentous algae, or forming gelatinous masses on various submerged plants. Cladophora is frequently covered with Cocconeis, an elliptically shaped diatom; Vaucheria is often covered with small forms. Diatoms frequently present as a brown, slippery coating on submerged stones and sticks, and may be seen to "stream" with river current.

The surface mud of a pond, ditch, or lagoon will almost always yield some diatoms. They can be made to emerge from the mud by putting black paper around the jar and letting direct sunlight fall upon the surface of the water. The diatoms, within a day or less, will come to the top in a scum which can be easily isolated and secured. Since diatoms form an important part of the food of molluscs, tunicates, and fishes, the alimentary tracts of these animals often yield forms that are not easily secured in other ways. Marine diatoms can be collected by direct water sampling, though benthic forms can be secured by scraping barnacles, oyster shells, and other shells.

The silicious shells of diatoms are among the most beautiful objects which can be examined with the microscope. To obtain perfectly clean mounts requires considerable time and patience, but once the material is cleaned, preparations may be made at any time with very little trouble.

GOLDEN ALGAE

The golden algae or chrysophytes are a large group of heterokont algae, found mostly in freshwater. Originally they were taken to include all such forms except the diatoms and multicellular brown algae, but since then they have been divided into several different groups based on pigmentation and cell structure. They are now usually restricted to a core group of closely related forms, distinguished primarily by the structure of the flagella in motile cells, also treated as an order Chromulinales. It is possible membership will be revised further as more species are studied in detail. They come in a variety of morphological types, originally treated as separate orders or families.

Most members are unicellular flagellates, with either two visible flagella, as in Ochromonas, or sometimes one, as in Chromulina. The Chromulinales as first defined by Pascher in 1910 included only the latter type, with the former treated as the order Ochromonadales. However, structural studies have revealed that short second flagellum or at least a second basal body is always present, so this is no longer considered a valid distinction. Most of these have no cell covering. Some have loricae or shells, such as Dinobryon, which is sessile and grows in branched colonies. Most forms with silicaceous scales are now considered a separate group, the synurids, but a few belong among the Chromulinales proper, such as Paraphysomonas.

Some members are generally amoeboid, with long branching cell extensions, though they pass through flagellate stages as well. Chrysamoeba and Rhizochrysis are typical of these. There is also one species, Myxochrysis paradoxa, which has a complex life cycle involving a multinucleate plasmodial stage, similar to those found in slime moulds. These were originally treated as the order Chrysamoebales. The superficially similar Rhizochromulina was once included here, but is now given its own order based on differences in the structure of the flagellate stage.

Other members are non-motile. Cells may be naked and empeded in mucilage, such as Chrysosaccus, or coccoid and surrounded by a cell wall, as in

Chrysosphaera. A few are filamentous or even parenchymatous in organization, such as Phaeoplaca. These were included in various older orders, most of the members of which are now included in separate groups. Hydrurus and its allies, freshwater genera which form branched gelatinous filaments, are often placed in the separate order Hydrurales but may belong here.

Algaculture

Algaculture is a form of aquaculture involving the farming of species of algae for purposes of producing food or other products that can be extracted from the cultivated species. The focus of this article is on the cultivation and uses of microalgae, (also known as phytoplankton, microphytes, or planktonic algae), and cyanobacteria.

Cultivation

Algae can be cultured in raceway-type ponds and lakes. Due to the fact that these systems are "open" to the elements, sometimes called "open-pond" systems, they are much more vulnerable to being invaded by other algal species and bacteria.

The number of species that have been successfully cultivated for a given purpose, (ie: as a food source, for oil production, or for pigments.), in an outdoor system, are relatively small. In open systems you do not have control over water temperature, and you have little control over lighting conditions. The growing season is largely dependent on location and, aside from tropical areas, is limited to the warmer months. Some of the benefits of this type of system are that it is one of the cheaper ones to produce - at the most basic you only need to dig a trench or pond.

It also has one of the largest production capacities compared to other systems, and depending on how large it's made. A variation on the basic "open-pond" system is to close it off, to cover a pond or pool with a greenhouse. While this usually results in a smaller system, (for economic reasons), it does take care of many of the problems associated with an open system. It allows more species to be able to be grown, it allows the species that are being grown to stay dominant, and it extends the growing season, only slightly if unheated, and if heated it can produce year round.

Algae can be grown in a photobioreactor. A photobioreactor is basically a bioreactor which incorporates some type of light source. While almost anything that it would be possible to grow algae in could technically be called a photobioreactor, the term is more commonly used to define a closed system, as opposed to an open tank, or pond. Because these systems are closed, when used to cultivate algae, everything that the algae need to grow, (carbon dioxide, nutrient-rich water and light), all must be introduced into the system. A pond covered with a greenhouse could be considered a photobioreactor.

Different types of photobioreactors include:

- Tanks provided with a light source.
- Polyethylene sleeves or bags.
- Glass or plastic tubes.

In most algal-cultivation systems, light only penetrates the top 3-4 inches of the water. This is because as the algae grow and multiply they become so dense that they block light from reaching deeper into the pond or tank. Algae only need about 1/10 the amount of light they receive from direct sunlight. Direct sunlight is often too strong for algae. In order to have ponds that are deeper than 4 inches algae growers use various methods to agitate the water in their ponds, exposing the algae below to light and keeping algae on the surface from being over-exposed.Paddle wheels can be used to circulate the water in a pond.

Compressed air can be introduced into the bottom of a pond or tank, bringing algae from the lower levels up to the top to receive its share of light. Aside from agitation, another means of supplying light to algae is to place the light in the system.

Glow plates are sheets of plastic or glass that can be submerged into the water of a tank, providing light directly to the algae at the right concentration. The odour that many people assosciate with bogs or swamps, or stagnant ponds that have been taken over by algae is due to the depletion of oxygen in the water caused by the death of algal blooms that have been left to decay. Often the oxygen is depleted to the point where it kills all the fish, resulting in an even worse smell. In a system where algae is intentionally cultivated, maintained, and harvested, this situation shouldn't arise, the air around an algal pond should actually be very fresh from all the oxygen produced.

HARVESTING

Algae can be harvested using microscreens, by centrifugation, or by flocculation. Alum and ferric chloride are chemical flocculants used to harvest algae. Chitosin can be used as a flocculant. The shells of crustaceans are ground into powder and processed to acquire chitin, a polysaccharide found in the shells, from which Chitosin is derived. Chitosin is commonly used for water purification.

The more brackish, or saline the water that the algae is being grown in, the more chemical flocculant will be required to induce flocculation. Harvesting by chemical flocculation is a method that is often too expensive for large operations.

OIL EXTRACTION

Chemical solvents-Algal oil can be extracted using chemicals. Benzene and ether have been used, oil can also be separated by hexane extraction, which is widely used in the food industry and is relatively inexpensive. The downside to using solvents for oil extraction are the dangers involved in working with the chemicals. Care must be taken to avoid exposure to vapors and direct contact with the skin, either of which can cause serious damage. Benzene is classified as a carcinogen. Chemical solvents also present the problem of being an explosion hazard.

Enzymatic extraction-Enzymatic extraction uses enzymes to degrade the cell walls with water acting as the solvent, this makes fractionation of the oil much easier. The costs of this extraction process are estimated to be much greater than hexane extraction

Expression/Expeller press-When algae is dried it retains its oil content, which then can be "pressed" out with an oil press. Many commercial manufacturers of vegetable oil use a combination of mechanical pressing and chemical solvents in extracting oil.

Osmotic shock-Osmotic shock is a sudden reduction in osmotic pressure, this can cause cells in a solution to rupture. Osmotic shock is sometimes used to release cellular components, such as oil.

Supercritical fluid-In supercritical fluid/CO2 extraction, CO2 is liquefied under pressure and heated to the point that it has the properties of both a liquid and a gas, this liquified fluid then acts as the solvent in extracting the oil

Ultrasonic-assisted extraction- Ultrasonic extraction, a branch of sonochemistry, can greatly accelerate extraction processes. Using an ultrasonic reactor, ultrasonic waves are used to create cavitation bubbles in a solvent material, when these bubbles collapse near the cell walls, it creates shock waves and liquid jets that causes those cells walls to break and release their contents into the solvent

Algae as An Energy Source

Biodiesel Production

small scale production experiments bear out then using algae to produce biodiesel may be the only viable method by which to produce enough automotive fuel to replace current world gasoline usage. Microalgae have much faster growth-rates than terrestrial crops. The per unit area yield of oil from algae,(estimated to be from between 5,000 to 20,000 gallons per acre, per year), is 7 to 31 times greater than the next best crop, palm oil(635gal).

Algal-oil processes into biodiesel as easily as oil derived from land-based crops. The difficulties in efficient biodiesel production from algae lie not in the extraction of the oil, which can be done using methods common to the food-industry such as hexane extraction, but in finding an algal strain with a high lipid content and fast growth rate that isn't too difficult to harvest, and a cost-effective cultivation system (ie, type of photobioreactor) that is best suited to that strain. Open-pond systems for the most part have been given up for the cultivation of algae with high-oil content. Many believe that a major flaw of the Aquatic Species Programme was the decision to focus their efforts exclusivly on open-ponds, this makes the entire effort dependent upon the hardiness of the strain chosen, requiring it to be unnecessarily resiliant in order to withstand wide swings in temperature and pH, and competition from invasive algae and bacteria.

The energy that a high-oil strain invests into the production of oil is energy that is not invested into the production of proteins or carbohydrates, usually resulting in the species being less hardy, or having a slower growth rate. Algal species with a lower oil content, not having to divert their energies away from growth, have an easier time in the harsher conditions of an open system. Research into algae for the mass-production of oil is mainly focused on microalgae, (which is a term generally referred to as organisms capable of photosynthesis that are less than 2 mm in diameter, including the diatoms and cyanobacteria), as opposed to macroalgae, (ie. seaweed).

This preference towards microalgae is due largely to its less complex structure, fast growth rate, and high oil content- (for some species). Some commercial interests into large scale algal-cultivation systems are looking to tie-in to existing infrastructures, such as coal power plants or sewage treatment facilities. This approach not only provides for the needs of the system, such as CO_2 and nutrients, which, if the facility were built independently, would have to be otherwise acquired, but in addition it remediates waste.

SVO

The algal-oil feedstock that is used to produce biodiesel can also be used for fuel directly as "Straight Vegetable Oil", (SVO). While using the oil in this manner does not require the additional energy needed for transesterification, (processing the oil with an alcohol and a catalyst to produce biodiesel), it does require a special engine designed for its use, or modifications to a normal diesel engine, whereas biodiesel can be run in any modern diesel engine, unmodified, that is designed to use ultra-low sulfur diesel, the new diesel fuel standard for the United States that goes into effect in the fall of 2006.

HYDROGEN PRODUCTION OF ALGAE

Algae can be grown to produce hydrogen. In 1939 a German researcher named Hans Gaffron, while working at the University of Chicago, observed that the algae he was studying, Chlamydomonas reinhardtii (a green-algae), would sometimes switch from the production of oxygen to the production of hydrogen. Gaffron never discovered the cause for this change and for many years other scientists failed in their attempts at its discovery. In the late 1990's professor Anastasios Melis a researcher at the University of California at

Berkeley discovered that by depriving the algae of sulfur it will switch from the production of oxygen (normal photosynthesis), to the production of hydrogen.

He found that the enzyme responsible for this reaction is hydrogenase, but that the hydrogenase will not cause this switch in the presence of oxygen.

Melis found that depleting the amount of sulfur available to the algae interrupted its internal oxygen flow, allowing the hydrogenase an environment in which it can react, causing the algae to produce hydrogen. Chlamydomonas moewusii is also a good strain for the production of hydrogen

BIOMASS

Algae can be grown to produce biomass, which can then be harvested and burned in the same manner as wood, to produce heat and electricity.

DYES AND COLORANTS

Feedstock

Algae can be dried and used as a feed for livestock, or fresh as food for fish and crustaceans.

MEDICINAL

Nutritional

There are many algae that are cultivated for their nutritional value, either for supplemental use, or as a food source. Spirulina(Arthrospira platensis) is a blue-green algae(cyanobacteria) that is quite nutritious, this species does very well in open systems and commercial growers have found it well-suited to cultivation. One of the largest production sites for Spriulina is Lake Texcoco in central Mexico.

Astaxanthin

Overexposing the algal-species Haematococcus pluvialis to direct sunlight will cause it to produce astaxanthin in defence. Astaxanthin is a carotenoid that has the ability to pass between the blood and the brain, carotenoids such as lycopene and beta-carotene can not. There are companies setting up large scale bioreactors for producing Astaxanthin.

"Astaxanthin is an antioxidant pigment found in microalgae such as Haematococcus algae. Its strong antioxidant action is linked to anti-inflammatory activity, and may have a role in strengthening immunity, slowing the aging of skin (age spots and wrinkles) and alleviating muscle fatigue. It is attracting a great deal of attention, not only in the health food market, but in the sports and beauty markets as well. "

Algal-oil

Algal-oil is very high in unsaturated fatty acids. Some UFA's found in different algal-species include:

Arachidonic acid(AA)
Eicospentaenoic acid(EPA)
Docasahexaenoic acid(DHA)
Gamma-linolenic acid(GLA)
Linoleic acid(LA)

PLASTICS PRODUCTION FROM ALGAE

Pollution Control

Much of the carbon dioxide that is released into the atmosphere is from the burning of fossil fuels for the production of energy. Coal power plants are a major source of carbon dioxide emissions. With concerns over global warming, new methods for the thorough and efficient sequestration of CO_2 are being sought out. Algae can be used to reduce coal power plant emmissions. By attaching an algae pond, or photobioreactor to a coal burning plant, the carbon dioxide produced during combustion can be fed into the algae system.

ALGAL CULTURES

ALGAL BLOOM

An algal bloom is a relatively rapid increase in the population of (usually) phytoplankton algae in an aquatic system. Typically only one or a few species are involved and the bloom is recognized by discoloration of the water resulting from the high density of pigmented cells. Although there is no officially recognized threshold level, algae are unlikely to be considered to be blooming unless more than 10,000 cells per millilitre occur. Algal bloom concentrations may reach millions of cells per millilitre. Colours observed are green, yellowish-brown, or red.

Cities and farms along the shores of lakes produce a variety of pollutants that get into the water. Improved laws and regulations have reduced the amounts of toxic chemicals entering lakes. However, even small quantities of toxic chemicals can enter food chains and create health hazards over the long term. Non-toxic chemicals also produce problems. For example, rivers and run-off from farmers' fields bring extra nutrients into a lake, allowing increased growth of algae in the water. The result is a population explosion of algae called an algal bloom. The algae cover the water with a film of green slime. Too many nutrients in fresh-water lakes and ponds can cause populations of algae to increase rapidly. The algae reduce the oxygen supply available to other organisms in the water.

The excessive growth of algae may disrupt higher links of the local food web. Algae that die and sink to the bottom stimulate growth of decomposers, especially bacteria. Decomposition can result in the depletion of oxygen in the deeper water layers, and these conditions may result in fish kills or replacement with less valuable species more tolerant of higher phosphorus and lower oxygen levels. Deoxygenation also may cause chemical changes in the mud on the bottom, lowering the redox value of the sediment, releasing chemicals and toxic gases. All these changes further accelerate the eutrophication of the aquatic ecosystem. Algal blooms may also be of concern as some species of algae produce neurotoxins. At the high concentrations reached during blooms, these may cause death if affected water is ingested.

ALGAL CULTURES

Algal blooms are monitored using biomass measurements coupled with the examination of species present. A widely-used measure of algal and cyanobacterial biomass is the chlorophyll concentration.Peak values of chlorophyll a for an oligotrophic lake are about 1-10 μg/l, while in a eutrophic lake they can reach 300 μg/l. In cases of hypereutrophy, such as Hartbeespoort Dam in South Africa, maxima of chlorophyll a can be as high as 3,000 μg/l (Zohary and Roberts, 1990; Bartram *et al.*, 1999).

Red Tide

The so-called red tide is an example of a naturally occurring estuarine or marine algal bloom. Red tide is caused by species of dinoflagellates, often present in sufficient numbers (thousands or millions of cells per millilitre) to turn the water red or brown.

Black Water

So-called black water is a dark discoloration of sea water, first described in the Bay of Florida in January 2002 Although fishermen in Florida complained and requested that the "government do something", scientists say that black water results from a non-toxic algal bloom, probably of diatoms. It dissipated within a few months by transport through the Florida Keys into the Florida Straits and by disruption by winds and wave action.

Water Treatment

Algal blooms sometimes occur in drinking water supplies. In such cases, toxins from the bloom can survive standard water purifying treatments. Researchers at Florida International University in Miami are experimenting with using 640-kilohertz ultrasound waves that create micropressure zones as hot as 3,700° C. This breaks some water molecules into reactive fragments that can kill algae

Dead Zone (ecology)

F Dead zones are hypoxic (low-oxygen) areas in the world's oceans, the observed incidences of which have been increasing since oceanographers began noting them in the 1970s. The term could as well apply to the identical phenomenon in large lakes. it reported 146 dead zones in the world oceans where marine life could not be supported due to depleted oxygen levels. Some of these were as small as a square kilometer, but the largest dead zone covered 70,000 square kilometers.

Causes of Dead Zones

Aquatic and marine dead zones can be caused by the process of eutrophication, triggered by an excess of plant nutrients (nitrogen and phosphorus) from fertilizers, sewage, combustion emissions from vehicles, power generators, and factories. In a cascade of effects, the nutrients trigger a bloom of phytoplankton at the bottom of the marine food chain, allowing zooplankton to proliferate. As phytoplankton and zooplankton die and sink below the photic zone where photosynthesis can occur, a bloom of natural bacterial degradation exhausts the water's dissolved oxygen.

Dead zones can also be produced by the natural event of river flooding. Large amounts of fresh water empty into the ocean forming a thick layer of fresh water atop the denser salt water, effectively forming a barrier between the ocean water and oxygen in the atmosphere. Remains of organisms found within sediment layers near the mouth of the Mississippi River indicate four hypoxic events before the advent of artificial fertilizer. In these sediment layers, anoxia-tolerant species are the most prevalent remains found. The periods indicated by the sediment record correspond to historic records of high river flow recorded by instruments at Vicksburg, Mississippi.

Effects of Dead Zones

Low oxygen levels recorded along the Gulf Coast of North America have led to reproductive problems in fish involving decreased size of reproductive organs, low egg counts and lack of spawning.

In a study of the Gulf killifish by the Southeastern Louisiana University done in three bays along the Gulf Coast, fish living in bays where the oxygen levels in the water dropped to 1 to 2 parts per million (ppm) for 3 or more hours per day were found to have smaller reproductive organs.

The male gonads were 34 to 50% as large as males of similar size in bays where the oxygen levels were normal (6 to 8 ppm). Females were found to have ovaries that were half as large as those in normal oxygen levels. The number of eggs in females living in hypoxic waters were only one-seventh the number of eggs in fish living in normal oxygen Fish raised in laboratory created hypoxic conditions showed extremely low sex-hormone concentrations and increased elevation of activity in two genes triggered by the hypoxia-inductile factor (HIF) protein. Under hypoxic conditions, HIF pairs with another protein, ARNT. The two then bind to DNA in cells, activating genes in those cells.

Under normal oxygen conditions, ARNT combines with estrogen to activate genes. Hypoxic cells in a test tube didn't react to estrogen placed in the tube. HIF appears to render ARNT unavailable to interact with estrogen, providing a mechanism by which hypoxic conditions alter reproduction in fish. It might be expected that fish would flee this potential suffocation, but they are often quickly rendered unconscious and doomed. Slow moving bottom-dwelling creatures like clams, lobsters and oysters are unable to escape. All colonial animals are extinguished. The normal mineralization and recycling that occurs among benthic life-forms is stifled.

Locations of Dead Zones

In the 1970s, marine dead zones were first noted in areas where intensive economic use stimulated "first-world" scientific scrutiny: in the U.S. East Coast's Chesapeake Bay, in Scandinavia's strait called the Kattegat, which is the mouth of the Baltic Sea and in other important Baltic Sea fishing grounds, in the Black Sea, (which may have been anoxic in its deepest levels for millennia, however) and in the northern Adriatic.

Currently the most notorious dead zone is a 20,000 square kilometer region in the Gulf of Mexico, where the Mississippi River dumps high-nutrient runoff from its vast drainage basin, which includes the heart of U.S. agribusiness, the Midwest, affecting important shrimp fishing grounds.

Other marine dead zones have appeared in coastal waters of South America, China, Japan, southeast Australia, and even in the waters of New Zealand, which have a reputation for being some of the planet's most pristine near-coastal marine environments.

Reversal of Dead Zones

Dead zones are not irreversible. The Black Sea dead zone, previously the largest dead zone in the world, largely disappeared between 1991 and 2001 after fertilizers became too costly to use following the collapse of the Soviet Union and the demise of centrally planned economies in Eastern and Central Europe.

Fishing has again become a major economic activity in the region. While the Black Sea "cleanup" was largely unintentional and involved a drop in hard-to-control fertilizer usage, the U.N. has advocated other cleanups by reducing large industrial emissions.

From 1985-2000, the North Sea dead zone had nitrogen reduced by 37% when policy efforts by countries on the Rhine River reduced sewage and industrial emissions of nitrogen into the water.

Germination and Growth in the Seedling

Seeds have a multicellular embryo with meristematic tissues for the first leaves and root already established oils for high energy food storage and to help prevent decay a seed coat to prevent desiccation and provide mechanical protection carbohydrates in cotyledons which when broken down will supply the energy needed for rapid uptake of water for germination.

Angiosperm seeds are an effective evolutionary adaptation allowing plants to weather rough conditions. If they germinate when the conditions are not right the species could perish. An Australian medical journal reported that a woman had suffered from persistent sore throat until doctors discovered a pear or apple seed had germinated in her throat and put down roots.

While it was reporting this peculiar plant performance, the journal went on to mention that doctors in south Africa had found a little boy with a chrysanthemum seed growing out of one eye and a British woman who had a tomato plant growing out of her upper gum. It seems a seed had gotten caught under her dentures and found the temperature and moisture conditions ideal.

Requirements for Germination

Dormancy may be broken when conditions favour plant growth. All seeds require:

- Adequate supply of water
- Favourable temperatures
- Presence of oxygen

Depending on the species some special conditions for seed germination are:

- Freezing to help open the seed coat
- Heat or smoke from fire
- Abrasive action of sand or grit to wear down the seed coat
- Hydrolyzing action of animal digestive system

How dormancy is Broken: After the uptake of water the plant hormone gibberellic acid is released. It acts on the aleurone layer found just below the seed coat to stimulated the enzyme alpha-amylase.

The amylase mobilizes the starchy reserves of the endosperm. The resulting sugars quickly supply the meristematic tissues the energy they need to begin cell division.

THE SEEDLING

In all seedlings a root develops first and grows down into the soil. Then the shoot must emerge from the embryo and search out light above the soil and leaf litter. When stems and leaves emerge above ground chloroplasts in them rapidly mature turning the plant green. Although there are differences from plant to plant the basic pattern is the same and is illustrated below.

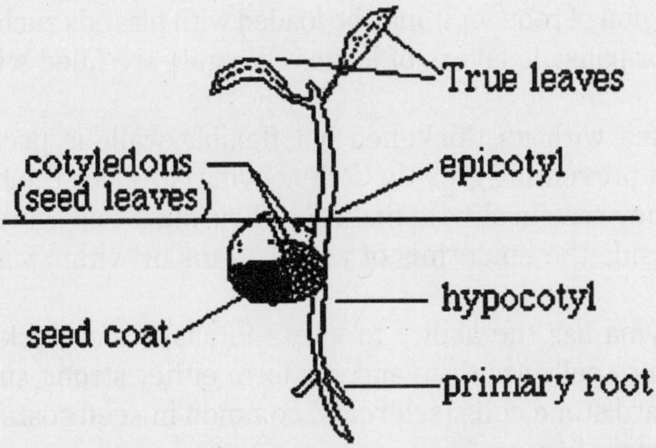

The epicotyl which emerges from a seedling is an example of primary growth.

Tissue Organization and the Plant

All plants have a special tissue that continues to divide called apical meristem which is generally found at the tip of roots or shoots. The apical meristem in turn gives rise to just 3 types of primary tissue which can divide and differentiate into all the other plant tissues. These primary meristems are:

- Protoderm
- Ground Meristem
- Procambium

Given the right environment each plant cell can develop into a new plant —the ultimate in cloning. Such cells and tissues are totipotent.

Protoderm matures to form epidermis. This flattened irregular layer covers roots, shoots, and leaves. Some epidermal cells specialize to become:

- Guard cells (always in pairs) to help in gas exchange, generally found in the lower epidermis of leaves.

Root hairs which emerge from the epidermal cells just above the root tip help increase the absorptive surface area of the root. Many epidermal cells produce a waxy water proofing, the cuticle of a leaf for instance.

- leaf epidermis produces a wax called cutin stem and root epidermis make subrin

Where secondary growth occurs the epidermis within becomes a special layer of cells called periderm. Periderm is responsible for the production of a waterproofing material called subrin.

Ground meristem differentiates into 3 important tissues:

- Parenchyma
- Collenchyma
- Sclerenchyma

Parenchyma cells are generally spherical or slightly irregular in shape with thin walls. Parenchyma forms a versatile tissue which can be used as storage in the cortex region of roots or it may be loaded with plastids such as chloroplasts in the spongy or palisade layers of leaves. Wounds are filled with parenchyma cells.

Collenchyma with its thickened but flexible walls is primarily used for support without preventing growth. Collenchyma cells lack the hardening agent called lignin. They remain alive at maturity, their thick walls create a supportive cylinder just inside the epidermis of young stems or within vascular tissue of leaf veins.

Sclerenchyma has the ability to very the shape and thickness of its cell walls. The mature cells are dead and can form either strong supporting fibres or extremely hard stone cells (sclereids) common in seed coats and nut shells. These cells contain lignin.

Procambium gives rise to the following vascular tissue:

- Primary xylem
- Primary phloem

VASCULAR CAMBIUM

Primary xylem has the dual roles of support and water transport, although the latter is by far the most important. When mature the lifeless tracheids and vessel elements form hollow, continuous, conduits from root to leaf

All vascular plants contain tracheids, long slender cells with tapered ends. Pits along the sides and at the ends of these cells line up permitting water to flow from cell to cell. The pits of pit pairs consist of thin primary cell walls only.

Only Angiosperms have vessel elements. These shorter, wider, cells align end to hollow end to form uninterrupted pipelines. The cell wall at the end of vessel elements is often missing, and pits dot their sides for lateral movement of water. Primary phloem is living tissue which is involved in the active transport and bulk movement of sugars and other nutrients. It is a complex tissue that has two major components: sieve elements and companion cells.

The backbone of phloem function is carried out by sieve tube members which when arranged end to end form sieve elements. Because the end walls of sieve tubes have pores materials dissolved in the cytoplasm can stream from cell to cell. Sieve tube members, while alive at maturity, lack a nucleus so their energy and metabolic needs are supplied by companion cells which lie adjacent to them. Thus companion cells also play a role in conduction of food.

PRIMARY GROWTH IN THE ROOT

Differentiation in the Young Root Tip

The root tip to be successful must accomplish 3 tasks:growth (root apical meristem) penetration of the soil (root cap and region of elongation) absorption of water and minerals (root hairs in region of maturation and mycorrhiza)

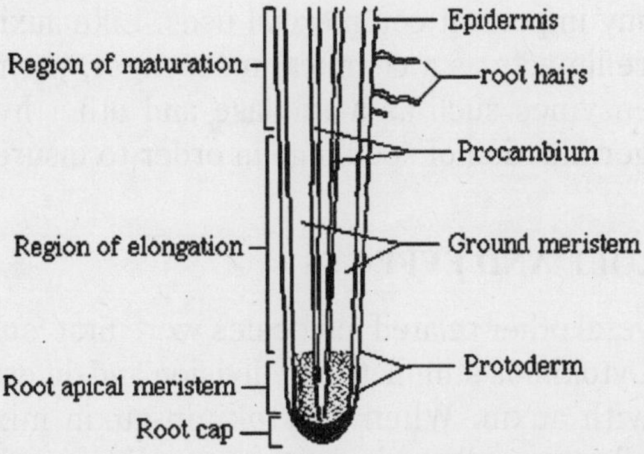

Primary growth involves the ability of a cell with a potentially rigid cell wall to: divide — root apical meristem elongate.

Roots can penetrate soil without destroying the delicate meristem cells because they create a protective cap. The turgor pressure helps elongate cells above the root tip which forces the tip deeper into the soil.

As the epidermal cells of the root tip mature many will produce hair-like extensions. These root hairs intermingle with soil particles and perhaps the hypha of fungi to create a large absorptive surface area.

THE STELE

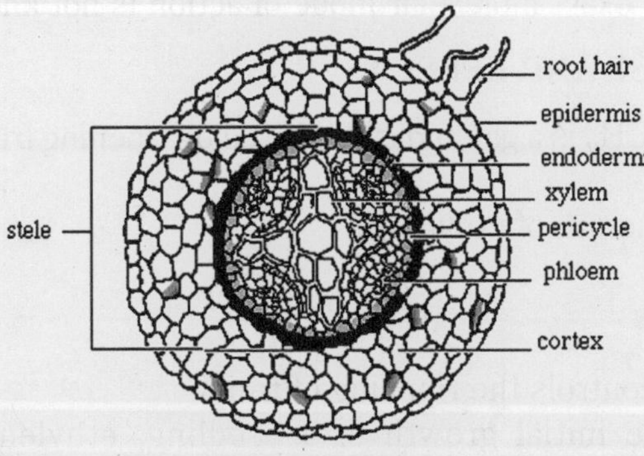

The epidermis is a single layer of cells which first produce root hairs and when mature the wax subrin. Cortex cells are thin walled and generally used for storage. The endodermis produces the water tight casparian strip. The stele is composed of conductive tissues such as xylem and phloem and gives the root support. The pericycle lines the outer edge of the stele and can produce lateral roots.

OTHER PLANT GROWTH HORMONES

GIBBERELLIN: ROLE AND EFFECTS

Originally gibberellin was found in a fungus, Gibberella fujikuroi, which caused rice seedlings to grow so tall and spindly that they fell over. Today gibberellin has many important commercial uses. Like auxin it promotes cell elongation. Gibberellin acts as a chemical messenger, hormone, to stimulate the synthesis of enzymes such as a-amylase and other hydrolytic enzymes important during germination of seedlings in order to insure release of stored nutrients.

CYTOKININS: ROLE AND EFFECTS

Zeatin and several other related molecules were first found in coconut milk and corn kernels. Cytokinins stimulate cell division and differentiation in plants; usually coupled with auxin. When a cytokinin-auxin mixture is added to undifferentiated cell mass (callus which forms around wounded tissue) either a root or a shoot will being development depending on which hormone is higher in concentration: auxins for root development and cytokinin for shoot development.

Plant cells exhibit totipotency—the ability to dedifferentiate and develop into any tissue necessary. Cytokinins retard senescence (aging); a process in which chlorophyll breaks down and various molecules and minerals are removed from deciduous leaves before they fall Protein synthesis stops when leaves are picked, but if they are treated with cytokinin they remain green, protein synthesis continues, and carbohydrates do not breakdown. Plant tRNA's contain cytokinins: but the exact molecular mode of action is not known.

ETHYLENE: ROLE AND EFFECTS

Ethylene CH_2CH_2 is a gas and is produced by ripening fruit from the amino acid methionine.

Molecular structure of ethylene

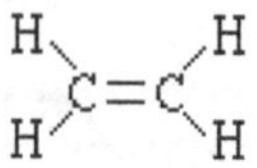

- Ethylene controls the ripening of fruits
- During the initial growth of a seedling, ethylene appears to be responsible for the hooked shape of the developing shoot. This helps protect the young tender leaves from the wear and tear of growing up through the soil.
- Ethylene may insure that flowers are carpelate (female) while gibberellin confers maleness on flowers
- Ethylene promotes the production of cellulases prior to leaf fall and abscission.

Commercially ethylene is used to artificially speed ripening of fruit prior to marketing. Tomatoes if picked red and ripe would spoil before reaching market, so most are picked green then reddened with ethylene rather than the slow natural way.

ABSCISIC ACID (ABA) ROLE AND EFFECTS

- ABA induces winter dormancy by suppressing mRNA production. Without mRNA auxin and gibberellins are no longer produced.
- ABA enters guard cells during periods of water stress, and brings about the outward transport of potassium ions (K+). Where salt goes so goes water so guard cells lose turgor pressure and collapse which, because of their unique shape, closes the stomata, preventing the further loss of water.

TROPISMS: PLANT GROWTH RESPONSES AND MOVEMENT

Plant Growth Responses (Cell Division Involved)

Phototropism - plants grow in response to light (nonreversible growth towards light stimulus. Light causes auxin to move laterally in the apical meristem. An unknown (yellow pigment) receptor absorbs blue light and helps transport auxin to the unlighted side where cells respond by elongating auxin is quickly inactivated by enzymes further down the stem.

Gravitropism - non-reversible growth towards gravity. Young roots tend to grow down into the soil thanks to the combinations of gravity and inhibitory effects of high auxin concentrations

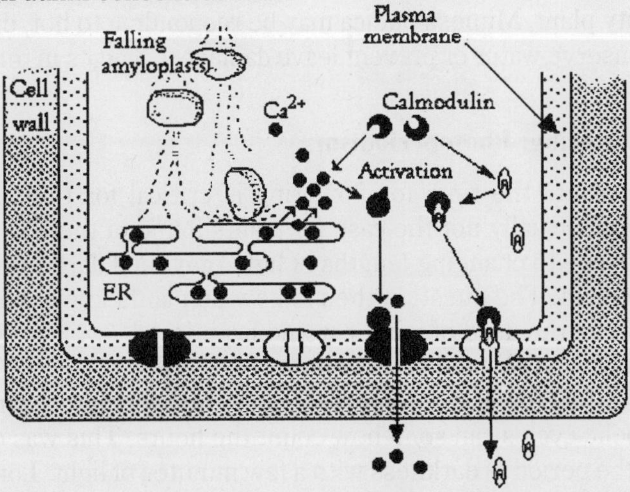

In a hypothetical cellular mechanism, falling (amyloplasts) statoliths in the horizontal root cause the release of calcium ions, which activate calmodulin, a well known enzyme activator. The activated enzymes, in turn start calcium and auxin pumps working in the nearby plasma membrane.

Both Ca2+ and auxin leave the columella cells and migrate to the lower margin of the root cap, whereupon the auxin begins its journey along the lower side of the root--its accumulation there leading subsequently to the inhibition of cell elongation. Thigmotropism - nonreversible growth towards touch stimulus. Special young slender branches called tendrils respond to touch by growing towards the contacting surface and entwining around it. Both auxin, light, and cell division are probably involved.

PLANT MOVEMENTS - NASTIC RESPONSES

"Sleep movements". In many plants, during the last glimmerings of day, leaves bend up and together vertically (like hands in prayer) This has been shown to reduce the loss of heat during the night. When light returns the leaves fold down again for maximum exposure to the sun. A special structure at the base of the leaf's petiole, called a pulvinus, contains motor cells specialized in pumping potassium ions into nearby tissues changing the turgor pressure. The result is the nastic movement.

Solar Tracking- "Heliotropism". This nastic response, often seen in sunflowers, causes the plant to act as if it were a astronomical tracking device aimed at the sun. In fact the flowers are often parabolic in shape, ideal for receiving the maximum amount of light energy. Growth is not involved in this fast and reversible response (which is repeated in an identical fashion day after day.)

Thigmonastic response. Some plants have the ability to respond quickly to touch by collapsing their leaves, often together (at the base of the petiole). The sensitivity plant, Mimosa pudica may be responding to hot, dry conditions in order to conserve water or prevent leave damage. Changes in turgor pressure is involved.

Light and Flowering: Photoperiodism

In many plants the time for flowering is critical for their reproductive success. (This is usually not the case for plants found in the tropics. Why?)

The response to changing lengths of light (day) and dark (night) is known as Photoperiodism. The question then is how do plants know when to bloom, since to flower in the dead of winter would be suicidal.

WHAT WE KNOW (and don't know) ABOUT PHOTOPERIODISM

The critical factor is the length of night (uninterrupted darkness). The plant's metabolic cycles can somehow count the hours. This was confirmed by interrupting the period of darkness with a few minutes of light. Long Day-Short Night plants (treated with a flash of light to breakup a long night) will begin to flower during winter. Those plants which need Short Days (actually a Long Night) to flower may be prevented from doing so by the same procedure, fooling them by apparently resetting their counting (clock) mechanism after the flash of light.The wavelength of the light is also critical. Red light (predominant during the day) is necessary to reset the clock which counts the hours of darkness. Far-red light will reverse the effects of red light, but has no direct effect by itself.

The chemical which detects and is changed by red or far-red light is called phytochrome. The form of phytochrome which absorbs red light is designated Pr, and the alternate form which is sensitive to far-red light is Pfr. The conversions between these two forms of phytochrome can be expressed by the following:

red far-red spontaneously
Pr ----> Pfr Pfr ------> Pr Pfr -----> Pr
lightlight given time

Pr is the more stable form and slowly accumulates spontaneously during the night (or is synthesized etc.). Pfr predominates during the day since sunlight has more red light than far-red.

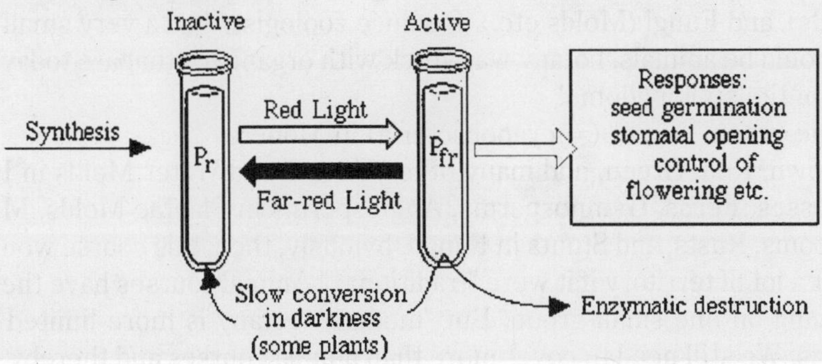

solutions of the two photoreversible forms of phytochrome. Absorption of red light causes the bluish Pr to change to the blue-greenish Pfr. Far-red light reverses this conversion. In most cases, it is the Pfr form of the pigment that triggers physiological responses in the plant.

Phytochrome is only an entraining mechanism, a link between the actual clock (mechanism unknown) and the hormone (yet to be discovered, but presumptively called florigen) which activates the flowering process. Many plants are simply not light sensitive which means that the change in concentration of other hormones such as auxin are involved in the flowering process. This area of botany is under active investigation.

WHAT IS A PLANT?

There is a $64,000 question. It seems the accepted definition has changed through history.

Long ago it was simple. If it moved around it was an animal (anima = movement) and if not, it was a plant (anchored). So scientist created the subdisciplines of biology called zoology (study of animals) and botany (study of plants). Of course popular culture decided that "animals" meant only the warm fuzzy or cute ones nursing babies (mammals). More-enlightened people figured out that most of the animals lack backbones and are really cold and prickly or cold and slimy (think invertebrates). In pop culture, plants are not really alive (of course this is wrong!). As we studied the plants and the animals more closely, scientists decided that our old ideas were really erroneous. There are so many organisms out there and their features intergrade finely. It became hard to distinguish organisms in one way, but clearly the two-kingdom system was inadequate. In particular, the group called plants was really a mixture of very-different organisms; it needed"to be divided. Enter the six-kingdom system. The Eubacteria and Archaebacteria are most unique (prokaryotic). The two kingdoms have organisms that lack nuclei or other membrane-bound organelles.

Protista is an artificial catch-all with intergrading forms and difficult organisms. Most are aquatic and are fundamentally unicellular. These are crowned by three well-developed kingdoms: Plantae (Plants), Animalia (Animals), and Fungi (Molds etc.). Because zoologists had a very small idea of what should be animals, botany was stuck with organisms that are today fournd in four of the six kingdoms!

Blue-green "Algae" (= cyanobacteria) in Monera

Brown, Red, Green, and many other "Algae" and Water Molds in Protista

Mosses, Ferns, Gymnosperms, Angiosperms in Plantae Molds, Mildews, Mushrooms, Rusts, and Smuts in FungiObviously, then, this course would have to cover a lot of territory if it were "traditional." Animal courses have the luxury of focusing on one small group. But "modern" botany is more limited than it once was. We still need to cover more than animal courses and thereby get the advantage of seeing a vast array of diversity in form and function. However eliminating all prokaryotes except cyanobacteria (because they are the closest relatives to chloroplasts), all algae except the green algae (because they are the closest relatives to true plants), and all fungi (they are not plants by any modern concept).

CLASSIFICATION AND TAXONOMY

It is important that biology as a discipline develop some orderly way of handling taxonomy. Here are the levelsof classification that are commonly recognized:

Kingdom (Eubacteria, Archaebacteria, Protista, Animalia, Fungi, Plantae)

- Division (Anthophyta = flowering plants)
- Class
- Order
- Family
- Genus
- Species Latin binomial (two names)
- Variety

Why do we use Latin? Well, Latin was selected as the universal language for taxonomy very early on, which means that you can pick up a journal in botany written in just about any language and somewhere in the title or abstract the taxonomic name will appear in Latin. You will instantly know whether or not this chapter is working with the taxon you prefer to study. Secondly, Latin is a dead language. Because it is no longer in use by people, the meanings of root words do not change with time. This gives the language some stability that would be demanded by a universal language. Thirdly, because the international bodies that determine nomenclature for taxonomy come to agreement on the names to be used, each organism has a particular name that is unique to that organism. Common names, while cheerful and sometimes cute, do not share these properties with Latin names. So there are at least 100 unrelated plants called "Black-Eyed Susans" world wide...if you use "Black-Eyed Susan" in your published article without any reference to a Latin binomial, other scientists world-wide would make different assumptions about the plant you worked with, leading to chaos in the literature.

Humans have been selecting various forms of particular taxa for use as food and so on. This unnatural selection process has produced plants that are unique but are still belonging to a single species. Good examples include the many cultivated subspecies of Brassica oleracea. The names below are all Latin subspecies or variety names that go with certain common vegetables. If you dissect the Latin roots of these namescapitata = cabbage

- Acephala = kale
- Gemmifera = Brussels sprouts
- Italica = broccoli
- Botrytis = cauliflower
- Caulorapa = kohlrabi

Of course human selection created more than just subspecies! Indeed each subspecies includes many cultivated varieties or cultivars of each subspecies. For example, a perusal of any seed catalogue

- Brassica oleracea capitata 'Late Flat Dutch'
- Brassica oleracea capitata 'Copenhagen Market'

You might notice that the cultivar name is NOT Latin and is surrounded by single quotes rather than typed in italic font.

Older Ways of Classifying Plants

Plant classification is one of the older pastimes in botany. Every culture, society, and religion has taken it as its duty to name and organize the plants in its area into some manageable arrangement. The arrangement at first was meant to facilitate communication between people about the plants under discussion.

The earliest organizations, naturally, were based upon the uses of plants by people. Obviously some plants were useful as vegetables, others were fruits, some make great spices for other foods, some were a good source of sugar, or perhaps starch. Oils could be extracted from others. Some were good for spinning fibres into thread for clothing. Yet others were ornamental. But as is true of any classification scheme based upon human uses, there are problems of several sorts. The uses of plants by one civilization might not be the same as those by another. Even among individuals in the same society, one person's fruit is another's animal fodder. Moreover there is the conundrum of what to do with plants that present several different uses. George W. Carver demonstrated many uses for peanuts. Henry Ford produced an automobile with body parts made of soybean products. But the worst problem with human-use classification is that it puts related organisms in different classification categories. There is no relationship between the classification system and the evolutionary pathway through which the diversity was obtained.

Other early attempts at classification focused upon the form of the plant. Thus one category would be the herbs; here we use herb in its botanical sense meaning non-woody plants. A second category would be woody plants. These plants have secondary xylem and would need to be subdivided into several sub-categories of form: shrubs, vines, and trees. Each of these categories would have to be subdivided into deciduous and evergreen. Of course whether an evergreen vine is closer related to an evergreen tree or to a deciduous vine is a very good question. And of course there are both woody and herbaceous vines. It is sounding messy. Also, we find that a natural grouping may have members in all of these "unnatural" categories.

Humans have also noticed that the seasonality of plants varies and so can be another way to divide up the plants into categories. Thus in one category we have annual plants which are planted in the spring, flower that summer, and die in the fall. Then we have the perennial plants which are planted in one year, grow vegetatively in the first year, overwinter, and then flower in each year thereafter. Nurseries like to sell annuals because they know customers will come back next year to replace them all. They sell an entire flat of annuals for a small price. Perennials are planted, and then never replaced, so nurseries place a premium price upon them.

However, there is a real "gotcha" among the perennials in some nurseries, however...biennials act like perennials in the first and second year, but die all the way to the ground at the end of that second year. You will have to replace those. Many seed catalogues are divided into annual and perennial categories. However even a casual gardener will be able to find plants of the same kind in both the annual and perennial categories. Certain species of a genus may be perennial, and other species of the same genus may be annual. It is not a natural division.

Another way of dividing the plants is obvious in nursery catalogues: by climate. Tropical plants originate in areas of the world between the tropics of Cancer and Capricorn. This part of the world is very warm year-round with frost never occurring in the year. This is explained as the sun passes directly overhead on at least one day each year, and may do so for much of the year.

Tropical plants generally cannot withstand any kind of frost. Subtropical plants are slightly more hardy and can tolerate light frost but not heavy freezes. Hardy plants can tolerate frozen periods of various durations and temperatures. The hardy plants are classified on the basis of hardiness zones. In North America these range from Zone 1 (extremely cold and long winters) to Zone 10 (no frost at all). Subtropical plants survive in zones 8-10. Some hardy plants can only survive in the warmer zones (7-10) while others are tough enough to survive in zones 1 and 2! In nursery catalogues the coldest zone for each plant is noted and gardeners are well-advised to heed those. Connecticut is zone 6 except for the northwest corner which is zone 5. Along the shore one might experiment with some zone 7 plants if a sheltered spot is available. However, again, close relatives can be tropical plants and extremely hardy plants. This is not a natural classification scheme. It is useful but not based upon genetics and evolution, the cornerstones of biology!

All of the possible classifcation criteria above create artificial groups of virtually unrelated plants AND separate out close relatives in what appear to be distinct classification categories. Juniper is an evergreen conifer that is a tree, holly is an evergreen dicot that is a tree. These criteria also artificially separate related plants. Spruce is an evergreen woody tree, larch is a deciduous woody tree but both are conifers. So we really needed something better; this was recognized by Linnaeus and others who established the foundations of Botanical Taxonomy or what is known as Systematics.

5

Methods of Mushroom Cultivation

Mushrooms are a source of high quality proteins, essential minerals and vitamins. Being low in starch content, they are good for diabetic patients. Mushroom is grown on agriculture residues like straw, leaves, and sticks which (after use) can be recycled back as organic manure. Mushroom cultivation demands no land but closed structure, which can even be a thatch-hut. it can be a remunerative cottage industry for rural women.

CULTIVATION AND TISSUE CULTURE METHOD

Select fresh mature mushrooms (straight from the bed), wash them gently under running water to remove surface dirt. Blot it dry. Wipe the surface gently with 70 per cent alcohol.

Using a sterile knife, cut a small slit at the bottom of the mushroom. Tear it into two halves and avoid touching the inner surface. Transfer a few pieces of tissue (Pseudoparen-chyama) from the centre of the mushroom to the media in plants.

Spore Culture Method

The surface of a mature (open) mushroom should be disinfected by wiping gently with 70 percent alcohol. Remove the spore and place the cap with gills downwards on a piece of sterile white paper. Cut tiny piece of spore print and soak it in a few mi. of sterile tap water.

Dip a sterile inoculating loop into this spore suspension and streak gently across the surface of an agar plate. Incubate the plate for a few days and a white mycelial (white thread like structure) mass resulting form the growth of germinated spores becomes evident.

Preparation of Spawn

Cook the grains (rye/sorghum/maize/wheat) in water until they are ready to burst. Drain off excess water. Mix in 2 percent (w/w) lime (calcium carbonate), fill loosely into glass bottles and plug with cotton. Sterilize them in an autoclave for 30 minutes at 1210C and allow it to cool. Inoculate the bottles with the prepared pure culture and incubate at 25^0 C for 10-15 days. When the whitish mycelium has spread over the Surface, the spawn is ready for use.

MATERIALS OF MUSHROOM PRODUCTION

For 100 kg Mushroom production - Dry paddy straw (not more than one year old) -100 kg; gram powder-2 kg; spawn bottles-l0; polythene *bag-i* kg; water and a shady structure.

PREPARATION OF SUBSTRATE

Use Dried paddy straw, 500 gms per bed (Hulled maizecob, sugarcane bagasse, cotton waste, paper waste, etc., can also be used instead of paddy straw). Cut the paddy straw into small bits of 3 to 5 cm. and soak it in cold water for 6-8 hours.

Take it out and then immerse in boiling water for 15-30 minutes and drain off excess water. Dry the straw in the shade for 2 hours and then use. Remove the spawn from the bottle using a small iron rod with hook on one side. Dip the iron rod in antiseptic solution before use.

Separate the spawn into four lots. Fill the first layer with straw bits upto 5 cm height in the bottom of the polythene bag. Sprinkle spawn over the entire surface of the straw layer. Add second layer with the straw bits upto 10 cm height and sprinkle spawn as before. Similarly, repeat the spawning for third and fourth layers of straw bits. Then, fill the final layer with straw bits upto 5 cm height. After spawning the beds keep the beds in the growing room for spawn running. Keep bags in the dark for a period of twenty days. White mycelial growth appears throughout the bed.

Cropping & Harvesting

After the spawn running period *(nearly* 15 days), remove the polythene cover and keep in the rack. The relative humidity is maintained above 75% by spraying water on the floor and gunny screens, but watering the beds is avoided for first two days.

After two days, sprinkle beds with water twice daily. Allow aeration and light for 20-30 minutes in morning and evening by opening windows of the growing room. First harvest can be made 24 to 26 days after spawning. Harvest fully-opened mushroom before the edge of the top rolls downward. Two to three harvests can be taken at 7-10 days interval. The yield is about 35 to 85 percent of the substrate (w/w).

HISTORY AND BACKGROUND OF MUSHROOM

FOLKLORE

Mushrooms, the plant of immortality? That's what ancient Egyptians believed according to the hieroglyphics of 4600 years ago. The delicious flavour of mushrooms intrigued the pharaohs of Egypt so much that they decreed mushrooms were food for royalty and that no commoner could ever touch them. This assured themselves the entire supply of mushrooms. In various other civilizations throughout the world, including Russia, China, Greece, Mexico and Latin America, mushroom rituals were practiced. Many believed that mushrooms had properties that could produce super-human strength, help in finding lost objects and lead the soul to the realm of the gods.

EARLY CULTIVATION

France was the leader in the formal cultivation of mushrooms. Some accounts say that Louis XIV was the first mushroom grower. Around this time mushrooms were grown in special caves near Paris set aside for this unique form of agriculture. From France, the gardeners of England found mushrooms a very easy crop to grow which required little labour, investment and space. Mushroom cultivation began gaining popularity in England with more experimentation with spawn and publicity in journals and magazines.

In the late 19th century, mushroom production made its way across the Atlantic to the United States where curious home gardeners in the East tried their luck at growing this new and unknown crop. However growers had to depend on spawn imported from England which, by the time it reached the U.S. was of poor quality. This industry text suggested that mushroom growing was perfect for florists. Since they grew flowers on benches, florists could just slide mushroom beds right under their flower benches and gain a profit in growing two crops in the area of one. Falconer also thought that mushroom growing was ideal for farmers who had access to growing their own manure and spawn.

At the time skilled labour was not a necessity of mushroom growing. It was recommended to house wives as well as a source of home income. Not only did Falconer's book develop target groups for which growing was suited. It also contained much practical advice on building beds for cultivation, the perfect growing temperature and where mushroom markets were developing.

The first producer of pure culture virgin spawn was the American Spawn Company of St. Paul Minnesota, headed by Louis F. Lambert, a French mycologist. He began the production of brick spawn and advertised it across the country as "Lambert's Pure Culture Spawn." It soon proved to be a very popular product. This spawn received a silver medal at the Universal Exposition in St. Louis in 1904. A measure of Lambert's success was that English spawn was soon being sold under the name "English Pure Culture Spawn."

By 1914, mushroom marketing began to play a much greater role in the industry. It was estimated in one publication that four to five million pounds of mushrooms were grown in the U.S. Cost to the mushroom grower was fifteen to twenty-five cents. At retail, the price was forty to sixty cents per pound. Marketing became very important with a popular theory being to bypass the middle man and aim directly at consumers. It was pointed out that attractive containers would move product and only good looking product would sell.

U.S. Mushroom Cultivation

In 1891, the first book on mushroom growing was published and it shed new light on the theory of cultivation. William Falconer, a mushroom grower and experimenter from Dosoris, Long Island agreed with the recommendations of agricultural journalists and compiled their theories in Mushrooms: How to Grow Them; A Practical Treatise on Mushroom Culture for Profit and Pleasure.

New Developments

At this time of budding industry developments, one problem continued to face farmers intent on perfecting the growing conditions–poor spawn quality. With such an important ingredient standing in the way of mushroom cultivation, the U.S. Department of Agriculture began work on manufacturing spawn as an alternative to the English product that had proven to be unreliable. In 1903, after much experimentation, two USDA scientists had produced the perfect pure-culture virgin spawn. The U.S. Mushroom Industry was finally freed from depending on imported spawn that had caused so many problems in the past.

Industry Growth

Moving from an easy back yard crop in the early days of cultivation, to a large money maker, the mushroom industry began to grow in certain parts of the country. Concentrated areas of industry growth were Long Island, Central Massachusetts, Chicago, Michigan and California. Southeastern Pennsylvania was (and still is today) the largest centre of mushroom production in the country. In 1924, the Pennsylvania Department of Agriculture boasted that 85% of U.S. mushrooms were grown in Pennsylvania. In 1930, the U.S. Census Bureau revealed that there were 516 growers in the U.S. and that 350 were in Chester County, Pennsylvania.

After 1930, the industry changed rapidly with better spawn production, the development of synthetic manure, and improvements of mushroom growing houses- hence the production of healthier crops. Organizations such as the Mushroom Growers' Cooperative Association were developed to assist and protect growers. In attempt to find better marketing techniques for mushrooms, the Farm Credit Administration became involved. Pennsylvania State University also became a major factor in the growth of the U.S. mushroom industry helping improve productivity dramatically in the 70's and 80's allowing growers to produce more and more mushrooms per square foot of growing area.

Development of AMI

One organization was developed to coordinate the actions of independent growers and act on behalf of the mushroom industry as a whole. The American Mushroom Institute was brought to life by the Chester County, Pennsylvania growers. Unfortunately, the first meeting for the development of the AMI was held December 4, 1941, only three days before the Japanese attack on Pearl Harbor. With America's involvement in World War II, this new mushroom growers' organization was put on hold. After the war was over and things began settling down for the nation, the American Mushroom Institute was finally officially organized with over 275 growers signed up to join. On January 14, 1955, AMI was legally incorporated as a non-profit organization.

Their goal was to promote the consumption of all cultivated mushrooms through research, advertising, publicity, merchandising, consumer education and government relations as well as to assist the industry in developing better growing and handling methods. Every broadcast medium was employed to promote mushrooms. Radio, television, magazines and newspapers told of different ways to eat them. Even produce stores and supermarket chains were displaying mushroom merchandising posters. The American public was learning that mushrooms weren't something just used to garnish steaks. The AMI promoted recipes that used mushrooms in casseroles, appetizers, salads and other combinations. These were the first organized marketing efforts of the American Mushroom Institute.

THE NATIONAL MUSHROOM GROWERS ASSOCIATION

In 1985, the National Mushroom Growers' Association was established in Illinois to promote the sale of fresh mushrooms on a national basis. They developed a newspaper and magazine promotion programme. In spite of its small budget, the programme was very successful in receiving coverage from national women's magazines and newspapers.

THE MUSHROOM COUNCIL

In 1990, the Mushroom Promotion, Research and Consumer Information Act was passed by Congress to strengthen the mushroom industry's position in the marketplace, maintain and expand existing markets and uses for mushrooms, and develop new markets and uses for mushrooms. In 1993, the Mushroom Council was established to carry out the direction of this act. The Council started out with a meager budget and a lot of inspiration about promoting mushrooms.

They began doing research to closely define the mushroom user which became the foundation for all of their communication efforts. Once the ground work was laid, a successful promotions programme began to shape.

Immediate targets for consumer communication were food editors of newspapers and magazines, TV and radio personalities, chefs and cookbook writers. Mushroom recipes went out to hundreds of venues each year — thus increasing consumer awareness and demands for literature on mushrooms. In 1996 the Mushroom Council made the pages of more than ten national women's magazines including Family Circle, Women's Day and Good Housekeeping.

Today, the Mushroom Council plays a very important role in the national promotion of fresh mushrooms through consumer public relations, foodservice communications and retail communications.

Many different venues are used in promoting fresh mushrooms to consumers such as working with professional chefs in developing and promoting new recipes, working with produce department managers to maintain the highest quality mushroom product for customers and sending out thousands of brochures each year to consumers hungry for new mushroom ideas. Thanks to the Mushroom Council, mushrooms have their own month to be honored and eaten. September is National Mushroom Month.

Today mushrooms are commercially produced in virtually every state. Pennsylvania, however, still accounts for 61% of total U.S. production, which in 2006/07 reached 827 million pounds.

From the caves of Paris to the dinner tables of millions of Americans, fresh mushrooms have come out of the dark and into a spotlight that's intensity is ever increasing.

THE HISTORY OF MUSHROOM CULTIVATION IN INDIA

People have been eating mushrooms for a long time. They used to go the forests and other wild places and based on their personal knowledge of edible and poisonous mushrooms they used to collect the edible ones. Even now some people collect wild mushrooms from the forest and eat them. Sometimes they turn out to be poisonous and have harmful effects. In India the 'Guchhi' mushroom was a delicacy known to the people prior to the 1950's. It came mostly from Kashmir. Around the early 1950's the government of Himachal Pradesh appointed Shri S.S.Jain as its first Asstt. Plant Pathologist and Mycologist for the state. He worked in the Wild Flower Hall in Chharabra, Shimla. He was touring the interior areas of Himachal to help the apple orchardists and the farmers control the diseases of apples, other fruits and crops like potatoes and wheat. He noticed the poor hand to mouth condition of the poor farmers in the hilly state of HP. He wanted to help them.

While staying with some farmers in interior areas he noticed that there were rotting twigs and branches of apple and other fruit trees and wheat straw in the barn along with cow dung and in the environmental conditions there were a profusion of mushrooms growing in the dark barns. This led him to think of using the waste material with the farmers for growing edible mushrooms. He searched the literature and found that edible mushrooms were being grown in France and Japan. He made a research proposal on growing of edible mushrooms and got the permission for the same from the state government and obtained the mushroom spawn from Japan and France and started a laboratory in Solan, Shimla Hills and started his research experiments on growing edible mushrooms of Agaricus and other species, in laboratory conditions simulating those found in Himachal Pradesh.

When he was able to grow the mushrooms successfully on substrate prepared from rotting apple tree twigs and branches, cow dung and wheat straw etc. he published the results through the magazine of the HP state Extension department. These results when publicized and brought to the notice of the farmers and the poor people people in the state led to dissemination of information and spawn to them and mushroom farming started in Himachal Pradesh. The laboratory established by Shri S.S.Jain, the pioneer of mushroom cultivation in India later became the only important centre for training in mushroom cultivation to farmers of Himachal Pradesh and other states as also the mycologists and plant pathologists from all over India. Mr. Seth and others actually worked with Shri Jain and later became important in the area of mushrooms. But it is a fact that Shri SS Jain in Solan successfully completed the first research project in India. Shri Jain left Solan in 1962-63 for Cuttack.

Shri S.S.Jain then became the OSD and set up the first campus and office of the HP Agricultural University in Solan and later joined Central Rice Research Institute, Cuttack Orissa under the ICAR as a Senior Scientist Plant Pathologist and retired from there in 1978 after having published over a hundred research papers and also having been the Chief Editor of the International Rice journal 'Oryza'. Shri Jain had also done a monographical study of the Stem Rot disease of rice and also discovered the bacteria Xanthomonas oryzae, which caused the Bacterial Blight disease of rice, and Dr Devdath did his Ph.D on this bacteria and disease. Before Shri Jain expired after prolonged coma in Apollo Hospital Delhi and a small nursing home in Baraut, District Baghpat (Meerut), UP, he had been an award-winning President of the Rotary Club Baraut for his excellent social service work also winning International citation from Rotary.

Now the mushroom cultivation in India is something I could not have imagined from what I saw in the laboratory in Solan. I found a huge factory in Maharashtra near Talegaon, Lonavala, where there were huge godowns in a factory where tray upon tray of white beautiful button mushrooms created a sense of wonder and awe in me and I began thinking of the experiments of my father in the small room in Solan way back in the 50's when I was a student and used to walk to his office nearby and have lunch with him in his laboratory. By chance the Mycologist in that factory turned out to be one who had undergone training in the laboratory established by my father in Solan. Mushroom cultivation has become a huge export oriented industry and large foreign exchange earning business and also profitable for small time growers. Many Universities and State departments of agriculture as also private people are giving training in growing mushrooms, which are mostly, exported and also used in many good hotels in a variety of culinary delights.

LOW TECH METHODS TO GROW MUSHROOMS

Mushroom growing is both a science and an art form, and many people have developed their own unique methods for growing mushrooms. Some of these methods are good, some are labor intensive, other methods yield high volume or special quality fruitbodies. Over the many years we have been working and consulting in this field we have seen many methods, everything from in-ground growing in Mexico and Africa, to the highest tech autoclave based method used by the multinational agribusiness giants. Most of the books published today start out with the assumption that you have plenty of money to go out and buy the most expensive of boilers and baggers and autoclaves. In our experience this is not usually the case, especially when you are just starting out. So we would like to offer the benefit of our experience here, and show you some low tech methods that work well and can be accomplished for a low cost.

The goal of any mushroom farmer should be: The Production of High Volume of Mushrooms at a Low Cost, in a reasonable amount of time without too much waste, contamination or expense in new equipment. Further goals should be to have a system simple enough that you can easily train common farm laborers to perform every step of the process. Since most agricultural workers are not college educated microbiologists, it is more practical to simplify the system as much as possible so anyone can be trained to do the work, rather than depending on the availability of trained personnel that already know all the concepts involved.

Let's define the goals a bit further. High Yield: Every common strain of mushroom can be grown at 100% Biological efficiency or more, and in fact has to be if the project is to be successful. Often in Oyster cultivation 200% BE can be accomplished. Even with Shiitake, Lions Mane, or Maitake, 100% BE is quite possible to achieve with even minimal attention to your process. So what is this BE and how is it measured?

Biological Efficiency, abbreviated BE, is the most important and the most basic of all measurements in mushroom cultivation. BE is the comparison of the dry weight substrate to the fresh harvested mushrooms. In other words, we all buy our substrate in the dry form, like sawdust as an example. If you buy 1000 pounds of dry weight sawdust, and you are able to harvest 1000 pounds of fresh mushrooms, that is a 100% BE. If you start with 1000 pounds dry straw, and harvest 1500 pounds of fresh Oyster Mushrooms, that is a BE of 150%. Keep it always in mind that 100% BE is the minimum acceptable yield of mushrooms for your farm to be commercially successful. How can we start with 1000 pounds substrate and end up with 1500 pounds of mushrooms? Keep in mind, mushrooms are mostly water, and this is being added to the substrate. So what you harvest is anywhere from 50% to 90% water. Make sure you measure the weight of your bags, and measure the weight of your harvested mushrooms, because without the BE figures for your process, it is impossible to optimize your production and accurately predict the yield ahead of time.

Examples of changes in BE would be the comparisons of two different strains of Oyster mushrooms: They are bothPleurotus ostreatus, but maybe one came from a high altitude, and one came from a tropical forest. These are definitely going to perform differently on your farm, and they will absolutely and without question have a different BE when you grow them side by side. So how do you know which is the best strain for your operation? There is only one way to know – and that is to run trials and see which one performs better. A strain that works GREAT for another grower might work terribly for you.

You have different climate, different sterilization methods, different substrate, different spawn, and most importantly you have different competitor organisms on your farm. These things cannot be factored in ahead of time, only trials on your farm will give you realistic BE figures. However, ALL commonly cultivated mushrooms will give you AT LEAST 100% BE, so if you are not getting that from your operation, then something is wrong.

The question comes up "What mushroom should I grow?" Most people who get into this trade want to start off growing Shiitake. Just like most kids learning to drive want a new Ferrari. Maybe it is more practical for them to learn to drive on the old family Ford, and move up to the race car as their skill level (and income!) increases. The same holds true of mushroom growing. It is a lot easier (and cheaper!) to start out growing Oyster mushrooms, then graduate to other species as both your skill level and available equipment grows. So the remainder of this discussion will focus on Oyster mushrooms for the most part, although the general concepts hold true for other species as well.

Based on the teachings of some of the common books available today, many people start growing mushrooms by soaking straw in hot water for a pasteurization step. This system works, although it is our least favorite of all the pasteurization methods for many reasons, all going back to the basic goals of "The Production of High Volume of Mushrooms at a Low Cost". The hot water method is difficult, costly and all too often does not produce 100%+ BE.

We prefer the cold pasteurization methods because we are lazy, cheap, and do not like wrestling large bundles of hot wet straw. And oh yeah, did I mention it always yields a higher BE?

COLD PASTEURIZATION METHODS

It is necessary to pasteurize or sterilize your substrate prior to inoculating it with spawn. The decision of pasteurizing verses sterilizing depends on both your substrate and your species grown. In general, if you have a low nutrient value substrate like paper, cardboard, or straw, then pasteurization is sufficient and far cheaper than true sterilization.

Sawdust varies in nutrient value, and can sometimes work well with a simple pasteurization step, and sometimes needs to be sterilized. For those that say Shiitake cannot be grown on pasteurized straw, I encourage you to look at this short Write-Up from a grower in San Diego who grows Shiitake quite successfully on straw. Since Straw works the best for Oyster Mushrooms, I will describe the process on straw: To do a cold pasteurization step, you will need to soak your straw in a solution that will hydrate the straw and kill off the majority of single celled organisms present. There are a number of things you can use for this, but by far the best is hydrated lime. The exact concentration is not critical, just a big double handful in a 55 gallon drum of water. Then submerge your straw, weight it down so it does not float, and wait 12 to 24 hrs. No longer, as you do not want to develop a bunch of lime resistant organisms, between 16 and 18 hrs is perfect. After soaking, drain the excess water, bag the straw and inoculate it with your spawn. Works great, the typical BE is 150% to 175%, and the quality of the fruit bodies are better than with the hot water method. Here are the tips on what lime to use:

Lime Types and Usage in Mushroom Farming

The term "Lime" is widely used in English, and the same term often refers to different materials.

So an explanation of different types of lime and their usage is in order:

1. *Calcium Carbonate lime*: This is limestone, marble, chalk, eggshell, coral or oyster shell. This material is used as a pH buffer in substrate and media. This type of lime does not cause a major change in pH, but acts as a buffer, that is it holds the pH steady as the mushrooms grow. This is the preferred amendment for substrate to ensure the substrate does not go too acid during the growth cycle. This type lime will not work for pasturizing!
2. Calcium Hydroxide or Hydrated Lime, also known as Quicklime. This is used for pH increase and is used primarily for pasteurizing or sterilizing substrate. Many mushrooms do not grow well if this type of lime is added to the substrate, and is therefore NOT used for amending the substrate, it is used only for pasteurization, through the mechanism of rapid pH change. Adding about 3 pounds of hydrated lime (a big double handful) to a barrel of water will yield a high pH, around pH 12-13, which is sufficient to pasteurize straw in about 4-6 hrs.

Hydrated Lime is made by burning limestone (calcium carbonate) which converts it to Calcium Hydroxide. The starting material can be any of the calcium carbonate sources, or the raw material source can by Dolomite Lime, which has a high magnesium content. When Hydrated Lime is made from Dolomite, it is often called "Builders Lime" and is used in the concrete industry. High Magnesium Lime or Builders Lime WILL NOT WORK for growing mushrooms. When you purchase Hydrated Lime for pasteurization of your substrate, make sure it is the High Calcium Lime (90% or more calcium - usually about 95-96% calcium). If the bag says high magnesium, or states a percentage of magnesium anything over 10%, do not use it. The high magnesium will stunt the mushroom growth.

THE WISDOM OF SIMPLICITY

If you have been intrigued by the prospect of growing your own mushrooms, perhaps after a bit of research you became daunted by the complexity and potential expense of "sterile techniques," "clean rooms," "laminar flow hoods" and "agar media." Or perhaps you are the more adventurous type. You plunged ahead with valiant attempts to isolate pure mycelial cultures, only to discover your beautiful white mycelium bursting into a rainbow of brilliantly colored fuzzy mold spores or sleek, shiny pools of bacterial contaminants. Take heart, you are not alone!

As chairman of the Cultivation Committee for the Mycological Society of San Francisco (MSSF), I have talked with many aspiring cultivators, and those are two of the most common roadblocks preventing progress in mushroom cultivation.

Recognizing the need to help overcome these challenges, the MSSF cultivation committee has periodically offered classes and workshops on how to grow mushrooms. I'm indebted to past teachers of these workshops — Sal Billeci, Dr. Joe Mandell, Fred Stevens and Don Simoni — for some of the methods I'm presenting here.

Goals and Objectives

The ultimate goal, of course is to start with one mushroom fruitbody and turn it into a bunch of fruitbodies. There are two mechanisms by which mushrooms can reproduce -- by spores (similar to the seeds of plants) or by propagation of tissue extracted from a fruitbody (similar to plant propagation through cuttings).

To get started we will use the latter technique, which is simpler and which offers a better chance of success for the beginning cultivator.

The objectives are:

- Start with a sterile piece of mushroom tissue and transfer it onto a nutrient medium (actually to a series of media in phases, which you will see as you read on),
- Create multiple copies of the culture,
- Avoid contamination of the culture,
- And, finally, create an environment in which the fungal organism is induced to fruit.

There are three distinct phases through which your culture will pass:

1. Sterile culture on an agar medium
2. Sterile culture on grain, also known as "grain spawn"
3. Fruiting on a pasteurized substrate

Now don't let the word "sterile" and a bit of terminology deter you. Yes, potential "contaminants" exist everywhere in the environment, in our homes, and even in clean rooms and laboratories. The objective is to prevent them from entering our cultures and competing for the "food" we intend for the sole use of our mushroom cultures. With a modicum of care and a little practice using some simple techniques, this process can be successfully mastered by anyone.

Phase I: Agar Preparation and Sterile Culture

Agar Preparation: The first step is to prepare the agar media. Derived from a seaweed, agar plus added ingredients is often used to start and isolate cultures. Microbiologists tailor agar media to selectively grow fungi by adding various nutrients, e.g. minerals, antibiotics, etc. Agar media also has the advantage that contaminants are usually detected easily on the media surface, and thus can be eliminated early in the growing process. Many types of agar can be used to grow fungal mycelium. Two of the more common are Potato Dextrose Agar

(PDA) and Malt Dextrose Agar (MDA). Premixed commercial preparations are available from a number of sources. Although these may be considered relatively expensive compared with preparing the agar "from scratch," if one has more money than time, they are a good choice.

For the do-it-yourselfer, two methods of preparation for PDA are given below. The first is a method I have used with success; it is based on that described in The Mushroom Cultivator, by Paul Stamets and J.S. Chilton. The second is a method successfully taught and used by the MSSF cultivation group. Try one or the other, or both, on for size. You will soon either adopt one that suits you well, or you will more likely incorporate some of these jewels of wisdom into your own method, which will undoubtedly evolve over time. The procedures are similar to making a recipe from a cookbook.

Materials Needed for both Methods

- Rolled cotton (available at pharmacies)
- Aluminum foil
- Test tubes with autoclavable screw-caps -- get from a medical or biological supply company, or scrounge. Labs sometimes discard these tubes in vast quantities. Call around.
- A small funnel for pouring the tubes
- Measuring cups
- A pressure cooker
- Two 1-liter small mouth bottles or pouring flasks

First, a note on the pressure cooker. This is probably the most expensive essential item for the home cultivator. If you already own one, and it is of sufficient size to contain the bottles or jars you wish to use, you are all set. If not, a new cooker may cost somewhere between \$75 - \$250, depending on size and where it is purchased. They may be found at most hardware stores.

It is also possible to start out frugally if you are on a limited budget by searching at thrift stores, garage and rummage sales, and flea markets. Good quality pressure cookers may often be found for \$15 - 35. Be sure to get the type with a pressure gauge, rather than the type with weight only and no gauge. In any event, always read and carefully follow the manufacturer's safety instructions for the cooker you choose.

PDA Method No. 1

Materials Needed:

- 300 grams potatoes
- 20 grams agar (from biological supply, health food store or Asian food market)
- 10 grams dextrose or other sugar
- 2 grams brewers yeast (optional)

Boil potatoes in 1 liter of water for 1 hour. Drain the potatoes, reserving the liquid. Mash the potatoes, apply butter, salt and pepper, and eat. Sour cream, chives and other amendments are optional. Once your appetite for mashed potatoes has been satisfied, place the potato broth with the agar, sugar and optionally the yeast in a bowl and mix thoroughly (a wire whisk works well, but stir, do not beat). Pour half of the mixture into each bottle or flask. Fill to only 1/2 to 3/4 of the vessel's capacity. Plug the mouths of the bottles or flasks with cotton and cover with aluminum foil. Put about 1/2 inch water in the bottom of the cooker and insert its rack and then the bottles or flasks. Place the cover on the cooker and lock it in place. Turn on the heat until steam is venting from the cooker. Allow to vent for a few minutes before closing the stopper or applying the weight.

Cook the vessels at 15 psi (250 degrees F.) for 15 minutes. Do not allow the temperature to exceed 250 degrees F. for any length of time as that will caramelize the media and ruin it. After the 15 minutes, turn off the heat and allow the pressure to return to normal gradually (about 45 minutes). Meanwhile, remove the caps from the tubes, place the tubes in clean tin cans or test tube racks, and place cans or racks of tubes and caps on a clean work surface in a dust-free location. Once the media vessels can be handled with a clean towel or oven mitt, remove them from the cooker to the location with the tubes. Swish the media around a bit to mix it thoroughly and remove the foil and cotton. Fill the tubes to about 1/3 their capacity, using the funnel.

Place the caps loosely on the tubes, place the cans of tubes in the pressure cooker (replacing used water in the bottom of the cooker, if needed), bring back to 15 psi (250 degrees F.) and cook for 30 minutes. Again let the cooker cool gradually until the pressure is back to normal.

Remove the cans of tubes and tighten the caps. Anchor or otherwise support the cans such that they are tilted at an angle and the agar media surface is "slanted" in the tubes, offering the maximum amount of surface area on the media for future mycelial growth. (Tubes prepared in this manner are often referred to as "slants.")

As the media cools, it will harden into a gel, and eventually the cans can be returned to their vertical position and the media will remain in place. If tubes will not be used immediately, they may be stored in Ziploc bags in the refrigerator for weeks to months. If stored, be sure to check for mold or bacterial contamination before use.

PDA Method No. 2

Materials Needed:

- 10 ounces or 284 grams potatoes
- 3/4 ounce or 21.3 grams agar
- 1/4 ounce or 8 grams dextrose — table sugar may be substituted

Step 1: Mixing the ingredients:

1. Wash the potatoes but don't peel. Slice into small pieces.
2. Cook potatoes until tender, in about a pint of water.
3. Strain and save the liquid. Discard the potatoes or save to eat later.
4. Mix the dextrose/sugar, potato broth and agar in a metal or Pyrexè bowl. Add water bringing the volume to one quart.

Step 2: Liquefying the Agar:

1. Place the agar mixture in a pressure cooker. Cover the bowl with a sheet of aluminum foil.
2. Bring the pressure up to 15 psi. Cooking 20 minutes will liquefy the agar mixture.
3. Let the pressure cooker cool slowly.

Step 3: Pouring agar into test tubes:

1. With an oven mitt and measuring cup, fill each test tube about 1/3 full. Tin cans can be used in place of test tube racks. Excess agar may be stored in a bottle, closed with a cotton or polyester fiber plug and sterilized later along with the test tubes. If test tubes are not available, small bottles can be used instead.
2. If the test tubes have plastic caps, the caps should be loose so that pressure can equalize during sterilization. Cotton or polyester plugs avoid this problem but require an aluminum foil cover to prevent wetness from condensation when the pressure cooker cools.

Step 4: Sterilizing the Agar:

1. Place the test tubes/bottles containing the agar in a pressure cooker and sterilize at 15 psi for 25 minutes. Do not count the time needed to bring the pressure cooker up to 15 psi.
2. Let the pressure cooker cool slowly. Sudden release of pressure will cause the agar in the tubes to boil violently, splattering the caps and plugs and creating a potential for contamination.

Step 5: Slanting the test tubes:

1. Wipe down with water or a 10% bleach solution the area where the test tubes are to be slanted. Work in a draft-free room.
2. With an oven mitt, slant the still hot tubes against a raised surface. Experiment beforehand with strips of wood or magazines to establish the correct angle.
3. After the agar has jelled, tighten the caps. Store the finished tubes/bottles in a cool, dust free area. A cover of aluminum foil will help to prevent contamination before the tubes are used.

Sterile Culture Technique

Once the agar tubes have cooled, it is time to use them to start mushroom cultures.

Materials needed:

- A scalpel or very sharp thin-bladed knife
- An alcohol lamp or propane torch with fuel and lighting device or matches
- Tin can(s) or test tube racks containing the "slants" and one extra can or rack for storing completed tubes
- Scalpel or knife rest
- Micro-porous tape (such as first aid tape from the pharmacy or medical supply) or masking tape (a lower-cost option)
- Spray bottle containing 10% chlorine bleach, 90% water solution (optional)
- Fresh, clean mushroom fruitbody or bodies (Oyster mushrooms are highly recommended for beginning cultivators)

Select a location in your home, office or lab which is as free as possible from dust and molds. Work on a sturdy table or bench after cleaning the surface with warm, soapy water and wiping dry. (You may also spray the area with the 10% bleach solution for extra protection and wipe with a clean cloth or paper towel.) Have all windows closed and minimize air circulation. (Ideally, try to carry out your cultivation activities first thing in the morning before particulates are kicked up by household activities.)

Have all materials ready and within easy reach. Place the culture tubes in the tin can(s) or racks. Light the lamp or torch. Sterilize the scalpel blade thoroughly in the flame, then place it on the stand. The stand can be fashioned from a wire coat hanger, a tin can, or from other appropriate household material. The object is to have a stable place to rest the handle such that the blade is suspended in the air near the flame to avoid contamination when not in use.

Start with a fresh, clean mushroom fruitbody. Although the outer surfaces of the fruitbody may harbor bacteria and mold spores, the internal tissue usually is free from contaminating organisms, assuming the mushroom is not waterlogged.

Tear or break off a section of the mushroom (do not cut with a knife, as the blade will introduce contaminants from the outer surface). Place the mushroom on the table with the clean surface facing sideways (not touching the tabletop). The object is to have one clean, exposed surface from which a very small piece of tissue may be extracted, and to keep that piece as pristine as possible while removing it to the culture tube.

Now you must organize your materials so that when the tube is opened, it is exposed for a minimal amount of time while you insert the piece of tissue. To decrease the probability of contamination, neither the tube nor the cap are allowed to rest on the table. This part is just a bit tricky. It is a good idea to practice the technique a few times with an empty tube before using real agar tubes and a mushroom. (The following assumes you are right-handed. Reverse directions if you're not.) Position your left hand with thumb facing downward and fingers facing forward horizontally. Place the tube in your left hand between middle and ring finger, placing the ring finger on top of the tube and the middle finger underneath it, with the tube cap facing toward the inward direction. Always keep the tube positioned horizontally so particulates from the air cannot fall into the mouth of the tube. The slanted agar surface, on which you will place the mushroom tissue, should be facing upward. Carefully remove the cap from the tube and place it between the ring and pinkie fingers, sideways, with the pinkie on the top and the ring finger underneath.

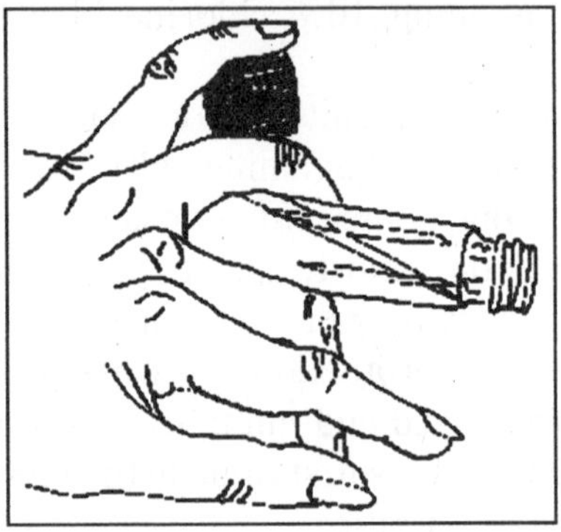

Fig. A Little Tricky, but Practice Helps

Using the remaining index finger and thumb of the left hand, firmly grasp a piece of the mushroom with the clean surface facing upward or sideways. With the right hand, grasp the scalpel firmly by the handle as you would hold a pen or pencil. Starting with the tip of the scalpel, cut a very small triangular shape from the clean inner tissue. Gently "tease" the tissue free and quickly place it just inside the lip of the tube. If necessary, tap the scalpel to free the tissue. Return the scalpel to its holder, and quickly replace the cap.

Now tap the bottom of the tube a few times on the heel of your hand to position the mushroom tissue on the surface of the agar.

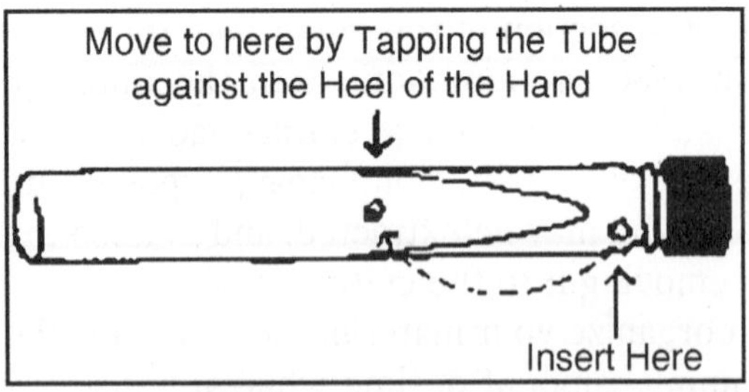

Fig. Move by Tapping

Voila! There is a very good chance you now have a pure tissue culture. Repeat these steps for the remaining tubes and mushroom material. It is important to make several tubes from each mushroom specimen, since even with the most careful technique contamination can sometimes occur. Between each tube inoculation, re-sterilize the scalpel or knife blade in the flame, and tear off a fresh piece of the mushroom exposing yet another clean tissue surface.

When finished with all of the tubes, tighten the caps, and tape each on the area where the tube meets the cap with micro-porous tape. Micro-porous tape allows the culture to "breathe" but prevents contaminants from entering the tube. Label the tubes with the date and contents using a permanent marker such as a Sharpie™ or use a small mailing label.

Finally, place the newly inoculated tubes in a low light location with moderate temperature (55 - 70 degrees F. is fine). In a few days to a week, the inoculated tissue will become "fuzzy" as the mycelium begins to grow. After several weeks, mycelium will completely cover the agar surface.

If you see mold (usually detected by its green, black or other colored spores) or bacteria (usually colored slimy material), you should discard the culture immediately and clean the tube and cap thoroughly with hot, soapy water. If possible, empty the contaminated tube in a location as far away from your other cultures as possible. For example, carry it straight outdoors to the garbage can. You may use a chopstick or other long thin tool to loosen the media; it should then slide out in one or a number of pieces.

PHASE II: STERILE TRANSFER TECHNIQUES AND GRAIN SPAWN

Sterile Transfer to Multiply Agar Tubes

Your agar tube containing mushroom mycelium is your "master culture." If you want to replicate the culture to increase your supply or to share with other cultivators, you should transfer the material from one tube into many tubes. This may be accomplished using the same process as during the initial sterile transfer described above. Instead of using a piece of tissue from a fruitbody, however, substitute a tiny piece of agar containing mycelium from the master tube. The transfer is made using a scalpel or another appropriate tool such as an inoculation loop, which can be purchased or easily made using a piece of thin wire attached to some sort of handle.

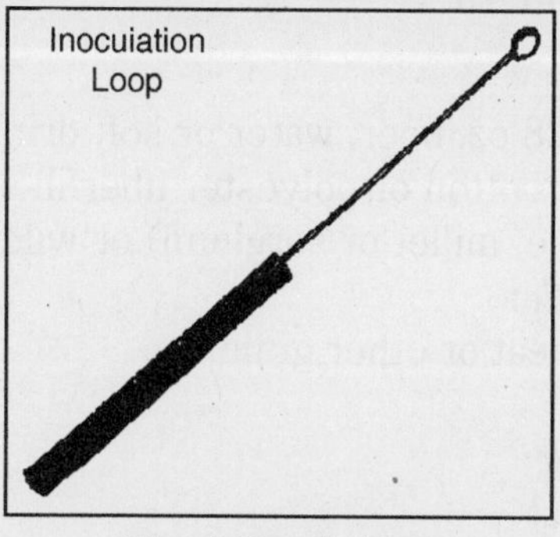

Fig. A Simple Tool

Again, be sure to sterilize the tool between each inoculation and minimize the amount of time the cap is off each tube. With a little practice, these transfers can be made quickly and with excellent success.

Sterile Transfer to Grain to Create Spawn

The next step in the cultivation process is to increase culture size by growing it on grain, often known as "grain spawn." Mushroom cultures grown on grain are easily proliferated by shaking and, as you will see, easily transferred by pouring from one container to another. As an example, one bottle of grain spawn can be used to inoculate many gallons of the final substrate. Grain is fairly inexpensive and readily available at health and animal feed stores.

It is readily colonized by many saprophytes, including Pleurotus species (Oyster mushrooms) andAgaricus species (button mushrooms and their relatives). Here are two equally viable methods. The first uses 12 or 28 oz. beer, water or soft drink bottles and wheat, although rye can be substituted. The second uses canning jars and wild bird seed. Although wheat, rye or bird seed can be used with either type of container, the two methods produce slightly different results. Experiment to see which works best for you.

The primary advantage of the bottles is that the opening is smaller, thus decreasing the chance of contamination during inoculation. That advantage can later be a disadvantage: It can be more difficult to remove the material from the bottles.

If you have, or later obtain, a laminar flow hood with a HEPA filter which minimizes contamination problems, canning jars are recommended because they are easier to work with. If you are careful, however, you can have good success using either technique, even without the hood.

Materials needed for both methods:

- A pressure cooker large enough to contain the bottles or jars
- Aluminum foil
- Grain (wheat, rye, millet or sorghum) or wild bird seed

Preparing Grain, Method No. 1

Materials Needed:

- Clear 12 oz. or 28 oz. beer, water or soft drink bottles
- Rolled cotton or cotton or polyester fiberfill such as pillow stuffing
- Grain (wheat, rye, millet or sorghum) or wild bird seed

Recipe for 12 oz. bottle:

- 1/3 level cup wheat or other grain
- 1/4 cup water

Recipe for 28 oz bottle:

- 1 level cup wheat
- 2/3 cup water

Guidelines for Grain and Bottles

An important consideration when preparing grain is that it will roughly double in volume when cooked or sterilized. A rule of thumb is to fill a container only one third full with dry grain. If using bottles, a breathable plug closure can be fashioned from rolled cotton (available at pharmacies) or polyester/fiberfill material (at fabric stores).

Cooking/Sterilizing

Grain can be cooked/sterilized in either one or two steps. The two step process produces a very uniform product, but if time is limited, cooking and sterilizing can be done simultaneously.

Two Step Process

After the bottles are loaded with grain and water, place them in the pressure cooker and cover with a sheet of aluminum foil. The foil will help to prevent the filter plugs from getting wet from condensate. Cook the grain in the pressure cooker by bringing the pressure up to 15 psi for seven minutes. Let the pressure cooker cool and use oven mitts to shake the hot bottles so that wet and dry grain is mixed. Place the containers back in the pressure cooker and sterilize by cooking at 15 psi for 45 minutes. Do not count the time required to bring the pressure cooker up to pressure. Then let the pressure cooker cool slowly.

Remove the bottles and shake once again to mix the grain. Cover the filter plugs with a paper dust shield. The will help prevent contamination. Store the bottles in a dust-free box or cabinet. Since the grain will gradually dry out over time, avoid making cultures too far in advance of your needs.

One Step Process

Proceed as above, but simply cook for 45 minutes at 15 psi. After the pressure cooker has cooled, but while the bottles are still hot, use oven mitts to shake the bottles to mix wet and relatively dry grain. Well-prepared grain is plump and shakes readily with a soft "ring." Overcooked grain prepared with too much water will often rupture. Then the starchy contents cause clumping, making it difficult to shake. Grain prepared with too little water will be flinty, shriveled and unsuitable for mycelial growth.

Spawn Method No. 2

Materials needed:

- Pint or quart sized canning jars
- Micro-porous filters fitting inside canning jar lid ring (Mycological supply or homemade)
- Aluminum foil
- Paint strainer bags (from hardware or paint store)

Grain Preparation

Measure one jar of grain for every three jars you will prepare into a paint strainer bag until the bag is half full. Use more bags as needed, keeping in mind that you can only cook as many jars at once as you have room for in the available pressure cooker(s). Tie the bag closed with twine, place in a large pot or the pressure cooker bottom (setting the lid aside), fill with water until just covering the grain, and boil for 30 minutes. Remove into the sink and rinse for a few minutes in cool water. This will remove some of the released starch so the grain will not stick together as readily.

Let drain for a few minutes. While draining, insert the canning jar lid filters into the lid rings. The filters may be obtained commercially, or may be cut out of appropriate fabric, such as a stiff, heavy polyester to fit the lid ring. Instead of using a filter, you may invert the disk that comes with the lid to provide for a small amount of air exchange, but the filtering capability of this method is questionable and a greater possibility of contamination may result. In the long-run, the commercially available filters or handmade filters are re-usable and probably worth the investment.

Cooking/Sterilizing

Untie the bag and fill the jars with grain. The jars should be 1/2 to 2/3 full. Place the lids with filters on the jars and seal loosely. Pressure cookers generally have some sort of rack to hold the jars away from the bottom of the cooker. Add water to the cooker so that it almost reaches the bottom of the jars when they are placed on the rack. (There should be at least a couple inches of water.)

Load the pressure cooker with the jars. Please see the manufacturer's instructions for the pressure cooker and observe all safety guidelines. Bring the cooker up to 15 psi (250 degrees F.) and maintain that pressure/temperature for 2-1/2 hours.

Turn off heat and let cool for 30 minutes. Carefully remove jars from cooker, using a clean potholder or oven mitt if the jars are still too hot to handle. Tighten the lids and cover with a small square of aluminum foil. Shake the jars vigorously to ensure that the grain is not sticking or clumping. Store in a clean, dust free location until ready to use.

Grain Amenities

For some mushrooms, the acidity of the media, as measured by Ph, can influence growth. Acidity can be adjusted by adding 1 tbsp of ground limestone ($CaCO_3$) to a gallon of water and using the latter to mix with the grain. Limestone may be purchased at a garden supply store (it's sometimes called "agricultural lime"). Agaricus bisporous (the button or Portobello mushroom) is an example of a species that benefits from adding $CaCO_3$ to the grain.

Transfer of Agar Culture to grain Bottle or Jar

One agar culture tube will inoculate several spawn bottles or jars, since only a tiny piece of mycelium is required for each jar. Generally you will follow the same sterile technique described for Phase 1. Again, have all of your materials and equipment in place before you begin, and practice the technique with empty jars and tubes until you are comfortable and can work quickly.

When working with grain bottles or jars it is helpful to have a cradle to hold the bottle or jar in a horizontal position while making transfers. This frees the hands for removing and replacing lids. A cradle can easily be fashioned from a quart or half-gallon milk carton by cutting appropriate-sized notches in it.

After you've removed the cap and lid of the agar tube and of the bottle or jar, use your inoculation loop to cut a small piece of mycelium from the agar. Place it just inside the jar and close both lids. Lastly, tap the jar against the heel of the hand until the mycelium is positioned on the grain.

Be sure to sterilize your inoculation loop in the flame between each inoculation, and to keep both tube and jar in a horizontal position at all times to prevent airborne particulates from entering.

Once the jars have been inoculated, date and label their contents using a permanent marker, such as a Sharpie, or use a small paper mailing label. You may also want to mark an arrow on the container closest to where the mycelium has been placed in the jar for further monitoring, as sometimes it is difficult to see until it begins to grow.

As the mycelium begins to grow through the grain, periodically shake the container to speed colonization and to prevent the grain from clumping into a thick mass. The mycelium of mushrooms like Oysters and Agaricus grows well at room temperature. Once a grain culture is fully colonized, it can be stored in a refrigerator to maintain its viability.

PHASE III: FINAL SUBSTRATE AND INDUCTION OF FRUITING

In the third and final phase of mushroom cultivation, culture size is increased again, but usually on a different substrate than grain, and environmental conditions are manipulated to stimulate fruiting. The choice of a fruiting substrate will vary, depending on the mushroom species. Here we will describe the substrate and conditions necessary to fruit Pleurotus species (Oyster mushrooms).

Oyster mushrooms are an excellent choice for the beginning cultivator as the mycelium grows quickly, competes aggressively with undesirable organisms, and fruits readily under a relatively wide range of conditions. Also, there are numerous varieties available in a rainbow of colors including white, brown, pink and yellow, and they are available from food stores or via mail order suppliers.

Fruiting Substrates

Oyster mushrooms can be fruited on a variety of materials, e.g. hardwood logs, straw, sawdust, corncobs, shredded newspaper, etc. Avoid sawdust and chips from softwood trees such as pine, as these inhibit the growth of Oyster mushroom mycelium. Substrate for growing Oyster mushrooms can be obtained from pet stores and feed stores where you can buy bales of straw or bedding material like aspen shavings. Lumberyards that saw hardwoods and tree-trimming companies are also good sources of growing material. Use your observation skills and creativity.

Preparing the Substrate

Oyster mushrooms can be grown on a pasteurized substrate, which is easier to prepare and less expensive than sterile media. The purpose of pasteurization is to leach out soluble compounds that might attract competing molds, and to kill insect pests.

A simple method for preparing pasteurized chopped straw or similar material is to soak it in steaming water for about 30 minutes, then drain and cool it. The moisture content will be about right when a squeezed handful just barely produces droplets of water between the fingers.

For larger quantities of substrate, or for logs, a 55-gallon metal barrel or drum may be placed on an outdoor type propane stove or burner. Filled with substrate and covered with water, it is brought to a boil and boiled for 30 or 45 minutes. The material is then strained (a homemade strainer is fitted inside the barrel before filling, made from wire mesh secured from a hardware store). Alternatively, fill paint strainer bags 1/2 full with the substrate, tie with twine, fill the barrel 1/2 or 3/4 full with these, add water and boil as described above.

Culture Size and Containers

Fruiting cultures can be of any size, but hobbyists would be wise to start small in order to perfect their technique. A half gallon milk carton lined with a plastic bag is a convenient hobby-sized culture. Another alternative is to purchase sleeves of clear plastic food containers and lids from a restaurant supply store.

These are inexpensive, sturdy, easy to transport, and offer good viewing of the culture's progress.

Enriching the Substrate

Yields can be improved by adding rich food sources like coffee grounds, cotton meal, rice bran, etc. However, these should be used with caution since they increase the likelihood of contamination. Typically these materials are added after initial spawn run, i.e. after the culture has already been fully colonized, to avoid contamination. Try using a chopstick to make small tunnels in the colonized mycelium from the top, then pour in the pasteurized additive, possibly using a funnel. This will help the mycelium to colonize it quickly, minimizing the opportunity for contaminants to get to it first.

Spawning

The more grain spawn used to inoculate a fruiting culture the faster it will be colonized, the less likely it will become contaminated, and the sooner it will fruit. Because pasteurized cultures can become contaminated with molds and bacteria, it's best to err on the side of excess spawn when inoculating your first cultures. A spawning rate of one cup of inoculum to a half gallon of substrate works well.

Shake the grain bottle or jar well before opening. Pour the grain spawn across the surface of the substrate or, with clean hands, mix the substrate and spawn together in a tray and then place in a bag or container. Tie the plastic bag loosely with a twist-tie, or simply fold over. With plastic lids, poke small holes after a few days to allow air exchange. Store in a dust-free location, possibly laying some paper or plastic loosely across the top to prevent dust and contaminants from entering.

Most strains of Oyster mushroom, especially those used commercially, grow quickly between 65 and 75 degrees F. and don't require a lowering of temperature to fruit. During spawn run, cultures should be maintained in a high humidity environment with low levels of air exchange. Oyster mushroom mycelium can endure fairly high levels of CO2. A culture the size of a half gallon milk carton should be fully colonized in about three weeks when grown at 70 - 75 degrees F.

Fruiting

A culture normally will not fruit until the substrate has been fully colonized. Small aggregations or coralloid growths on the surface indicate that a culture is beginning to fruit. Although lowering the growing temperature is usually not required for fruiting, a small amount of indirect light is believed to be important.

For normal-shaped fruitingbodies, more air exchange is required than during spawn run, but high humidity must still be maintained. To achieve this condition, cultures are often kept in a plastic bag with holes for aeration or in a box or terrarium with a moistened tray of gravel. If using container and lid, once the mushrooms begin to "pin" (i.e., begin to form tiny masses resembling miniature fruitbodies), the lid may be removed, and then the aforementioned plastic bag with holes or a terrarium may be used to maintain humidity.

It can take as long as a week for an Oyster mushroom to grow from a pinhead to maturity depending on temperature and humidity. When the cap margin starts to turn up, the fruitingbody is past its prime, so by this point you ought to have either used it to start another series of culture, or to have eaten it for supper.

Fig. Successful Fruiting

SIMPLIFIED AND LOWER COST METHODS FOR MUSHROOM CULTIVATION

A universal need for all mankind is food. Reliable supplies of low cost, nutritious and readily available food for the population are the key items to a stable and healthy society. Until this basic need is met, poverty alleviation and economic development cannot take place. To achieve the goal of stable, healthy societies, mankind over most of the world has evolved from hunter-gatherer societies to agriculture, aquaculture and animal husbandry for production of their necessary foodstuffs. These farming endeavors have developed over centuries of social evolution, originally utilizing local plants, fish and animals as the raw inputs to make these production systems work. However, in the last century or two this has begun to change, with plants and animals from far and wide being imported into new locals as superior candidates for cultivation over the plants and animals found originally in that region.

This chapter presents the first step in this expanded cultivation scheme, which is Mushroom Production through utilization of low cost, technologically simple methods that can be easily implemented at the village level in any country, be it highly developed or the poorest country on earth. Our main goals in this research were to develop real world solutions to the problems faced in implementing these cultivation systems.

It is easy to imagine a system where if you have capital resources of X, Y and Z, you can achieve great results, but what about those peoples that do not have even the minimum resources of X, Y and Z? It is to these people that we have directed this research. Using only agricultural waste which we have diverted from the disposal stream, and commonly available tools, equipment and supplies, we will show in this chapter how anyone of even the most modest means can produce an economically viable crop of wholesome and nutritious food with which to feed their families and villages; the surplus of which can be exported to others as a cash crop. At the end of the first step of this diversified agriculture program the remaining mushroom waste and spent substrate can be utilized as nutritious fodder for goats, cattle, horses, camels, sheep, and fish farming. This allows a third crop, where initially there was only one.

To put this in perspective, consider the typical case of a cassava farmer in Africa: the cassava is grown and harvested, and the plant waste and cassava peeling are dumped back on the ground. The next crop to be harvested by the farmer is the next crop of cassava. By implementing the system proposed here, the cassava peelings and plant waste becomes the feedstock for the mushroom cultivation which yields a second crop (mushrooms) in approximately three weeks, long before the next crop of cassava would be ready for harvest.

The waste materials from the mushroom production then becomes a highly nutritious feed for animals and fish such as goats, cows and Tilapia, producing a third crop, where there originally had been but the one primary crop. This increase in efficiency with little additional input of labor or capital has perhaps the greatest potential for elevating the economic status of the rural farmer across much of the world.

Mushroom cultivation represents a very basic natural process, that of fungal decay. In nature the primary function of plants is to act as the molecular assemblers. Plants take simple compounds like water and carbon dioxide and using sunlight as energy they assemble these into complex organic compounds such as proteins, carbohydrates and fats.

Upon the death of these plants, the fungi move in and become the molecular disassemblers. That is to say the fungi take these complex compounds like the proteins and carbohydrates and disassemble them back into the simple compounds which we started with, those compounds like water and carbon dioxide. As these simple compounds are released back into the environment by the fungi, they become available as raw materials for the next generation of plant growth.

In the process of running through their life cycle, the fungi reproduce as do all other living things. In the case of fungi, the reproductive organ can take several forms. One particular class of fungus, the basidiomycetes, produce a large fleshy reproductive organ (or fruitbody as it is named) which we know more commonly as the mushroom. Many of these mushrooms are choice edibles, and in fact represent very high quality nutrition, often equaling or exceeding meat in quantities of protein and essential amino acids.

The basis of mushroom cultivation is the breakdown of cellulose. The cell wall structure of virtually all plants is a fibrous structure composed of cellulose and hemicellulose, surrounded by a structural compound called lignin. The lignin wraps around the cellulose fibers like plastic wrap. This makes for a very strong structure, allowing a tree to stand upright for hundreds of years. The cellulose and hemicellulose are made of sugars which are great sources of food, but these are protected by the lignin wrapping, which is a very stable compound and difficult to breakdown. Only a few organisms can breakdown the lignin and utilize it as a food source, thus exposing the underlying cellulose and hemicellulose for food use by other organisms. The best known and most effective of these lignin-breakdown organisms are known as the white rot fungi, of which the Oyster mushrooms are the prime examples.

Oyster mushrooms are a closely related species complex, comprised of many species within the genera Hypsyzygus and Pleurotus. Many different species of Oyster mushrooms have evolved in different locations around the world, and have become more or less specialized in degrading different raw material cellulose sources (the substrate) at different temperatures, oxygen and light levels (acclimatization).

A word on species and strains: There are many "Oystereque" mushrooms, with similar growth characteristics, morphology (looks) edibility, flavor, shelf life, etc. Some of these are very closely related to each other, even being given the same genus and species name such as the common tree oyster, Pleurotus ostreatus. However, not all Tree oysters are completely alike even though they may have the same genus and species name. For example, one may be growing on an oak tree in the temperate region, while another is growing on an oil palm in the tropics. These two mushrooms may be the same species but they are different strains. They will each have evolved different enzymes to degrade their respective substrates, and they will have different growth parameters having to do with the temperature, humidity and sunlight intensity in the regions where they evolved.

To try to grow the northern climate oak mushroom on a palm tree in the tropics would yield very disappointing results. Then there are other Oysteresque mushrooms that are not quite as closely related, but which still share most of the characteristics with the other Oyster mushrooms. These have names like Hypsyzygus ulmarius, the Elm Oyster, so named because it usually grows on Elm trees. This does not mean that it will ONLY grow on Elm trees, just that this is the usual native habitat.

Choosing the best species and strains of Oyster mushrooms is the single most important step in creating a successful mushroom cultivation program. There is no one strain that is any better than another strain, it all just depends on the climate and the substrate and the cultivation methods used. The only way to determine the best species and strain for growing in any particular situation is to experiment with several types of Oyster mushrooms and record the production time and yield for each strain in that particular circumstance. Generally, a local strain of mushroom isolated from nearby the growing location is the best choice, as this local strain will have become acclimatized to the local environment and the local competitors for the food source.

For the purpose of this research, we choose several strains of oyster mushrooms to indicate the possibility of different strains being utilizable in this cultivation method, and to show that the results obtained in this research were not strain specific. It should not be assumed that the results shown here will necessarily transfer 100% to another location, strain or substrate, although the general guidelines shown here will certainly apply. Anyone desiring to follow this methodology of mushroom cultivation is encouraged to experiment with different species and strains, and to include locally acclimatized strains whenever possible.

In some cases such as northern climates, it is not unusual to see several different species or strains cycled through a farm over the course of the year as the seasons change. This way, one can assure the maximum yield both in the hot season and in the cold season by selecting appropriate strains according to prevailing temperature.

Likewise, it is often preferable to cycle different strains through the farm seasonally depending on substrate availability. If you have a strain that grows well on straw, you should use that one at grain harvest time when straw is readily available, but if your only substrate is river grass or cassava peeling, by all means be flexible in switching to strains particular to these substrates so as to maximize your harvest.

The most basic concept of mushroom cultivation is that we need to produce an environment in the substrate that is selectively preferential to the growth of our target species of mushroom, and less amenable to other types of microorganisms and pests.

This involves sterilization (completely killing any other organisms that are present in the substrate which would compete with the mushrooms for utilization of the substrate as food) or pasteurization (killing off the majority of competitive organisms). Mushroom cultivation in more developed countries has evolved from an art into huge agri-business by way of the most innovative production technology and biotechnology available. This usually means the use of large pressure chambers and high temperature steam for sterilization of the substrate.

However, steam and the associated equipment are both expensive and technically demanding in terms of training and education. There are many less-developed areas of the world where access to these technological advances are either unavailable or too expensive to utilize. In those areas it would be much more practical to look at ways to pasteurize rather than sterilize the substrate, and by methods other than the use of steam. This can be done in many ways, such as using commonly available substances such as soap or hydrated lime to pasteurize the substrate. The mechanism by which these chemicals work is through the rapid change in osmotic pressure, causing the majority of microorganisms present to burst from the change in osmotic pressure. Once the substrate is treated in this way, and the majority of the microorganisms are killed, the substrate is suitable for the introduction of our target species, the Oyster mushroom.

Another potential treatment method is simply to soak the substrate in water for a week or so. This results in a rapid bloom of the bacteria that rapidly consume the readily available simple sugars. Once the sugars are exhausted after about 5 days of fermentation, the bacterial bloom dies off due to lack of any more simple nutrient availability. This leaves a substrate which is more selective to the higher cellulose degrading organisms such as Oyster Mushrooms. This is the simplest method of substrate preparation; however it is the least effective of any methods we have tried. It does work, but the chemical treatment methods are much more effective and not considerably more difficult or expensive.

MATERIALS AND METHODS

Substrates Used

Unshredded wheat straw (whole stalk), shredded wheat straw (reduced in a hammer mill to small particle size). Wheat straw is the post-harvest stalk of Triticum aestivum and Ground maize cob (Zea mays) were used as the substrates. All straw was obtained from S&W feeds, Carson City, NV. Ground maize cob obtained from Benson's Feed and Tack, Carson City, NV

Species and Strains Used

Although there are several fungi capable of white rot degradation, the following genus and species in the Tricholomataceae family were used in this experiment: Pleurotus pulmonarius AX strain, Hypsizygous ulmarium ELM 1 strain, Pleurotus D'jamor PDJ strain, Pleurotus Sajar-caju PSAJ strain, and Pleurotus ostreatus TL strain. All cultures used in this research are commercially used in mushroom farming, and were all provided by Aloha Medicinals Inc. of Carson City, Nevada.

Spawn Used

Spawn is the seed stock which is used for inoculating the substrate for mushroom production. The spawn for this research was generated under the usual conditions which will be known to anyone familiar with spawn production. Spawn is generally grown on grain or sawdust. The spawn substrate used for this project was White Sorghum grain, which was cooked with a measured quantity of water to achieve a moisture content of approx. 55%.

A small amount of ground oyster shell was added to control the pH. Approx. 400 g oyster shell per 100 kgs of dry grain was added, along with 102 kgs of water, which resulted in a pH of 7.0. After cooking with water and oyster shell, gypsum was added to keep the individual grains of sorghum separated and reduce the possibility of anaerobic conditions forming in the spawn. The cooked, gypsum treated grain was filled into glass jars of 1 liter size until jars were approx. ¾ full, which measured an average of 454 g of weight per bottle. The top of the jars were covered with a filter made of thick paper and a metal lid screwed over the filter paper. The metal lid has a single central hole drilled, measuring approx. 25 mm in diameter. This allows gas exchange of the spawn substrate while eliminating the introduction of any foreign organisms during the spawn grow-out period. These filled, capped jars were then sterilized in an autoclave at 17 psi steam pressure (approx. 1.2 bar) for a period of 2.5 hours. After the sterilized jars were cooled overnight, a small piece of inoculum of the appropriate strain was introduced into each jar under sterile conditions, as will be known by anyone familiar with the art of spawn making.

The inoculated grain jars were then grown for 10 to 12 days before use, which resulted in completely colonized grain of only the target species of mushroom. The short growth period of 10 to 12 days means there was some percentage of unconverted grain starch remaining in the spawn, which becomes a nutritional amendment to the final mushroom substrate, raising the nutrient level of the straw without raising the risk of contamination. The concept of using young grain spawn as both inoculum and nutrition amendment is important to the simplified cultivation process presented here.

Simply adding a rich nutrient amendment to an otherwise poor nutrient base like straw or maize cobs will increase the risk of contamination by unwanted competitive organisms, which would lead to reduced yield of the target mushroom fruitbodies. By using the grain spawn as mentioned here, the grain is already fully colonized by the target mushroom strain, which in turns eliminates all competitors within that jar, while still providing a rich nitrogen and starch base for quick initial growth in the final fruiting substrate for the mushroom organism.

Spawning

Spawning is the term used in inoculating the final substrate in order to grow mushrooms. In this research 2.73 kg (6 pounds) of treated straw was added to a growing bag for each strain and treatment type. All treatment methods and strains were done in triplicate time three, or in other words a minimum of nine bags were used for each treatment method and strain, insuring statistical reliability of the data generated.

The growing bags in this research were spawned with the addition of 454 g (one pound) of spawn per bag, added alternately with a layer of straw as the bags were filled. The layers of straw were approx 50 mm thick, then a sprinkling of spawn, before adding the next layer of straw, followed by another layer of spawn until the bags were full. This represents a spawn to substrate ratio of approximately 1:6. This is considered a heavy rate of spawning, but any spawn ratio from 1: 5 up to 1: 40 can be used. It is generally considered that the higher the rate of spawning (the more spawn used per bag) the better the results in consistency and yield, while a lower rate of spawning (less spawn per bag) is cheaper, but with a correspondingly lower yield of mushrooms per bag. The spawn rate chosen in farm practice is usually a reflection on the price and availability of spawn in the local region. Whenever possible, it is always better to use more spawn rather than less.

Growing Bags Used

Clear polyethylene bags of approx 300mm × 700mm × 2 mil thickness were used as growing containers for this experiment. The bags were filled with substrate and spawn, with the excess air squeezed out and the bags sealed by twisting a short metal wire or twine around the top. After 48 hours post-filling, the bags were ventilated by puncturing repeatedly with a knife blade. The knife blade was 20 mm wide and ventilation holes were made on a random spacing approx 100 mm apart all over the surface of the bag. These holes both allowed gas exchange and a place for the mushrooms to erupt from when fruiting.

Substrate Treatments

A control batch of substrate for comparisons was made by first soaking the whole straw for 24 hours in barrels of water, and then draining the straw by setting it on racks and covering by newspaper for two hours resulting in approx. 60% moisture content in the straw.

The straw was then treated with steam at atmospheric pressure for 2 hours (maintaining ~100 degrees C). The bins were then tipped so that the excess water condensed from the steam treatment could drain off for two hours while cooling. This straw was filled into growing bags and spawned in an identical fashion to the experimental bags and grown in the same room and under identical conditions throughout the trial period. This method of steam treatment is well known in the mushroom industry and is well understood by anyone familiar with mushroom farming, and so forms the comparison for all other substrate treatment methods. Straw treated with steam had no detergent, lime or other substances added.

A second method of substrate conditioning used was the natural fermentation of one bale of wheat straw (approx. 30kgs) in water. The straw was soaked in barrels containing 200L of clean tap water for seven days with no other substances added.

The straw was removed from the barrels after seven days and placed on racks for two hours, covered with newspaper in order to drain excess water and allow outgassing and aeration of the straw. As can be imagined, after soaking this straw for seven days in the sun it was a stinky, sticky mess, but actually had a fairly low microorganism load due to exhaustion of the simple sugars and other readily utilizable food sources upon which the first generation of decay organisms depend.

This fermented straw was layered with spawn (454g bottle) in four plastic bags for each test strain until a total weight for each bag was 2.73kg (6 pounds). When the bags reached the proper weight, they were compressed by hand and tied off using segments of light metal wire.

These bags were kept outside in the shade for two weeks of observation and then moved inside a climate controlled grow room. While this treatment method did produce edible mushrooms, it did not result in commercially viable quantities of mushrooms being produced. This method of substrate preparation is not very effective when compared to the other methods utilized in this trial, and the data on this treatment method is not shown in the results.

Two more bales (approx 30kgs ea) of unshredded wheat straw were soaked in several barrels containing 200L water each and treated with one of the following substances for 24 hours:

1. 355 g per 200 L water of hydrated lime (Calcium hydroxide) (Rockwell Lime Co. Manitowoc, WI, 54220)
2. 710 mL per 200 L water 5% Chlorine Bleach (Clorox Co. Oakland, CA 94612
3. 240 g per 200 L water Tri Sodium Phosphate (Savogran Co. Norwood MA 02062)
4. 120 g clothes washing powder per 200 L water (Tide brand clothes washing detergent, Procter & Gamble, Cincinnati, OH)

After soaking in the above solutions for 24 hrs, the straw was laid out on top of large metal racks and drained in the same manner as the fermented straw.

Shredding Straw and Maize Cobs

When shredding the straw for the second portion of this experiment, dry straw was run through a 50 HP Jacobsen Hammer Mill for shredding. The shredded straw was soaked and processed in the same fashion as the regular unshredded straw. The bags used for the shredded straw were hand compressed more tightly than the regular straw after inoculating each, due to the smaller particle size of the substrate. This additional compression was possible due to the increased density of the shredded straw.

In addition, a drum containing 118 mL bleach was mixed with 200L water and filled with shredded maize cob. Once the maize cob was sufficiently drained of excess water by straining through small mesh chicken wire, the material was filled into identical growing bags and spawned. For spawning, bottles containing 454 g of spawn were mixed intermittently with the maize cob in a manner similar to that described for the straw. The maize cob substrate treated with bleach resulted in yields very similar to the straw substrate. The outcome for the maize cob substrate is not shown in the results simply in the interest of conserving space. It is notable that the information regarding straw can be used representatively with maize waste and with most other cellulosic agricultural waste. All of the shredded straw bags and the maize cob bags were compressed by hand and tied with string, wire or rubber bands. They were then placed on racks in the same growing room as the unshredded straw bags. The growing room was maintained at a temperature of 21 to 25 degrees C. Humidity was maintained with the use of humidifiers from 80% to 90% RH. Light cycle was natural shaded light, approx 12 hrs per day of light and dark.

Picking Procedure

Mushrooms were checked daily and picked at specific times in their growth cycle: eg. When the edges of the caps were flattening out or showing signs of slight curling. The mushrooms were picked, weighed, and put into a refrigerator. Pictures and weights were taken and recorded at each harvest.

Observation Sheets

Observations were taken bi-weekly in order to note the difference and similarities between growth for each treatment method and individual strains. Characteristics noted included the amount of straw sprouts, bacterial contamination, competitive mold contamination, mycelial growth, number of primordial formation (not yet matured fruitbodies called pins) and fruitbody clusters, and the weight of the harvest.

Bag Disposal

Bags that became overly contaminated by bacteria or competitive molds, were weighed using an accurate lab scale (Yamato Corp. Colorado Springs, CO 80906) and disposed of immediately to avoid contamination of the remaining bags. After the remaining bags passed two months of active harvest, they were also weighed and disposed of.

Biological Efficiency Equations

Biological efficiencies (BE) for each strain were attained by dividing the average fresh weight totals for that treatment method and strain type by the average dry weight of substrate and multiplied by 100. This resulting figure is given as a percentage, with the figure of 100% BE equal to 2.5 kgs of dry weight substrate yielding 2.5 kgs fresh weight mushrooms. In this study, it was not uncommon to see BE's in the neighborhood of 150% to 180%.

RESULTS AND DISCUSSION

In the third world, protein-energy malnutrition (PEM) affects 500 million people and kills 10 million annually. To avoid ill health from certain diseases, and death caused from nutritional deficiency, daily dietary protein consumption is essential. In countries that suffer from widespread protein deficiency, available potential food sources are generally full of plant fibers from the terrestrial biomass. Wood and straw make up about 80% of this biomass, which equates to approximately 800 billion tons worldwide, an extremely large reservoir.

Although four stomached mammals (ruminants) such as cows, goats and sheep produce the enzymes necessary to digest the cellulose present in plants such as grasses and hay, they are unable to directly utilize the biggest cellulose reservoir; woody biomass, which remains untouchable because of the lignin that is structurally wrapped around the cellulose. Lignin is a complex chemical compound found in plant cell walls that is cross-linked with other cell wall components, and acts as an integral part of the secondary cell walls of plants and some algae.

As a biopolymer lignin is unusual because of its assorted composition and lack of defined primary structure. Lignin is most commonly noted for support through strengthening of wood (xylem cells) in trees and is generally only degradable by highly specialized enzymes associated with very few organisms.

Therefore, this molecule is also generally associated with reduced digestibility which helps defend against pathogens and pests. In order to make use of this cellulose, the lignin first needs to be stripped, leaving the cellulose exposed. As mentioned above, all forms of White-rot fungi such as Pleurotus ostreatus are able to oxidize the non-digestible lignin, setting free the cellulose and giving both the fungi and other potential consumers such as ruminants ready access to this significant nutrient reservoir.

As will be seen from the graphs and illustrations, this experiment clearly shows the practicality of raising mushrooms on low cost (or free) agricultural waste such as straw, that has been treated by simple and low cost methods such as soaking with a bit of lime or washing powder. In fact, the yields seen with these low cost treatment methods were not substantially different than the results seen in the steam treated control group. The biggest factor in yields was the density of the substrate, with shredded straw showing considerably higher yields than was seen with unshredded straw.

The results of this experiment show no significant difference between the mushroom marketability and characteristics such as fruit body, yield, taste, color, smell, etc. of the low cost, low technological methods described here as compared to the high technological steam-based methods used with most mushroom farming in developed nations today.

Implementing a low tech mushroom farming approach and low cost mushroom production program in West Africa would have compounded benefits for the region. This process would allow the use of secondary crops to be produced in a short time frame from agricultural by-products derived from the primary agriculture of the region. These and other potential raw materials are currently being disposed of as waste. After the production of edible or medicinal mushroom crops from these agricultural wastes, the residual mushroom substrate represents a bioconversion from the non-nutritional cellulosic material to an edible fungal matrix of high protein content which can be utilized as nutritious cattle and goat fodder as well as feedstock for tilapia farming. This diversified approach results in three successive cash crops from any agricultural operation where previously there had been only the one primary crop.

The oyster mushroom complexes used for this experiment appear to be one of the best candidates for production in the West African climate. These mushrooms are primary lignin/cellulose decomposers, and can grow on almost any plant material available for substrate including banana waste, coffee residue, sugar cane baggasse, cassava peelings, paper or cardboard waste, river grass, sawdust and nearly any other agricultural waste.

Other studies have been successful in revealing semi-low-tech methods for recycling agricultural waste and utilizing white-rot fungi for the production of fodder for ruminants which showed fast and reliable conditioning of the substrate. This type of study was also able to help in the revitalization of the local grazing areas from being arid and dry from overgrazing to lush and green. To take this idea one step further, we explored different detergent solutions as a method for pasteurization to find which will provide the best protection against outside contamination, while still producing the highest yield. In this way farmers will be able to produce less expensive fodder for their herds while having a commodity that will feed and support their families as well as raise their economic status and maintain the land. This type of technique has been shown to decrease the land needed to maintain healthy livestock and increase employment availability. An addition to these obviously stated facts, the first harvest of oyster mushrooms occurs only 18 to 21 days after inoculation. There are few other crops that can give the farmer an income in less than 3 weeks.

Many of the graphs employed are calculated using the values found for biological efficiency (BE). BE is the amount of growth determined by a comparison between the dry weight of substrate compared to the fresh weight of the mushrooms harvested. BE indicates the total percentage of growth in terms of mushroom fruitbodies. The figures below indicate a significant increase in BE when using shredded straw as the substrate for all the treatment types. In fact, substrate density appears to be the greatest determining factor in mushroom yield, at least for straw substrate and the strains of oyster mushrooms tested in this experiment.

MUSHROOM GROWING METHODS

Wild mushrooms have been collected by people for food since before recorded history. Once is it was figured out which ones were the best to eat, it made sense to try to grow them to have a more reliable supply. It is difficult to trace the origin of mushroom cultivation but some species have been grown in Asia for many centuries. Mushroom growing techniques have been developed to a high degree of technical efficiency. Proven methods exist for growing many kinds of edible or medicinal species. Mushroom growing kits can be purchased ready to be set up in an appropriate environment to produce mushrooms. Commercial quantities of mushrooms are grown using large scale production methods.

MUSHROOM SPAWN PRODUCTION

Before mushrooms of any kind can be grown, a quantity of starting material called "spawn" must be produced. Spawn is a mass of the white fuzzy mycelium of a fungus grown in a form suitable for inoculating a large quantity of substrate on which the mushrooms will be produced.

Grain spawn is prepared by inoculating sterilized rye grain in jars, with a pure mushroom culture. The mycelium grows throughout the grain and the spawn is expanded to any desired quantity by using the colonized grain to inoculate still other jars of grain. This type of spawn is convenient for starting mushrooms that grow on compost.

Dowel spawn is produced by inoculating bags of sterilized wooden dowels. When fully colonized, the dowels are driven into holes drilled in logs, or the dowels can be added to bags of wood chips or sawdust. Dowel spawn is used to culture wood-inhabiting species.

Growing Mushrooms in Compost

The classic commercial button mushroom (Agaricus bisporus) and the larger portabello (which is the same species allowed to get well past the button stage) is grown in tremendous quantities on mushroom farms around the world. Since the natural habitat of its wild predecessors was in pastures, this species is grown on a substrate of horse manure compost. Other species grown on compost include Coprinus comatus (shaggy mane) and Lepista nuda (blewit).

A number of variations on preparing mushroom compost exist, but two basic formulas are:

1. Horse manure, cotton seed hulls and gypsum, and
2. Chicken manure, wheat straw and gypsum.

The compost is inoculated with spawn and the spawn is allowed to completely colonize the substrate. Production of mushrooms is initiated after a peat-based casing layer is spread on top of the colonized compost.

Growing Mushrooms on Logs

Reishi mushrooms (Ganoderma lucidum), shiitakes (Lentinula edodes), hen of the woods (Grifola frondosa), lion's mane (Hericium eriniceus), and oyster mushrooms (Pleurotus spp.) are all wood-decomposing fungi suitable for log cultivation. Most species prefer hardwoods such as oak, maple, elm, cottonwood and alder, although a few species do well on conifer wood. The logs are best cut just before winter when their sap content is highest. Logs should be allowed to age for a month or two, but should be used before they dry out. Any size of log should work, but if the logs are big and heavy they take longer to colonize and are difficult to transport and handle. A length of about a meter and a diameter of 10-30 cm is good.

Dowel spawn is driven into holes drilled into the logs. The holes are sealed with wax to keep the inoculum from drying out before the fungus can become established in the wood. Under the best of circumstances, the logs will not be ready to produce mushrooms for six months, with a year or more being more typical. When the logs begin to show mycelia on their cut ends, they are close to being ready for fruiting.

Fruiting is initiated by soaking the logs in cold water for twenty-four hours followed by further incubation in high humidity with fresh air. Logs can produce flushes of mushrooms periodically for several years.

Growing Mushrooms in Sawdust and Wood Chips

Mushroom species that do well in logs can also be grown in bag culture using the same kinds of woods. Dowel inoculum, liquid spawn, or grain spawn is added to bags of sawdust and/or wood chips and thoroughly shaken. After two or three weeks, once the mycelia have grown throughout the wood, the bags are shaken again and incubated an additional couple of weeks until the substrate is fully colonized.

Then the bags are usually slit open and soaked in cold water for a day and incubated in high humidity with fresh air exchanges. Mushrooms are produced in as little as five weeks with this method and two or three flushes, with rest periods in between, can be achieved before the nutrients are used up in the wood.

Many mushroom growing kits come with a substrate of sawdust and/or wood chips. These kits are easy to set up since they come ready to produce mushrooms in a relatively short time once the correct conditions are established. Other methods of mushroom growing exist for research or laboratory purposes, but the methods outlined above are the ones usually used by both commercial and hobby mushroom growers.

CULTIVATION TECHNIQUES OF MUSHROOMS

Many mushrooms species can be cultivated at home on a small scale using various methods. Since home cultivation is mainly a hobby, the expectations of the grower and the results of cultivation can vary widely. However, the experimentation alone can lead to a satisfactory experience. Commercial cultivation, on the other hand, requires reproducible, quantitative results. A steady supply of mushrooms is necessary in order to enable the commercial grower to satisfy the demands of potential customers such as food distributors, supermarkets and restaurants. Since commercial cultivation usually involves a substantial investment, the cultivation techniques must be dependable with consistent results so that a marketing strategy can be formulated and a reasonable return on the investment realized. Successful commercial cultivation techniques have been determined for a number of edible and medicinal mushroom species which have a high potential for commercial production.

These techniques fall into three categories:

- Indoor cultivation on sawdust, straw, compost, or other suitable substrates
- Outdoor cultivation on hardwood logs
- Outdoor cultivation in beds of straw, wood chips or other suitable materials

INDOOR CULTIVATION

Many mushroom species are cultivated indoors in climate-controlled buildings. Certain species require adherence to strict parameters for successful cultivation, others are more forgiving. The buildings necessary for indoor cultivation can therefore vary in sophistication and cost depending on the species of mushroom that is cultivated. Substantial investment will likely be required for cultivation of sensitive species such as Maitake (Grifola frondosa) or Enoki (Flammulina velutipes), where temperature control and humidity are critical. Others, such as Oysters, can prosper in insulated greenhouses with less sophisticated controls.

Indoor Tray Cultivation

The Agaricus species, which includes the common White Button mushroom, is the most widely cultivated species of mushroom using this technique. There are two commonly-cultivated species – Agaricus bisporus and Agaricus bitorquis. There are also various strains of these species which exhibit different characteristics such as the colour range which varies from pure white to light brown. The substrate for this species normally consists of animal manners and cereal straw which have been well composted, meaning that high temperatures were achieved during the composting process. The elevated temperature is necessary in order to kill the many fungal competitors present in the original substrates.

The compost is then placed in long trays in a climate-controlled building and inoculated with spawn of the desired strain. The trays are kept in the dark while the mycelium from the spawn runs through the compost. At the appropriate stage the substrate is "cased", which means covered to a depth of about 3 to 5 cm with casing soil that is in general a mix of peat moss and limestone. The casing protects the colonized substrate from drying out and provides a stimulus for the mushrooms to form. After approximately two weeks of casing, the mushrooms start forming and are picked continuously until the nutrients in the compost are used up. These mushrooms have been traditionally picked at a fairly young stage, hence the name "button". A recent development in North America has been to allow these mushrooms to fully develop to maturity. At this stage they may reach 3 to 6 inches across the cap. They are sold as Portobello or Portabella mushrooms, and have become very popular.

Tray cultivation has been used with varying degrees of success for the production of Oyster and Shiitake mushrooms using pasteurized straw and sawdust respectively. Paddy Straw mushrooms are also produced on trays in temperate zones, whereas they are normally cultivated in outdoor beds in tropical climates.

Indoor Bag or Bottle Cultivation

Bag cultivation is used worldwide to grow Shiitake (Lentinula edodes), Oyster (Pleurotus ostreatus), Buna-Shiimeji (Hypsizygus tessulatus) and Maitake (Grifola frondosa). Substrates suitable for this method include hardwood sawdust, aged softwood sawdust, wood chips, cereal grain straw, corn cobs, various types of hay, and sugar cane bagasse depending on the mushroom.

The above materials are supplemented for nitrogen with such ingredients as bran from rice, wheat, spelt and other grains, cornmeal, cottonseed meal and many more at a ratio of approximately 1 to 4. Other additives may include sugar, molasses, gypsum, limestone, and other organic and mineral materials.The ingredients are mixed together dry, then water is added to obtain a 60 to 70% moisture content level. The wet mix is then bagged in autoclavable bags made of high density polypropylene or polyethylene and with filter patches.

The bags are then sterilized by steam in an autoclave or retort for 1 to 4 hours at 15 pounds per square inch (psi) or 1.1 kg/cm^2. Alternatively, atmospheric pressure steam sterilization can be done for 12 to 18 hours. This has the effect of destroying competitor fungi or bacteria which may be present in the substrate. However, for certain species, such as the Oyster mushroom, pasteurization rather than sterilization of the substrate is sufficient since this particular mycelium can tolerate or even benefit from some bacterial action.

Following sterilization protocols, the bags are inoculated with the "spawn" (mycelial culture grown out on grain or sawdust) of the desired species in a very clean or sterile environment. The bags are sealed but some form of gas exchange must be provided, either via a breathable patch on a specially-made bag or simply by closing the bag around a wad of cotton. The mycelium will run through the substrate in a few weeks, binding the mix together into a solid block. The bags are then moved to a grow room which has a high humidity level — approximately 80 to 90%.

In order to initiate "fruiting" (mushroom production; also referred to as "flushing"), the plastic bag is either removed for Shiitake or simply perforated for Oyster mushrooms. Within a few days the mushrooms begin to form on the blocks, or through the holes in the bag.

The mushrooms will be ready to pick within days or weeks depending on the species. After picking, the blocks should be left to rest for a period of time after which they can be soaked in water to initiate a second fruiting, or flush. Three to four flushes are possible with Shiitake.

Other species may only have one flush. Typically the nutrients in the blocks are used up within a couple of months. Mushroom yields vary from 30 to 50% of the original weight of the substrate.

A variation of this method is the use of autoclavable bottles instead of bags. This method is used extensively for Enokitake, Shiimeji, and Hiratake mushrooms. Specialized equipment for mixing, filling and inoculating bags or bottles, and for emptying bottles, is available, mainly from Japan.

Growers who are not prepared to set up sterilization and laboratory facilities can purchase bags or bottles already grown out on substrate and ready to fruit. All the grower has to do is provide suitable conditions in growing rooms or outdoor areas and carry out the final stage of the process, which is fruiting and harvesting. There are suppliers of ready-to-fruit bags or bottles in North America, Europe and Asia.

Set-up costs for a cultivation enterprise vary widely depending on factors such as the geographical location (due to the heating and/or cooling of buildings), desired species, whether a laboratory and sterilization facility is required, availability of ready-to-fruit substrates, and the degree of sophistication of grow-out areas, among others.

The first step in setting up a mushroom-growing enterprise is to carry out a survey of the mushroom species that can successfully be marketed in a given geographical region. This is essential since the species chosen will, to a very great extent, determine the setup that will be required.

OUTDOOR NATURAL LOG CULTIVATION

Commercial cultivation on hardwood logs such as Oak, Sugar Maple, Ironwood, Beech, Birch, Hickory and others consists of inoculating logs from the limb-wood of freshly cut trees (preferably winter cut) with mycelium of the desired mushroom species. Softwood logs are not suitable because of the fungal inhibiting substances in their resin. Logs can be cut just as the leaves finish turning colour. This is the point of highest concentration of sugars and vitamins in the wood. However some experts recommend cutting during winter freeze-up because the bark is tight and will adhere to the log better. The logs should be left in long lengths in the bush, off the ground, until spring. At this point they can be cut into shorter lengths of 3-4 ft. The mycelium is introduced in the form of spawn - a blend of sawdust and nutrients with the mycelial culture growing through it, or wooden dowel plugs which have been impregnated with the mycelium. Suitable size logs vary from 7cm to 20cm in diameter and from 1m to 1.3 m in length.

Oak is the preferred species for a commercial operation because it gives consistent results but other hardwoods can be used as mentioned above. Inoculation involves drilling a series of holes into the logs (average 50 holes per log) and introducing the spawn into the holes. The holes are filled with either dowel plug spawn or sawdust spawn. Sawdust spawn works faster than wooden dowel plugs. Wooden dowel plugs can be placed in the holes by hand and tapped in with a hammer; loose sawdust spawn requires an applicator. The holes are then sealed with molten wax. The wax prevents the spawn from drying out and dying. A newer method uses sawdust plug spawn with a styrofoam backing. These sawdust plugs are inserted manually into the pre-drilled holes and no wax is necessary as the styrofoam cap acts as a vapour and contamination barrier. Sophisticated machines that insert loose spawn into the pre-drilled holes and seal them with a piece of styrofoam in a single step are available. These machines are expensive but for a large commercial operator who is inoculating thousands of logs, the machines can save on labor costs.

Log Management

Following inoculation the logs are stacked outdoors under a natural or artificial canopy that provides from 65 to 85 % shade. Various stacking patterns are suitable such as this 'log cabin' method. The logs are monitored for moisture content (by weighing pre-selected logs), and are watered as necessary.

The mycelium takes off from the spawn and runs through the sapwood. Depending on climatic conditions and strains used mushrooms will start to fruit out on the logs after 6 to 18 months.

When the logs start producing they are stood up in various configurations such as in rows between posts or trees. This arrangement facilitates picking and keeps the mushrooms away from dirt and insects. The logs will produce mushrooms for an average of five years without further inoculation as long as the mycelium is kept alive.

A consistent mushroom production schedule is achieved by watering or soaking the logs in tanks. Soaking stimulates (forces) the mycelium in the log to fruit, producing mushrooms a few weeks later. The logs are then allowed to rest for 6 to 8 weeks before soaking is repeated. Managing a population of logs on a staggered soaking schedule can yield a consistent supply of mushrooms.

More recent developments in log cultivation techniques make use of shaded greenhouses to house the logs. Humidity can be controlled much better, inoculations can start sooner, and fruiting seasons can be extended.

The methods described above are used for Shiitake cultivation where a supply of hardwoods (preferably Oak) are available. Outdoor and indoor cultivation make Shiitake the second most cultivated species after the White Button mushroom in terms of quantity of mushrooms produced. Outdoor cultivation is dependable, with a 90% success rate.

A top-quality product is produced from this method, which, if not sold fresh, can be dried to top grade dried mushrooms. The same technique can be used for Wood-ear, Nameko, Kuritake, Tremella, Maitake, and Reishi. Variations in log management include partially burying the logs in the ground in the bush or under artificial shade and having the mushrooms pop up from the ground. This is done for Nameko and Reishi. Depending on your location and availability of logs outdoor cultivation can be the method requiring the least investment for startup. If a woodlot is available as a growing area, no artificial shade is required. If you can cut your own logs a major cost is eliminated.

Typical startup costs for an outdoor Shiitake mushroom operation in Southern Ontario for a person who has the woodlot but must purchase Oak logs are as follows:

- Logs at approximately CDN$ 150.00 per bush cord plus delivery translates to approximately $ 2.00 – $3.00 per log
- Spawn at $ 1.00 to $ 3.00 per log, depending on inoculation rate
- Drills and drill bits can add $.50 – $ 1.00 per log
- Labor costs for inoculation will be approximately $ 1.00 per log
- The total is approximately $ 5.00 to $7.00 per inoculated log

The lifespan of a log (with a single inoculation) averages 5 years. The rule of thumb is 1 year for each 1? diameter. With proper management the mushroom production can be at least ½kg. (1lb) per log per year. If you use shaded polyhouses and "force" the logs you can get up to 4 lbs. per log per year but for a shorter lifespan. In total you can expect approx. 3 kgs. or 6 – 8 lbs. from each log.

Outdoor Sawdust Substrate 'Log' Cultivation

A variation on the natural log cultivation method is the use of 'logs' made up from substrates such as those described in the indoor bag cultivation section. This technique is used extensively in China for the production of Shiitake mushrooms. Once the blocks have been fully colonized with mycelium and have matured, they are placed outdoor in partially buried grow houses and fruited. This technique is only suitable for certain areas with the right climatic conditions such as high humidity and temperature.

Outdoor Bed or Mound Cultivation

This method is suitable for species that require an outdoor stimulus to produce mushrooms and involves the preparation of outdoor trenches or mounds located in a shady area. Substrates such as wetted straw, sawdust, wood chips and compost are used. The substrates are sequentially layered in the trenches or mounds with spawn added between layers. A final layer of soil is placed on top. These beds or mounds are then thoroughly watered and are then covered with cardboard or plastic until the mycelium has run through the substrate. At this point the cover is removed and mushrooms will start to come out. Species that are cultivated using this method include Paddy Straw mushrooms in warm Southeast Asian countries and Blewit and Stropharia in Europe. Some success has been achieved with Oyster mushrooms in Ontario using a variation/ adaptation of this method.

Marketing

Thorough research into the markets available for your product is a must before any investments are made. Marketing requires the investigation and consideration of a number of factors which will likely be peculiar to any geographic location.

The sale of the fresh mushrooms locally is the most desired venue as processing and delivery costs can be kept to a minimum. Potential customers include food distributors who supply restaurants, wholesale vegetable distributors at local wholesale food terminals, supermarket chains, restaurants, gourmet food retailers, and farmer's markets. If you are considering a large operation, distributors and chains are your best bet as they already have a distribution network set up.

Careful attention must be given to pricing and profit margins. The price structure for sales to a distributor or large chain would be lower than for direct sales to restaurants or retailers, but the volume would likely be higher. These kinds of marketing decisions will impact both profitability as well as the level of production required (i.e., labor costs will be higher). For smaller operations, establishing a relationship directly with some upscale restaurants and retailers or attending farmer's markets may prove successful. Your personal attention to these groups will be rewarded with a higher price for your mushrooms.

Markets for a quality fresh product may be found in wholesale food distribution centers of major cities. Demand for exotic mushrooms is usually higher in larger urbanized areas with access to a market that is willing to pay the extra costs for high quality mushrooms. As of 2008 farmers markets are becoming popular due to the "shop local" movement and may provide opportunities to market directly to customers and realize full retail price. Adding mushrooms to a current line of market products will enhance your bottom line.

A number of mushrooms can be marketed in a dried, pre-packaged format. Shiitake is one such mushroom. Large quantities of dried Shiitake are brokered through Japan which is one of the largest producers, importers and exporters of dried Shiitake in the world. Their grading system is very strict and access to this market is extremely difficult. Local markets for dried mushrooms are likely to prove more successful. A potential vast market is in the production of value-added products using dried mushrooms. Attending food trade fairs can be useful in determining which products to produce and for contacting potential customers.

Government agencies may be available to provide assistance in promoting your product. Contact your state or provincial department or ministry of agriculture as well as government economic development agencies.

TECHNOLOGY FOR MUSHROOM CULTIVATION

Mushroom cultivation is increasingly becoming popular because it not only meets the dietry requirements but also adds to the income, especially of growers with insufficient land. It is considered to be a very rewarding and fascinating hobby for the retired persons as well as house-wives who can grow mushrooms in small boxes or other containers while attending to household chores. Today, mushroom cultivation faces less difficulties provided the grower will follow simple rules of growing.

It is really amazing that a small quantity of spawn when planted in suitable growing medium can, within almost six weeks, grow into a highly profitable crop inside a room, where no other crop would grow. Moreover, mushrooms have more uses in modern culinary cuisine than any other food crop. Mushroom cultivation is carried out indoor in any room, shed, basement, garage, etc. which should be well ventilated. However, paddy straw mushroom can be grown outside in shady places also. Of the many mushrooms only three kinds namely button mushroom (Agaricus bisporus), straw mushroom (Voluariella uoluacea) and oyster mushroom (Pleurotus sajor-caju) are suitable for cultivation in India.

Button mushroom is grown in winter. The most suitable temperature for the spread of the mycelium is 24-25°C, while 16-18°C is essential for the formation of fruit bodies.Higher temperature is harmful and low temperature retards the development of both mushroom mycelium and fruit bodies. Paddy straw mushroom can be grown around 35°C. Temperature should not go below 30°C or above 40°C for more than 4-8 hours during growing period. In northern India it can be grown from April to September but the most suitable period is from middle of June to middle of September. Dhingri (oyster mushroom) grows best between 22-28°c. It is grown in northern plains from October to March. The methods of cultivation of these three mushrooms are given separately.

BUTTON MUSHROOM

The cultivated mushroom is grown on special composts, which are of the following two types, synthetic and natural.

COMPOSTING YARD

The compost should be prepared on well cleaned concrete or pucca floor, which should be at a higher level so that the run-off water does not collect near the heap.Composting is usually done in the open, but it has to be protected from rain by covering it with polythene sheet. It can also be done in a shed with open sides or a large room to shelter it from rain.

Synthetic Compost

The following ingredients are required for 15-16 trays of size 100cmx50cmx15cm: Wheat straw (chopped 8-20cm long)-250kg, Wheat/rice bran-20kg, Ammonium sulphate/calcium ammonium nitrate-3 kg, Urea-3kg, Gypsum-20kg The straw is uniformly spread over the compo sting yard in a thin layer and wetted thoroughly by sprinkling water. All ingredients such as wheat bran, fertilizers, etc. except gypsum, are mixed thoroughly in the wetted straw, which is finally heaped into a pile. The pile, 1m high, 1m wide and length adjustable, can be made with hand or stack mould. The straw should be firmly but not compactly compressed into the mould.

It is essential to open the entire pile and remake it a number of times according to the following schedule:

- Mixing of material and making pile - 0 day
- 1st *turning*: 4th day
- 2nd *turning*: 8th day
- 3rd *turning*: 12th day, add 10 kg gypsum
- 4th *turning*: 16th day, add 10 kg gypsum
- *Final turning*: 20th day, spray 10 ml malathion in 5 litre water (any other available pesticide like DOT, BHC, lindane can also be used)

At each turning water should be sprinkled to make up the loss of water due to evaporation. If it is desired to add molasses, then 10 kg molasses diluted 20 times with water should be poured over the straw mixture during the first turning. Sixty kg chicken manure, if available, can also be added at the time of start of pile.

Natural Compost

The following ingredients are required for 15-16 trays of size 100cm × 50cm × 15cm:

Horse dung-1000kg, Chopped wheat straw-300 to 350kg, Gypsum-25kg Poultry manure-100 to 110kg (or 3kg urea). It is prepared from pure horse dung (dung of other animals should not be admixed), which must be freshly collected and should not have been exposed to rain. Chopped wheat straw is mixed with horse dung, urea or poultry manure.

The mixture is uniformly spread over the compo sting yard and water is sprinkled over it so that the straw becomes sufficiently wet. The manure is then heaped in a pile as for synthetic compost. After 3 days when the manure in the heap gets heated up due to fermentation and gives off an odour of ammonia it is opened. The process is repeated 3 or 4 times after an interval of 3-4 days. Twenty five kg gypsum per tonne is added in two instalments at the 3rd and 4th turning. At the final turning 10 ml malathion diluted in 5 litres of water is sprayed into the manure.

Compost Filling in Trays

The compost when ready for fil1ing and spawning has a dark brown colour and no trace of ammonia. There is no unpleasant odour but it smells like fresh hay. The pH is neutral or near neutral. The compost should not be too dry or too wet at the time of filling in the trays, which can be determined by the palm test. For this purpose a small quantity of compost is taken into the hand and pressed lightly, if a few drops of water ooze out of the fingers then it is of right consistency. If relatively dry then the water should be made up by sprinkling.

If too wet, the excess water should be allowed to evaporate. The prepared compost is now filled in trays, which may be of any convenient size but depth should be 15-18 cm. A standard size of tray is 100 cm × 50 cm × 15 cm. The trays should be made of soft wood and provided with the pegs at the four corners so that they can be stacked one over the other leaving sufficient space (15 cm) between the two trays for various operations. The trays are completely filled with the compost, lightly compressed and the surface levelled.

Spawning

Spawning means sowing the beds with the mycelium (spawn) of the mushroom. Spawn can be obtained from Mushroom Laboratory, Y. S. Parmar University, Chambaghat, Solan; National Mushroom Research and Training Centre, Chambaghat, Solan (H.P.) at a nominal cost. Small quantity of spawn is also available from the Division of Mycology and Plant Pathology, Indian Agricultural Research Institute, New Delhi.

The grain spawn is scattered on the surface of the tray bed which is covered with a thin layer of compost. Spawning can also be done by mixing the spawn with compost before filling it in trays. Five hundred gram spawn is sufficient for five trays of standard size. After spawning, the compost surface is covered with old newspaper sheets, which are wetted by sprinkling water to provide humidity but no water is directly added to the compost during spawn running.

The trays after spawning are stacked vertically one over the other in 4-5 tiers. One metre clear space may be left in between the top tray and ceiling.

There should be about 15-20 cm space between the two trays. The room should be maintained around 25°c. The humidity should be built up by frequently watering the floor and walls. The room may be kept closed as no fresh air is needed during the spawn run. White cottony mycelium spreads and permeates through the compost. Eventually the compost surface gets covered with the mycelium. It takes 12-15 days for complete spawn run. Low temperature prolongs the spread of the mycelium.

Casing

After the spawn run is complete as is evident by white cottony growth, the surface of the compost is covered with 3 em layer of casing soil. A suitable casing soil can be prepared by mixing equal parts of well rotten cowdung (finely crushed and coarsely sieved) and garden soil. The casing material should possess high water holding capacity, good pore space and pH should not be lower than 7.4. The casing material is sterilized to kill insects, nematodes and molds. Sterilization can be accomplished either by steaming or by treating with formalin solution.

For one cubic metre of casing soil, hqlf litre for formalin (40%) diluted with 10 litre of water is sufficient. The casing soil is spread over a plastic sheet and treated with formalin by sprinkling. The treated soil is piled up in a heap and covered with another plastic sheet for 48 hours. The soil is turned frequently for about a week to remove all traces of formalin which can be tested by smelling. After casing, the temperature of the room is maintained at 25°C for further three days, after which it must be lowered to below 18°C. At this stage lot of fresh air is needed and, therefore, the growing room should be ventilated by opening windows etc.

Cropping and HalVesting

The first flush of the pin heads become visible 15-20 days after casing or 35-40 days after spawning. Small white buttons develop 5-6 days after pin head stage. The right stage of harvest is when the caps are still tight over the short stem. In case the buttons are allowed to mature further, the membrane below the cap will rupture and the cap will open up in umbrella-like shape. Such mushrooms are considered to be inferior. Harvesting is done by holding the cap with forefingers slightly pressed against the soil and twisting it off. The soil particles and mycelial threads clinging to the base of the stalk are chopped off. Mushroom can also be harvested by cutting off with a sharp knife at soil level.

Yield

The average yield of 3-4 kg per tray is considered normal. However, if compost is carefully prepared, spawn is reliable and temperature is favourable, then a yield of 5-6 kg per tray is possible. Partial or complete failure may also happen due to negligence.

Storage

The mushrooms are best consumed fresh. Storage in refrigerator for a few days is possible if they are placed between moist paper towel.

PADDY STRAW MUSHROOM

Paddy straw mushroom, also called Chinese Mushroom (Volvariella spp.) is grown in South-East Asia. This mushroom is dark in colour and is very delicious. It is usually grown in raised beds, which are laid in open, exposed or shady places over which temporary sheds are built to protect the beds from direct sunlight and rain. If the beds are made indoors in a well ventilated room which is not very dark, the production of mushrooms is more steady and reliable.

Spawn of Paddy Straw Mushroom

Spawn of this mushroom is made on grains of cereals or millets and is called grain spawn. Sometimes it is also made on soaked chopped paddy straw. This spawn is called straw spawn. Spawn may be available locally from the Division of Mycology and Plant Pathology, Indian Agricultural Research Institute, New Delhi-llOO12 at nominal price. Spawn is also available from Plant Pathology or Microbiology Departments of the respective Agricultural Universities. One bottle of spawn is sufficient for one bed.

Substrate or Bedding Material

This mushroom is usually grown in paddy straw. Thoroughly dried and long paddy straw is preferred. The straw is tied in bundles of about 8-10 cm in diameter. The bundles are then cut to a uniform length of about 70-80 cm, and are soaked by immersing them in a tank of water for about 12-16 hours. Later, excess water is allowed to drain off.

Procedure for Making the Bed

Make a 15-20 cm high foundation of soil or bricks. The size of the foundation is slightly larger than that of bed.

The foundation must be strong and firm to hold the weight of the bed:

- Make wooden bamboo frame of the size of the bed foundation and place it on top of the raised bed foundation.
- Place four bundles of soaked straw side by side on the wooden/bamboo frame.Over these, place another set of four bundles similarly but the loose ends on the opposite side. These 8 bundles consitute the first layer.

- Scatter grain spawn about 8-12 cm from the edges of the first layer. If straw spawn is used, small pieces of the size of the thumb are planted about 4-6 cm deep and about 10-15 cm apart along the edges. Dust the spawn with powdered gram/arhar dal powder or rice/wheat bran.
- Now place a second layer of eight bundles across the first layer and spawn it as before.
- Again place a third layer of straw bundles across the second layer and spawn it all over the surface.
- Finally cover with a fourth layer of four straw dundles. Press it lightly.
- Completely cover the bed with a transparent plastic sheet, taking care that the plastic sheet is not in contact with the bed.

Care of Bed

Remove the transparent plastic sheet after the mycelium has thoroughly permeated the straw. This would take about a week at 35°c. If the surface of the bed gets dry then it should be watered lightly by a sprayer at least once a day.

The Crop

Mushrooms begin to appear within 10-15 days after the beds are spawned and continue to do so for about a week or 10 days. The total yield of the bed is about 2-2.5 kg. The mushrooms should be picked when the volva (cup like veil) just breaks to expose the mushroom inside. The mushrooms are very delicate and must be consumed fresh. If stored in fridge they can stand for 2-3 days. These mushroo~s can easily be air dried in sun or shade.

OYSTER MUSHROOM

This mushroom is simple to grow and has excellent flavour and texture. It is very popular in many countries particularly in South-East Asia, where cultivation of button mushroom (Agaricus bisporus) is not possible under natural climatic conditions. Besides its delicious taste, it is known to be very nutritious and is recommended to control obesity and is reported to be beneficial for diabeties. Its low fat content makes it an ideal diet for the blood pressure patients.

Substrate or Growing Material

It can be cultivated on a wide ranging of cellulosic farm wastes or other materials. Cereal straws, banana pseudostems, waste paper, cotton waste are particularly suitable.However, paddy straw is the most commonly used material.

Oyster Growing in Polythene Bags

Paddy straw is chopped into small pieces 3-5 cm long. It is soaked in water for. about 8 hours, after which water is squeezed out.- About 200 g grain spawn (half litre bottle) is mixed thoroughly with about 5-6 kg of wet chopped straw (= lY2 kg dry straw).- A polythene bag 45 cm long and 30 cm diam. is used. It is perforated with 2 mm diam. holes, about 4 cm apart, all over the surface. Spawned straw is filled about 2/3rd of the capacity of the bag and mouth tied.- The bags containing spawned straw are placed in shelves in the growing room (RH. 80-85% and room temperature 24-26°C).

Growing in Rectangular Blocks

A wooden tray mould (50cm × 33cm × 15cm) without bottom is required. A one m2 piece of transparent polythene sheet is spread so that it forms the bottom of the tray mould and also lines the sides from inside. The loose edges hang out from the tray mould.- Fill in the wetted chopped paddy straw (as described for bag method above) to make a 5 cm thick bottom layer. Scatter spawn uniformly. Lay another 5 cm thick layer over the top of bottom layer and spawn it in the same way. Finally, lay the third layer (final layer) over the second and repeat spawning. Spawn is covered with more wetted straw to bring it in level with the top of the mould. Compress it firmly by hands or a board. Two hundred grams (Yz litre bottle) spawn is sufficient for two blocks.- Fold the loose hanging edges of plastic sheet over the straw block, fasten with a string. Remove the rectangular block from the mold.

Spawn Run

The spawned bags/blocks are placed on shelves in the growing room where RH. is maintained at 80-85 per cent and room temperature is about 24-26°C.- Spawn run is complete in about 10-12 days and is indicated by white cottony mycelium which permeates throughout the straw. As a result the straw becomes compact and does not split when handled. At this stage, polythene covering is removed by cutting it open in case of bags and by untying the polythene sheet in case of blocks. In case of bags, the straw gets compacted in the form of a cylinder.- The cylinder/blocks are neatly arranged on shelves and gently watered at least twice a day.

The Crop

First mushrooms appear about 18-20 days after spawning. Two or three flushes appear at an,interval of about a week. Dhingri should be harvested when the cap starts becoming folded. Harvesting can be done by cutting with a sharp knife or by twisting it off with fingers from the substrate. Dhiilgri is best consumed fresh. It can be dried in sun or in a mechanical drier, and stored in polythene bags.

Yield

About one kg mushroom can be obtained from 1Yz kg dry straw (=5-6 kg wet straw).

THE WILD EDIBLE MUSHROOMS

The next two lectures will be concerned with the mushroom eating. The term "mushroom" is used here in a broad sense as used in popular mushroom guides and include members of the Basidiomycota that produce fruiting bodies and some members of the Ascomycota as well.

The latter, as you should recall, produce asci and ascospores during sexual reproduction while mushrooms and other members of the Basidiomycota produce basidia and basidiospores. In addition, not all members of the Basidiomycota are mushrooms with stalk, caps and gills. There are coral fungi, polypores, puffballs and boletes, just to name a few. Boletes are very similar to mushroom in appearance. The most obvious difference being that they have pores, as the polypores, instead of gills.

However, instead of having fruit bodies that are leathery to woody, boletes are fleshy like mushrooms and there are a number of highly desirable species, with respect to their edibility.

Obviously, there are species such as the *Pycnoporus sanguineus* and *Geastrum indica* that cannot be eaten because of the texture of their fruitbodies and the powdery masses of their spores, respectively, but they have traditionally been included in popular mushroom guide books since one of the goals of the authors of such books is also to show the reader the diversity that exists in mushrooms.

Although the topic of eating mushroom is inevitably tied to mushroom poisoning, we will cover this topic in a later lecture. This is a very fascinating topic because there are many people that are intrigued by eating wild mushrooms and will go out, daily, and forage for them in spite of the possibility that they may inadvertently eat a poisonous mushroom.

Despite this danger, the number of people, in this country, consuming wild mushrooms increases each year. However, while incidents of mushroom poisonings continue to occur each year, sometimes with fatal results, mushroom poisoning is still not a major problem in the United States. Even with an increase in the number of people going out to collect mushrooms, there is no evidence that the number of incidents of mushroom poisonings have increased in recent years.

The practice of eating mushroom probably began during the hunting and gathering period, in our prehistory. They were collected along with fruits and berries, as well as other plant material that could be consumed.

Also, like plants the gatherers learned which ones were edible and which were poisonous and if there were other uses for mushrooms, *i.e.*, medicinal or religious uses. However, unlike plants, mushrooms must have been have been shrouded in mysteries since unlike the plants, there was not an obvious way in which they could be reproduced. Even much later, during the dark ages, mushrooms became more firmly embedded in the mythology of the supernatural. Many myths arose because of their seemingly supernatural characteristics and strange habitat. Their growth was rapid and they seem to suddenly appear overnight as if from nowhere.

Thus, their origin appeared to be magical. They sometimes formed circular patterns or "rings" as in *Chlorophyllum molybdites* where the grass is greener *inside* the ring. Some glow in the dark. Many have bizarre shapes, and are ephemeral.

They became part of the lives of fairies, elves and witches. This concept is evident even today in common name of mushrooms, such as "The Fairy Ring Mushroom" for *Marasmius oreades* and "Witches Butter" or Fairy Butter for *Tremella mesenterica*.

Because so little was known about mushrooms, many misconceptions concerning the edibility of mushrooms have developed. Some of the more common ones are briefly discussed below:

- There is a fool-proof test for distinguishing edible from poisonous mushrooms. The most common ones that can be heard are that "a poisonous mushroom will turn silver black while it is being cooked", "if you can peel the cap of the mushroom it is safe to eat" and "observing which mushrooms foraging animals consume will tell you which species are safe to eat". While there are some generalizations that can be made within certain groups of mushrooms, there is no fool-proof test that can be used for all mushrooms. Those species of mushrooms that are edible are known to be edible because someone at one time had tried it and discovered it to be safe to eat.
- Most mushrooms are poisonous. Of the thousands of species known, perhaps 60 or so are poisonous, and of these only a handful will be fatal if consumed (these numbers will vary depending on your source). However, it does not require a lot of mushroom to poison a lot of people. In Europe, of those that die mushroom poisoning, probably 90% die as a result of mistaking *Amanita phalloides* (Death Cap) for *Amanita calytroderma* (Coccora). The remaining species, however, are not necessarily good to eat. I am using edible here to means non-poisonous and not necessarily good to eat. Thus, an edible mushroom may have a strong bitter, peppery or some other unpleasant taste, be bland or have no taste at all.
- There are a large number of people that die from mushroom poisoning each year. "Large" is somewhat ambiguous here. If we are talking about the number of people that go out collecting for mushrooms each year, in this country, then the number of people that die as a result of mushroom poisoning is few relative to that number.
- Poisonous mushrooms must taste bad. Many mushrooms that are non-poisonous may have a very bad taste. The opposite can also be

true. *Amanita phalloides* is said to have a quite pleasant taste, but is one of the most deadly species of poisonous mushrooms.

- You can be poisoned by touching a poisonous mushroom. As deadly as some toxins may be, touching the mushroom is harmless. The harmful toxins in mushrooms must be consumed in order to harm you.
- Collecting mushrooms for consumption is unsafe and even experts have died from picking the wrong mushrooms. This last misconception is the one that continues to be perpetuated by the news media every year. Probably every year, you can read a headline that goes something like "Expert Mushroom Hunter Dies From Eating Deadly Mushroom". However, upon closer examination of such a story it is often the case that the person that died is far from being an expert. Even those who are avid collectors that have been foraging for wild mushrooms, for only a short period of time, are unlikely to die from mushroom poisoning, if they have even had a minimum of training in the do's and don'ts of mushroom collecting and if common sense is used. There are a number of species that are very good to eat that cannot be mistaken for other species. If collectors stick with those species, mushroom poisoning is highly unlikely. New species can be tried through interaction with other collectors who have eaten other species.
- Species determined to be edible are always safe to eat. Whenever a new species is tried for the first time, even if it is one that is highly regarded and is said to be being very tasty, it is best to be cautious. Try only a few bites and wait 24 hours before consuming more. There are a vast number of compounds that occur in wild mushrooms that may cause adverse reactions, when consumed by a few individuals, but are safe for the general public. There is also the possibility of an allergic response to a particular species. Other precautions that should be taken:
- When preparing mushrooms for a meal, always inspect the mushrooms to determine if it is firm and fresh. Bacterial and fungal decomposition may be taking place in old mushrooms.
- Avoid eating raw mushrooms. Many edible species have toxins that are heat sensitive and will be rendered harmless by cooking the mushroom. Also, the cell wall of mushrooms is composed of chitin, which the human digestive system cannot break down. must also be degraded by heat so that we can absorb nutrients within cells. If the cell walls remain intact, the nutrients in the mushrooms will simply pass through our digestive system. Cooking will break down the cell wall and release its contents, which are digestible.

- Mushroom are of no nutritional value. Although mushrooms will never be one of the world s staples, it has been a food supplement in various cultures. Nutritionally speaking, mushrooms fall between the best vegetables and animal protein source. Their protein content may vary anywhere between 15-40% of *dry* weight (keep in mind that mushrooms are more than 90% water). However, all essential amino acids are present in mushrooms, as well as water-soluble vitamins and all the minerals that our bodies require are present. A generous serving of mushrooms (0.5 lb) of fresh mushrooms provides approximately 70 kcal.

EAT MUSHROOMS FOR NUTRITIONAL VALUE

While some people may hype the nutritional value or even medicinal value, of mushrooms, as the reason for consumption, that does not appear to be the main reason for going out to collect wild mushrooms. Presently, there are a good number of species that have been cultivated. Many of these have had research carried out indicating their high nutritional value as well as medicinal value. So, why not eat only cultivated mushrooms rather than taking the risk of possibly being ill eating an unsafe mushroom or perhaps even being fatally poisoned? As far as individual collectors are concerned this seems to be an impossible question to answer. Perhaps, it is the thrill of the hunt, or the idea of going back to nature and collecting non-cultivated food. There are probably numerous reasons as to why an individual would participate in collecting and eating wild mushrooms. However, if we look at this question from a cultural perspective, according to R. Gordon Wasson, the father of ethnomycology (in its simplest definition it is the study of the relationship between people and fungi), how readily a person will collect and consume wild mushroom is dependent upon the culture in which they were raised.

Wasson believed that cultures could be divided into two categories, with respect to mushrooms: 1.) Those that are mycophilic may regard mushrooms as the epitome of gastronomy, and 2.) Those that are mycophobic may despise and regard them all as being poisonous and would surely cross the road just to stomp the life out of them. With the exception of possibly a single species that has been cultivated (*Agaricus bisporus*), those cultures that belong to the latter category will usually have nothing to do with mushroom. While Wasson truly believed that there was this dramatic like or dislike of mushrooms, Benjamin (1995) has pointed out that there are cultures that neither like nor dislike mushrooms, with respect to their use as food. In a few cultures mushrooms have even been regarded as being magical and utilized in religious ceremonies.

The observation that different cultures may vary as to their attitudes concerning mushroom consumption was made in the late 1920s by R. Gordon Wasson and his wife Valentina, during their honeymoon in the Catskill

Mountains, in New York State. A somewhat humorist recount of this observation was made by Wasson, in 1968, in his book, Soma, The Divine Mushroom of Immortality:

We had been married less than a year and we were off on our first holiday, at Big Indian in the Catskills. On that first day, as the sun was declining in the west, we set out on a stroll, the forest on our left and a clearing oil the right. Though we had known each other for years we had never discussed mushrooms together. All of a sudden she darted from my side, with cries of ecstasy she flew to the forest glade, where she had discovered mushrooms of various kinds carpeting the ground. Since Russia she had seen nothing like it. Left planted on a mountain trail, I called to her to take care, to come back. They were toadstools she was gathering, poisonous, putrid, disgusting. She only laughed the more: I can hear her now. She knelt in poses of adoration. She spoke to them with endearing Russian diminutives. She gathered the toadstools in a kind of pinafore that she was wearing, and brought them to our lodge. Some she strung on threads to hang up and dry for winter use. Others she served that night, either with the soup or the meat, according to their kind. I refused to touch them.

After this incident, the couple talked about the difference in their attitudes toward mushrooms. Upon further conversations with close friends, from various cultures, they discovered a relationship between their ethnic identity and their like or dislike concerning mushrooms. Their fascination with this topic would turn out to be a life-long interest and would lead to the origin of a new field of study, ethnomycology. Some of the attitudes of different cultures are briefly summarize, below, from Benjamin (1995), to demonstrate some of these differences.

Anglo-Saxon

Historically, the British and their colonies, *i.e.*, English-speaking cultures were widely recognized as being mycophobic or at least not very interested in consuming mushrooms. It is also true that these cultures are for the most part reliant upon domesticated food rather than foraging for wild food. Those individuals in these countries that may have interest in collecting wild mushrooms as well as plants for consumptions are usually the descendents of Asian, and Slavic cultures of Eastern Europe and Russia. In the United States, such people are usually looked upon as being eccentric.

Mushrooms were often closely associated with toads, snails, snakes spiders and witches, in Western mythology. Although witches have never been a very popular figure in any country or cultures in which they have been recognized, it was only in some countries where a deep-seated fear of them existed. This aversion may be related to the association of witches with the devil. It was primarily in the Germanic, Celtic and Anglo-Saxon countries such as Germany,

Switzerland, Scotland, England, and subsequently America that witches were actively and systematically persecuted.

The origin of this mycophobic attitude for any country is unknown. However, that it is a deep seated response can be seen by its expression in the English literature for centuries: They are all very cold and moist and therefore do approach unto a venomous and mothering faculty and ingender clammy and cold nutriment if they be eaten. – John Gerard, Herball or Generall Historie of Plantes (1597) But whatever dressing one gives to mushrooms, to whatever sauce our apiciuses put them, they are realy good but to be sent back to the dung heap where they were born. – Louis de Jacourt, Champignon (1753)

It is tempting to say that such writing may have shaped the attitudes of the English. However, this is unlikely since most people of that time had no access to such writings and few would have been able to read it if they had. It is speculated that the sexual connotations associated with some mushroom led to their rejection by a prudish Anglo-Saxon culture. Undoubtedly, the phallic aspects of many mushrooms did not go unnoticed. Certain species were also regarded as aphrodisiacs. While this may be a reasonable argument when applied to the period in history of Victorian England, it does not explain the mycophobic attitude of its earlier and later history nor does it explain this same attitude in America

One particular mushroom was even regarded as being obscene to the morals of Victorian England, *Phallus impudicus*, one of the many species of stinkhorns. However, it is obvious that it is not the particular species, but rather the resemblance of any species of stinkhorn to the male organ that upset Victorian England.

In our native woods there grows a kind of toadstool called in the vernacular The Stinkhorn (though in Latin it bears a grosser name). The name is justified for the fungus can be hunted by scent alone, and this was Aunt Ettys great invention. Armed with a basket and a pointed stick, and wearing a special hunting cloak and gloves, she would sniff her way through the wood, pausing here and there, her nostrils twitching when she caught a whiff of her prey.

Then with a deadly pounce she would fall upon her victim and poke his putrid carcass into her basket. At the end of the days sport the catch was brought back and burnt in the deepest secrecy on the drawing room fire with the door locked – because of the morals of the maids!! – Gwen Raverat, Period Piece (1952)

An example of myco-phobism can even be found in a children's story of Babar the Elephant, by Jean de Brunhoff. Babar the Elephant is a famous series of French children's stories that first came out in 1931. The first part of the story tells of how Babar was orphaned and then became king after his mother was killed by a hunter and his father died from mushroom poisoning: When Babar was young and lived in the jungle, his mother was killed by a hunter.

The hunter was just about to capture Babar, when Babar quickly ran away into the forest and managed to escape. After a few days, he came across a town with many big buildings. There he met a kind old lady who adopted him. In this town, Babar bought clothes, was educated about the ways of humans and quite enjoyed himself. The years passed by quickly and it wasn't long before 2 years had elapsed since Babar had run away from the forest. He began to miss the jungle and longed to play with his relatives and friends. One day, as he and the old lady were walking down the street, they saw 2 young elephants running along. They turned out to be Celeste and Arthur (Babar's cousins). On seeing them, Babar ran and embraced them. He told his cousins of his life in the town and he decided to return to the forest with his cousins. Meanwhile, Arthur and Celeste's parents were very worried about their children's disappearance. They organised a big search party. A little bird, who had seen the children in the town, reported back to them and told them where their children were located. Then Arthur and Celeste's cross mothers journeyed to the town to fetch them. Once there, they were very relieved to have found them, but scolded them for running away. Babar, Arthur, Celeste and the cousins' mothers decided they'd all travel back to the jungle together. After Babar and the old lady exchanged tearful goodbyes, he packed all his belongings and returned to the jungle.

Once Babar arrived, he was joyfully greeted by the elephants, but became quite shocked when hearing of an illness that had struck the King of the elephants. The king of the elephants had eaten a poisonous mushroom, become seriously ill and then died shortly before Babar had arrived. Cornelius (one of the older elephants) and some other elephant elders had been discussing who should become the new King before Babar had unexpectedly arrived. Cornelius suggested that Babar should become the king and the other elephants agreed. Babar gratefully accepted the honour and said he would marry his childhood sweetheart, Celeste.

While writings for adults may not shape the culture's attitudes towards mushrooms, they may be key element to how the children view the natural world as well as how they interpret the story they hear.

Russia and Eastern Europe

If the British are the consummate mycophobic society then Russia and Eastern Europe are at the opposite end of the spectrum. Culturally, the people in these area are Slavic, which is a branch of the Indo-European language family that includes Bulgarian, Belorussian, Czech, Macedonia, Polish, Russian, Serbo-Croatian, Slovak, Slovene, Ukrainian and Wendish. Mushrooms play a very important role in the Slavic culture

Here mushroom collecting and eating is a favorite past time. This passion for mushroom that was first demonstrated by R. Gordon Wassons wife, Valentina, shocked Wasson when he first observed her collecting mushrooms, but this was what began Wassons life-long quest into different cultures attitudes towards mushrooms.

Africa

Africa appears to be generally mycophilic. There are some some regions, such as Nigeria, where mushrooms are a part of everyday life as food, charms and remedies in traditional medicine. The most sought after species is the one belonging to the genus *Termitomyces*, a genus associated with termite nests.

There are also myths concerning mushrooms that are unique to this part of world. One is that if you feed a mushroom to a chicken, if it eats the mushroom, and nothing happens, it is thought to be edible, if it eats the mushrooms and vomits, it is edible with caution and if it rejects the mushroom, it is thought to be poisonous. Although there does not appear to be any logic to this myth, it should be kept in mind that this is also the case for those myths concerning mushrooms in Western cultures.

Neighboring Malawi is also a region of mushroom gourmands, whose women have knowledge of the edible and poisonous species, locally. Over 60 edible species are recognized in this area, predominantly belonging to the genera *Amanita*, *Cantharellus* and *Termitomyces*. The responsibility of foraging and identifying the mushrooms fall upon the women. Mushrooms also sold commonly sold on roadsides as well as in town and village markets.

In rural Zambia and Zaire, mushrooms are widely consumed during "hunger" months from late November through early April. A survey in the 1950s demonstrated that they were second only to caterpillars, as a food source. Although, there were a wide variety of edible species, the species selected for eating were restricted. *Termitomyces* was favored, but species of *Lactarius*, *Russula*, *Cantharellus* and *Amanita* are also eaten. Thus, consumption of mushrooms appear to be out of necessity for food than as a preferred food item. So Africa is not entirely mycophilic. Some parts of Africa will not readily eat the variety of species just mentioned, but still may have other favorites. While such areas will consume few species, they will still more readily consume the species that they recognize, unlike in Anglo-Saxon cultures.

India

The Indian subcontinent appear to be mostly mycophobic or at least do not eat mushrooms to the extent of the Asian continent. Most consumption seems to be in mountainous area of Kasmir and the Himalayas. Mushroom consumption is also common in Bengal. However, it generally appears that India lacks a culinary interest in mushrooms. One reason for this lack of interest may be that India was once a former British Colony, but it appears that this lack of interest predates British presence in India. Despite the lack of gastronomic interest in mushrooms, in the district of Kerala, mushrooms are utilized in religious ceremonies. In this district, burial monuments, resembling mushrooms, built between 2000-1000 B.C., are worshiped by present day people who also consume some of the hallucinogenic mushrooms in this area. It is believed that these mushrooms enable the user to communicate with their gods and the dead.

Native North American Cultures

Despite the large number of species in North America, Native Americans appear to be mostly mycophobic. Most tribes have never used mushrooms as a significant source of food, and in some case are not used at all in their diets. However, Native Americans did use mushrooms for other purposes: medicinal, and spiritual, religious and ceremonial. Puffballs were often used as a means to stop bleeding or to dry out wounds. This particular use is not unique and has been used for this purpose on all continents.

Puffballs are also used for healing umbilical cord, in poultices, and a lotion to treat various skin conditions. Puffballs that grew in fairy rings were used for spiritual, religious and ceremonial purposes. They were used as incense to ward off evil spirits. No evidence that any hallucinogenic fungi were employed in any ceremonies as they were in Mesoamerica.

Europe

European cultures are generally fond of mushrooms, with the exception of Holland, which has an attitude similar to that of England, but to a lesser degree. Species favored by each country varies. The Italians prefer the porcini (*Boletus edulis*) and white truffle (*Tuber alba*); the Germans and Swiss, the chanterelle (*Cantharellus cibarius*) the Catalonians, the delicious milky cap (*Lactarius deliciosus*).

Switzerland has developed a very sophisticated level of collecting and marketing of wild mushrooms. During spring and fall seasons, there is a space reserved for sale of wild mushrooms in open-air markets. Markets are under supervision of an inspector, a local mycologist responsible to the state government. Only truffles can be sold outside of these markets. All other mushrooms must pass the scrutiny of the inspector.

The most important job of the inspector is to confirm the identities of the mushrooms that are being sold by the amateur collectors, and minors are not permitted to sell mushrooms. There is a list of approximately 54 species that are sanctioned for sale. Other duties include surveillance for occasional deceptive practices. For example, soaking mushrooms in water to increase weight. Boletes must be cut in half so that customers can see the quantity of "animal protein" inside.

In Finland, people were indifferent towards mushroom eating, except in the southeastern parts of the country closest to Russia and its Slavic influence.

These areas had gathered mushrooms for centuries. However, during World War II, mushroom usage, as food, increased because of the scarcity of other food stuff. Years later, the Finnish government, along with several other organizations, launched an education training programme.

More than 1600 advisers, and 50,000 pickers were trained from 1969 to 1983. This effort was coordinated by 22 inspectors who provided the expertise in identification. Every citizen in Finland had the rights to pick mushrooms and berries on any land, provided no damage was done. By 1979, an estimated 72% of the population was picking mushrooms. Income from sale of mushroom, locally and abroad, is untaxed and the income generated during a good year can be substantial. Thus, the mycophilic attitude of the people was acquired shortly after WWII.

South America

For the most part, mushrooms are not a major food item for the majority of the people native to this continent. In studies carried out, few indigenous people have been identified who utilize fungi as part of their diet. Among those people who eat mushrooms, there is usually an elaborate taxonomic system for edible mushrooms and allows the distinction between species that look almost identical to the untrained eye. However, the inedible ones are just said to be "no good". This practice seems to be similar to other cultures whose mushroom knowledge is based on folk tradition.

Asia

Although the attitudes of different Asian cultures may vary, they are generally mycophilic. China has longest tradition in collecting mushrooms, not just for consumption as food, but also as an important role in traditional medicine. However, unlike most other mycophilic societies, species of mushrooms that are used in China for food and medicine are generally cultivated since the localities where these species can be collected are a long way from the densely populated agricultural areas where most people live.

Thus, the Chinese do not go out to collect mushrooms as other mycophilic cultures such as the Slavic and most Europeans, but instead have become very skillful in the cultivation of the various species that they utilize. Poisoning, for this reason, is not very frequent.

Japan is similar to China in having cultivation play a large role in supplying mushrooms for the diet and medicinal needs of its people. Once again, it is because of the population living in a limited space. However, the Japanese differ from the Chinese in that they have a great enthusiasm for mushroom collecting. One species in particular, the Matsutake, *Tricholoma matsutake*, is a particular favorite. Unfortunately, in the last few decades, this species has come under pressure and is now in a serious decline due to nematode infestation in their pine forests. Taking advantage of this situation, Korea and the Pacific Northwest, from Oregon to British Columbia have developed a lucrative business supplying the Japanese market with a similar species of mushroom, *Tricholoma magnivelare* (=*Armillaria ponderosa*), which is found in those areas.

Medicinal value of mushrooms is also valued in China and Japan. *Ganoderma lucidum*, called the Ling Chi, in China and Reishi, in Japan is a highly regarded herb that was first documented in China approximately 2000 years ago. Today, it is cultivated in both countries. Another highly regarded group of fungi in Asia are the stinkhorns. Unlike in Anglo Saxon cultures, the phallic appearance here represents the power of enhancing potency and performance, and in China, *Dictyophora duplicata, D. indusiata* and *D. multicolour* are species of stinkhorns that are cultivated for this reason.

Australia

Australia, because they were originally settled by Anglo-Saxon settlers, brought with them their mycophobic attitude. Few mushrooms are used and picked by the white populations there. The aboriginal use of fungi is also limited to a few species and do not make up a significant part of their diet.

TYPES OF EDIBLE MUSHROOMS

Mushrooms add something special to every meal and, when you are familiar with the different types of edible mushrooms, you can eat a different mushroom every night.

Exotic mushrooms add a distinct flavour to your cooking, but unless you are a mushroom expert, known as a mycophagist, you are safer getting your mushrooms from the store. Some types of poisonous mushrooms look just like some types of edible mushrooms.

While wild mushrooms may look wild and taste wild, it is better to play it safe when choosing your mushrooms. There are three types of mushrooms available at most markets: common, exotic, and wild. The common mushroom, known as a button or champignon mushroom, is the most readily available.

CULTIVATED EXOTIC MUSHROOMS

Cultivated exotic mushrooms are grown in a controlled environment and are very safe to eat. They are sometimes more expensive than button mushrooms but, as an occasional treat, are worth the extra money and effort to find them.

Shiitake Mushrooms

Shiitake mushrooms are also known as Black Forest mushrooms or Golden Oak mushrooms. They are mostly cultivated in Japan, China, and South Korea, but are also available from Australia and North America. When fresh, the mushrooms colour ranges from light golden brown to dark brown. They have a wide cap with a firm fleshy texture, but the stems are very tough and are either chopped very fine for sautéing or saved for making stock. These mushrooms are also available in a dried form.

The dried shiitake can be stored for a long time and can be revived just by soaking in water. The water that the dried shiitakes have been reconstituted in can then be used to make a delicious mushroom sauce. As far as exotic mushrooms go, the dried shiitake is a good way to add great flavour to your recipes while saving money.

Oyster Mushroom

The oyster mushroom, also known as Pleurotte, gets its name from the way it looks and not its flavour. Oyster mushrooms are light tan or cream coloured with a large, fan-like cap and a short stem. These tender mushrooms have a delicate flavour, so they are best prepared simply so the flavour isn't overpowered.

Enoki Mushrooms

These mushrooms are also known as Enokitake or Enokidake. They have a tiny white cap on a long slender stem. They grow in bunches from a single base, so you will be selecting clusters rather than single mushrooms. You just trim off the base and give the Enoki a quick rinse to prepare them. Enoki have a crisp texture and a fruity, sweet flavour and are very good raw in a salad. If you are using them in a cooked dish, add them last to keep the texture and flavour.

Cremini Mushrooms

Cremini mushrooms are fun and tasty. Cremini are the slightly more mature variation of the common button mushroom. Use these wherever you would use button mushrooms but expect a deeper, richer flavour.

Portobello Mushrooms

The most mature version of the Cremini mushroom these mushrooms can grow up to 6-inches wide, are great for grilling or stuffing, and can be used as a substitute for meat in some recipes.

WILD MUSHROOMS

Wild mushrooms bring more flavour to every recipes, but only use wild mushrooms that you find at the market. Eating wild mushrooms that you find in the actual wild can be dangerous.

These four wild mushrooms are available fresh as well as dried. You may want to consider the dried variation because they keep longer, have a more intense flavour, and the water that you use to restore them can be used to make a mushroom sauce.

Morel Mushrooms

Morels can be found in several varieties including black, golden, and white. Morels look like a conical sponge and are completely hollow inside. They are available fresh in the spring or dried all year long. Use them in butter or cream-based sauces.

Bolete Mushrooms

Also known as Cep, or Porchino, this brown capped mushroom has a light coloured bulbous stem and a cream coloured flesh. The smooth meaty texture of this mushroom has a rich earthy flavour and is best sautéed with butter and garlic. Bolete are usually available in late summer and early fall.

Chanterelle Mushrooms

Chanterelle are also known as Girolle. They are yellow to orange in colour and look like an inverted umbrella. This mushroom has ridges rather than gills. Chanterelle have a rich woodsy flavour and aroma and are best when sautéed with butter and a little garlic. Chanterelle are available during the summer and fall.

Black Trumpet

These mushrooms are related to Chanterelles, but are black in colour with much thinner flesh.

TRUFFLES

Truffles are very rare and only found in southern France and Northern Italy. Black truffles can cost as much as $130 to $390 dollars per pound while white truffles can cost as much as $1350 to $2700 per pound. Black truffles grow only on the roots of oak trees while white truffles can be found growing on the roots of oak, hazel, poplar, and beech trees.

Specially trained hogs and dogs are used to find these delightful fungi. In general, truffles are shaved onto a meal just before serving. A more cost effective way of getting the rich, earthy flavour of these mushrooms into your food is to look for truffle oil, which is usually olive or vegetable oil infused with truffles.

DIFFERENCE BETWEEN EDIBLE AND NON-EDIBLE MUSHROOMS

EDIBLE MUSHROOMS: POPULAR AND READILY

Most of us are familiar with the most popular and readily available edible mushrooms because we purchase them from the super market. You may not go out foraging for mushrooms, but many people do collect, identify and eat wild mushrooms, an activity that can prove to be deadly. Therefore, only a person who is extremely knowledgeable and has been trained in the identification of mushrooms should pick and consume wild mushrooms—due to the number of poisonous species.

Toxins

The difference between edible and non-edible mushrooms is the toxins that are present in the non-edible or poisonous mushrooms.

The fungi produce these toxins naturally, and the toxins that are present in a poisonous mushroom cannot be removed or made non-toxic by any process, including freezing, cooking, canning or any other process—these toxins are inherent to the mushroom.

Four Categories of Toxins

Mushroom toxins can be divided into four specific categories. Protoplasmic toxins/poisons destroy cells, and eventually lead to organ failure. Neurotoxins are compounds that cause various neurological symptoms like excessive sweating, convulsions, hallucinations, coma, depression and a spastic colon.

The third category, gastrointestinal irritants, produce nausea, vomiting, diarrhea and abdominal cramps. Disulfiram-like toxins only produce symptoms if alcohol is consumed (within a 72-hour window after consuming the mushroom); the person will experience an acute toxic syndrome, which is short-lived.

Edible Species

People can have different reactions to the foods they eat, and this holds true when consuming mushrooms. What is "edible" for some may not be "edible" for everyone. There are approximately 250 North American edible species of mushrooms.

These include Agaricus arvensis or horse mushroom, Agaricus campestris or field mushroom, Boletus badius or bay bolete, Boletus edulis or penny bun, Calocybe gambosum or St. George's mushroom, and Coprinus comatus or shaggy ink cap, to name a few.

Poisonous Species

Of the 10,000 North American species of mushrooms, 250 are edible and approximately 250 are poisonous, leaving an extremely large number of species that we know little about (edible or non-edible). Poisonous mushrooms, if ingested, can cause death within three to six days, which is why it is extremely important to seek treatment immediately.

Symptoms are breathing problems, dizziness, vomiting, diarrhea and dehydration. Some poisonous species are Amanita phalloides or death cap mushroom, Amanita virosa or the destroying angel, Amanita muscaria or the fly Agaric, and Cortinarius rubellus.

How to Avoid Poisoning

To avoid poisoning you must be extremely knowledgeable in mushroom identification when collecting mushrooms, as there is no simple cut-and-dry method, or list of identifiable characteristics between an edible mushroom and a poisonous mushroom.

If you do collect mushrooms, never mix the edible mushrooms with those that are non-edible, and only eat the edible mushrooms that are not damaged and are in good condition. Always refrigerate your edible mushrooms, and remember the number one rule: if you are in doubt about whether a mushroom is edible or non-edible, discard it.

6

Uses of Fungi

Many species produce metabolites that are major sources of pharmacologically active drugs. Particularly important are the antibiotics, including the penicillins, a structurally related group of ß-lactam antibiotics that are synthesized from small peptides. Although naturally occurring penicillins such as penicillin G (produced by Penicillium chrysogenum) have a relatively narrow spectrum of biological activity, a wide range of other penicillins can be produced by chemical modification of the natural penicillins. Modern penicillins are semisynthetic compounds, obtained initially from fermentation cultures, but then structurally altered for specific desirable properties. Other antibiotics produced by fungi include: griseofulvin from Penicillium griseofulvin used to treat dermatophyte infections of the skin, hair and nails; cyclosporins, commonly used as an immunosuppressant during transplant surgery; and fusidic acid, used to help control infection from methicillin-resistant Staphylococcus aureus bacteria.

Widespread use of these antibiotics for the treatment of bacterial diseases, such as tuberculosis, syphilis, leprosy, and many others began in the early 20th century and continues to play a major part in anti-bacterial chemotherapy. In nature, antibiotics of fungal or bacterial origin appear to play a dual role: at high concentrations they act as chemical defense against competition with other microorganisms in species-rich environments, such as the rhizosphere, and at low concentrations as quorum-sensing molecules for intra- or interspecies signaling.

FUNGAL PRODUCTS

Primary production concerns the commercial production of fungi (eg mushrooms) and fungal products (eg soy sauce) that are consumed by humans or other animals. Industrial production concerns the processes of fungi that may be exploited in any industrial (eg antibiotic production) or production (eg biocontrol of pests and diseases) activity. In principle, we aim to examine the variety of ways fungi can be of indirect benefit to humanity. The chapter includes selected examples; it does not cover the topics in detail.

Weed Control

Plant pathology is an extremely large field of study. The goal of the subject is to understand the field well enough to reduce the damage caused by (mostly fungal) pathogens on economically important plants and plant products. The subject of plant pathology can be turned around, by applying the principles of plant pathology to increase the damage being done to economically important weeds.

The basis of the biological control of weeds relies on pathogens attacking the weedy host. The aim of using disease is to poison, or reduce rate of growth, fecundity and general vigour of the host to thereby reduce the weeds competitive position in the crop. In perennials, longevity is also likely to be reduced. Over time, the population of the weed is reduced to a size where it no longer has an economic impact. The pathogen does not eliminate the weed, as that would also eliminate the pathogen.

In fact, the use of fungi to reduce weed populations is extremely difficult. Weedy plants are usually aggressive competitors, growing away from their natural control agents, in an environment that suits their requirements. It is within this severely different environment that we attempt biocontrol of weeds using fungi.

Natural Pathogens or Classical Biocontrol

Early attempts to control weeds took pathogens of the plant from the native habitat of both and introduced the pathogen to the isolated weed population in the new environment. The aim was to establish the pathogen on the weed in the new environment. The approach is called the Classical Strategy. In this approach, the pathogen is released, and the population of pathogens slowly increases. The result is a subsequent reduction in the population of the weed.

In only a few instances has the classical approach been successful. The most useful fungi have been rusts. This is due to their capacity to form specific associations with their host, they are capable of air-borne dissemination and they can cause epidemics in ideal conditions.

In Australia, the rust fungus Puccinia chondrillina has been successfully established on Skeleton weed (Chondrilla juncea), formerly a major weed of wheat throughout the eastern states. The weed has been reduced to tolerable densities. Interestingly, the initial development of control was associated with a change in the genetic structure of the remaining plant population, with weed genotypes tolerant of the fungus surviving.

In Australia, release of rust fungi for control of blackberries, Parthenium grass, Mimosa pigra and Rubber vine (Cryptostegia) has also been moderately successful. Use of natural pathogen populations to reduce populations of weeds has a number of problems, including:

Problems

The pathogen must be highly specific and the host highly susceptible. The pathogen must be sustained at levels that enable control of weeds should conditions change to favour the weeds. The pathogen must be able to mutate or change genetically to infect new races of weed. Other control measures should be available (chemicals, insects, mechanical) should the usual variation in environment result in an explosion in the size of the weed population and suppression of the pathogen. Lack of a commercial incentive for organizations to find new biocontrol agents. This last problem, in particular, has led to the use of other approaches to weed control.

INUNDATIVE APPLICATION OF PATHOGENS

In recent decades, the search for fungi that can control weeds when applied in an inundative manner (fungal products are also known as Bioherbicides) has led to several commercially attractive products. Effective bioherbicides must have: Specificity to the weed and no other plant. Virulence and pathogenicity to kill the weed when it is among its crop companion. Reproduction in artificial conditions, forming stable and long-lived reproductive units. Capacity to be delivered to control the weed.

Note that this process aims to kill the weed, much in the way a chemical herbicide functions. The pathogen does not survive in the crop after application. The control agent is just another commercial product. In Australia, the fungus Colletotrichum orbiculare has been developed to control Bathurst bur (Xanthium spinosum). The weed forms spiny seed casings that can damage the feet and contaminate wool of sheep. The pathogen was first discovered in the early 1980s. The development of the bioherbicide took over 10 years. One major stumbling block was the close relationship between the weed and an important crop species, sunflower. However, specific races were found, and subsequently used to establish the effectiveness of control.

Spores of the fungus are suspended in vegetable oil, sprayed on young seedlings and considerable damage develops with 2 weeks. Constraints to greater use of mycoherbicides include: Biological – genetic variation in host, host resistance to infection, interaction with other microbes on the phylloplane, host range, genetic stability of the fungus.

Environmental – germination requirements, especially temperature and moisture, spore formation and survival. Commercial – small returns. The future development of biocontrol of weeds requires an increase in our fundamental understanding of potential control agents and their interaction with hosts. While the future of biocontrol appears greater in currently underdeveloped countries, social pressures to reduce use of artificial chemicals because of their unintended effects, and costs of production, will increase pressure to seek alternative approaches to chemicals in developed countries.

Nematode Control

Fungi utilise many different sources of carbon, some living and some dead. The observation that nematodes can be colonised by fungi, led to the possibility that the interaction might be used to control pest nematodes of plants and animals. Closer examination of the interaction has led to the belief that only partial control of nematodes is possible, and that in a practical sense, populations of pests may be suppressed not eliminated. In the case of plant and animal parasitic nematodes, fungi may be used to regulate nematode populations in conjunction with other processes.

Pest Nematodes

Nematodes found in plants, animals and soil use various sources of organic energy. Plant parasites feed from plant cells, either by selecting and then establishing a single feeding site. This approach is usually known as sedentary feeding. Other plant-parasitic nematodes are migratory, moving from site to site on the root, rarely using a single cell for a long time. Migratory plant parasites can be further categorised as internal and external feeders, meaning that they migrate within or outside the root.

Animal parasites of agriculturally important ruminants tend to have a life cycle whereby colonisation of the host is followed by a soil phase. Larvae are consumed by grazing animals along with grass. The larvae establish a feeding location within the digestive system of the host. Adults lay eggs within the intestines, and the eggs fall to the ground with the faeces. Eggs subsequently hatch in the faeces and larvae move to blades of grass, where they are consumed.

Fungi that colonise nematodes fall into several different categories, each of which requires a different approach if it is to be used successfully to control pest nematodes. The fungi range from highly specialised nematode parasites, to generalists with no specific host. Some fungi may colonise organic matter and incidentally also colonise eggs, larvae or adult nematodes. Other fungi may trap larvae or adults and colonise their body, using the body as the primary source of nutritients.

Plant Parasitic Nematodes

Endoparasitic fungi are generally highly specific to single species or genera of nematodes, and are difficult to culture in the absence of their host. Hirsutella, Meria, Nematophthora and Nematoctonus have all been suggested as ideal biocontrol agents because of the specificity for economically important nematodes. However, the difficulty of culturing them and lack of experimental support for their activity has limited any commercial development.

Nematode trapping fungi utilise either sticky pads or constricting rings to immobilise nematodes. The fungus then digests the internal organs of the nematode and form reproductive cells. The main fungi in this group are

Arthrobotrys and Duddingtonia. These fungi are found in soil where they live on readily-available organic carbon, in the absence of nematodes. They are generalist feeders, being able to trap and digest many nematodes, not just those that infest plants. Thus, the fungi are rare, unable to exist at sufficiently high densities to control nematode populations, and parasitic on beneficial as well as pest species of nematode. Because of these difficulties, exploration of the use of trapping fungi in the field has largely ceased.

Parasitic fungi have been isolated from either eggs, larvae or adults nematodes. Several of these have been found to preferentially parasitise the nematode and thereby reduce the size of the nematode population. Research efforts in some crops are now concentrating on using fungi found locally to control local pests.

The general approach is to go to locations where nematodes have reached high densities. Parasitised individuals are extracted from soil, the fungi cultured and then tested as parasites of the pest nematode. The theoretical basis for this approach is that high densities of nematode will enable fungal parasites to increase population size dramatically. Thus by selecting fungi adapted to the host (nematode) and environment, effective and usually inundative inoculation will enable the control of the pest nematode. Biocontrol in problem fields has been achieved in a surprising number of cases. The species used include Paecilomyces lilacinus, a fungus that appears to have a wide range of potential hosts, both insect and nematode, and yet isolates have a degree of specificity for host.

Animal Parasitic Nematodes

Sheep and cattle have traditionally been drenched to control parasitic nematodes. However, nematodes are now developing resistance to the common active ingredients in drenches. Drenches using different active compounds are not being developed because the economical returns are low for the investment. Thus alternative approaches are being sought.

In one approach, nematode trapping fungi are placed into nutrient-rich faeces. As nematodes hatch, they become colonised by the fungus. The fungus Duddingtonia has been inoculated to the dung via a drench to the animal. Sclerotia of the fungus pass through the intestine, hatch in moist dung, and effectively reduce larval densities IN SOME CASES.

While further research is needed before commercial adaptation, this is one example where understanding of the biology of fungi, lateral thinking and some judicious research, has led to the development of a new approach to a serious problem.

Biocontrol of nematodes by fungi appeared to be possible when nematode trapping fungi were first discovered. However, it has taken some considerable time before their potential could be realised. Successful control relies on having sufficiently high densities of fungus where high densities of the pest are maintained. The requirement for high densities of agent to control the pest applies across all biocontrol efforts.

Plant Pathogen Control

Plant disease is caused by abiotic and biotic factors. While other chapters deal with the way fungi improve plant health directly, this chapter will ask the question "Can fungi be used to reduce the effects of biotic causes of plant disease?"

As most biotic causes are fungi, we will concentrate on fungus to fungus interactions. The topic of fungus to fungus interactions has been reviewed in depth on many occasions from many different points of view. This chapter will not attempt to summarise this work. Rather we will dip into some of the stories.

The interactions between fungi can be manipulated to reduce the damage caused by one fungus on the plant. Essentially, these interactions include: Direct hyphal parasitism Antibiotic interactions Competition for resources The mechanisms are not mutually exclusive and all may be evident within one interaction at the same or different times.

DIRECT HYPHAL PARASITISM

Mycoparasitism is where one fungus derives its nutrition from another without any benefit in return. The interaction can be where the parasite is biotrophic or necrotrophic. Biotrophs tend to have highly specialised relations with specific hosts. The parasite is extremely difficult to manipulate, and as a consequence few have ever been used in biocontrol.

Necrotrophs tend to kill the cell or tissue from which they feed. They tend to have a broad host range, and their activity commonly includes excretion of antibiotics. For instance, the necrotroph Trichoderma has the capacity to parasitise a great number of fungi, including pathogens Rhizoctonia, Scerotinia, Fusarium and Verticillium, and the Straminopiles

Pythium and Phytophthora. Trichoderma grows towards hyphae of other fungi where it binds following branching. Trichoderma releases hydrolytic enzymes that digest the walls of the host prior to penetration. Trichoderma is also thought to release a range of toxins that reduce any response from the host to invasion.

For biocontrol to be successful, the parasite must be placed in an environment favourable for its competitive success. Most soils are oligotrophic. Thus, the introduced fungus is unlikely to be favoured against isolates that are adapted to the local conditions. By judicious increases of organic amendments at inoculation, biocontrol can be attained followed by a decline in the population size of the parasite as the amendment degrades.

Antibiotic Interactions

Antibiosis can be defined as the inhibition of one organism by metabolites of another. Metabolites that kill another organism tend to be called toxins. Thus, dilute solutions of toxins may inhibit and high concentrations of antibiotics may kill.

Most fungi produce inhibitory metabolites. For instance, Gliocladium produces a diketopiperazine that kills Pythium probably because of coagulation of proteins in the cytoplasm. Use of the fungus on seed appears to reduce seedling damping-off as much as a common seed treatment. Volatile pyrones produced by Trichoderma appear to reduce damping off caused by Rhizoctonia. The potential to exploit these metabolites in biocontrol appears feasible.

The role of antibiotics in biocontrol is still unclear. Some spectacular experiments, when repeated in the field, have resulted in enormously variable outcomes indicating that other mechanisms may be important.

Competition for Resources

Competition is the active requirement for resources in excess of those immediately available to two or more organisms. The resource most commonly considered to be important for fungi is organic energy. Thus much competition exists for access to energy for growth and maintenance. However, competition between fungi is extremely difficult to quantify. The mycelium is constantly changing. Resources within the hyphae are constantly being shifted between different parts, with the consequence that over very short periods of time the nature of what is being assessed changes.

However, competition for nutrients can take place between survival propagules, germinating units and mycelium in any arrangement. The result of competition is stasis of the less successful competitor. Inoculum density does not decline and the potential to initiate disease remains. Thus the application of competition by itself to biocontrol appears limited.

Success

Some examples of fungi used against fungal pathogens with some success:

Agent	*Pathogen*
Chaetomium	Venturia
Coniothyrium	Scerotinia
Cladosporium	Botrytis
Gliocladium	Rhizoctonia
Trichoderma, Penicillium	Rhizoctonia
Scerotinia, Fusarium, Verticilium	*Puccinia*

Trichoderma viride, Trichoderma harzianum, Trichodermapolysporum and Trichoderma hamatum (all Deuteromycota) have all been formulated for commercial use against soil fungi. These fungi are common inhabitants of heated soil where the almost sterile medium is conducive to their explosive colonisation following germination of ascospores of the fungi. Over time other fungi colonise the soil, competitively excluding the species of Trichoderma. Thus their obvious presence in soil is short-lived. They colonise the soil, interact with the pathogen but then generally decline. For instance, Trichoderma can be used in potting media to reduce seedling diseases, but the fungus is unlikely to be successful in field soils.

While laboratory and glasshouse experiments indicate fungi have a role in biocontrol of plant pathogenic fungi, the results of field application of a single fungus have been discouraging. While it appears that some agents may have a place in integrated approaches to reducing plant pathology, we do not appear to understand the basis of interactions between fungi. Until this understanding is improved, use of fungi in biocontrol is likely to remain a dream.

Mushroom Production

A huge range of edible fungi are cultured. Agaricius bisporus is cultivated widely in western countries. In addition, Shii-take (Lentinus edodes), Straw mushroom (Flammulina velutipes), Oyster mushroom (Pleurotus ostreatus), and Chinese Black mushroom (Auricularia polytricha) are cultured in various Asian countries, and increasingly in western countries. The truffle industry is economically important in southern Europe and production is expanding in the southern hemisphere. Finally, edible mushrooms are picked from the wild, especially in Europe. Many of the fungi are eaten fresh, but there is also a market for dried mushrooms and canned truffles.

Common or Button Mushroom

The identity of the fungi grown as commercial or button mushrooms in western countries is not altogether clear. The generally accepted name is Agaricus bisporus. However, breeding over the years appears to have resulted in the infusion of some genetic material from various sources.

The commercial mushroom is grown in a process that varies only slightly around the world. Essentially, straw is mixed with stable and poultry litter, mixed with appropriate minerals and then composted. Spawn is added after the compost has cooled after the second phase. The compost is then laid out, usually in purpose built sheds in which the temperature is held at around 24 degrees. A layer of inert, alkaline material is added as a casing over the surface. Once mycelia reach the surface of the beds, the temperature of the shed is reduced to induce fruiting. Fruiting then proceeds over the next few weeks.

Commercial mushroom production is a highly intensive industry. Understandably, pests and diseases are a major problem. The principles of control do not differ from other intensive production systems.

Shii-take and Log Culture

Shii-take is a highly prized mushroom in Japan. The range of variation in flavours and odours make this a valuable ingredient in the usually delicately flavoured Japanese recipes. The range comes from the genotype of the spawn, selection of logs used to grow the fungi, and conditions under which the logs are then incubated. Shii-take is commonly grown on logs of deciduous trees. The logs are inoculated with spore suspension placed over the logs or by overgrowth from pre-inoculated fragments of timber. The base logs are not sterilised so the process has many potential hazards.

After extensive period of colonisation of the logs (laying), the induction of fruiting takes place when the logs are stood upright and the temperature drops. After a period of maturation, the temperature increases. Fruiting then takes place in spring, and continues for some years after.

Cropping takes place in the open, and traditional approaches continue to be used in Japan. Most of the crop is sold fresh, though some is dried, packaged and then sold throughout the year. The dried crop is most commonly exported from Japan. Artificial approaches to cropping have been developed. These use controlled temperatures, and selection of particular strains. In Australia, cropping more commonly uses these artificial environments and selected strains.

Auricularia and Pleurotis are also grown on timber logs. In Australia, Auricularia grows naturally on plum, fig and kurrajong trees, indicating a potentially huge potential host range for culture. Most Auricularia appears in Australia in the dried form, often sourced out of China.

Culture on Organic Waste

Straw mushrooms are traditionally grown on straw from the rice paddy. The bundles of straw are drenched and then stood upright. The bundles are usually inoculated with spawn. Each bundle can produce fresh mushrooms for several weeks.

More recently, bundles of straw have been mixed with cotton or other organic waste, or cotton waste used exclusively. These alternatives often produce more mushrooms, probably due to the improved N:C ratio.

ECTOMYCORRHIZAL FUNGI

Fruiting bodies of a number of ectomycorrhizal fungi are consumed. The Black truffle (Tuber melanosporum) is perhaps the most highly prized of these fungi. Other truffle species of Tuber, Terfezia and Tirmania are also eaten. Field collections of morels (Morchella sp) and chantarelles (Cantharellus cibarius) are also highly prized. The latter fungi are found naturally in Australia, and sometimes appear in the shops in autumn. Truffles have been harvested in European countries for many centuries. True Black truffles are found mainly in France, and other species of Tuber in France, Spain, Italy and many countries surrounding the Mediterranean sea. The Black truffle is the most highly prized, because of its very strong flavour.

Culture of Black truffles is indirect, in that trees that support the fungi are managed with the aim of maximising the production of the fruit bodies. The soil of the truffiere tends to be alkaline, calcareous, and well drained. Oak or Hazel trees are inoculated in the nursery. Saplings are plants out in the truffiere. The trees are then pruned to encourage lateral growth. After about seven years, the fruit body appears in late winter to spring. The vegetation above the zone of fruiting tends to die off. Pigs or dogs have been used to detect ripe truffles, though more biochemical methods are being developed. The truffiere continues to produce abundantly for around 30 years.

In New Zealand and southern Australia, truffieres have been established. The inoculum and trees have been imported mostly from France. Initial harvests of truffles took place in New Zealand in about 1993, and in Australia in 1998. The truffles are being produced in the autumn of the northern hemisphere, and it is expected that large profits will ensue from the out-of season availability of this highly prized food.

Problems of production are legion. In Europe, many traditional truffieres are showing signs of reduced production. The reasons are unclear. In Australia, other ectomycorrhizal fungi are known to colonise the oak and hazel host trees. Further, the climatic conditions are not exactly the same as in Europe. Thus we might expect selection of characteristics of fungi that benefit local conditions over time.

Conclusion

A small but active market exists for edible fungi, in Australia for most fungi, and overseas for truffles. The systems to culture the fungi are well known and in most cases highly controlled. The product is nutritious and well regarded. That Australia has developed a culture where the product is purchased in shops is perhaps of considerable importance. Unregulated picking from the field tends to reduce the crop, and among the Australian fungi are many that are toxic or hallucinogenic.

Insect Control

Most people who garden are aware of insects that have been apparently colonised by microbes. The small discoloured carcasses of larvae or adults are unearthed as you dig through soil. One group of organisms that parasitise insects are the fungi. Insects are common pests of agriculture and horticulture. Several fungi, like those seen in parasitised garden insects, have been used to control insect pests. A diverse array of fungi associate with insects, and the nature of the interaction is also diverse.

Biocontrol of insect pests can take one of several different approaches. The most common approaches are inundative application and classical control.

Fungi Used Inundatively

The Fungi

Several species of fungus have a reputation for biocontrol of insects. The fungi come from all divisions. The most commonly used ascomycetous fungi include the genera *Beauvaria*, Metarrhizium, Tolypocladium and Verticillium. These fungi have been applied inundatively, in the manner of a chemical application, with the goal of killing most, if not all, insects.

The most historically important of the commonly used fungi is Beauvaria bassiana. The Chinese have a long history of successfully culturing silkworms. However, when the industry was taken to Europe, cultures of insects were devastated by Muscadine disease. The causative agent in turn was determined as Beauvaria bassiana, a fungus that grew on and in larvae. Determination by Bassi that the infectious agent was a fungus, was one of the first examples where germs were shown to cause disease. The fungus has since been used successfully to control the larvae of many Lepidopteran insects.

Metarrhizium, also an asexual stage of an Ascomycete, has been successfully used to control spittlebugs in South America, Rhinoceros beetle in the South Pacific islands, and in Australia to control locusts.

Lecanicillium lecanii infects sucking insects such as aphids and scale insects. Some plant-sucking aphids also have the nasty habit of transmitting viruses, thus control of the insect and the viruses it transmits are both possible. The fungus is used to control sucking insects mostly in greenhouses where humidity is high. Tolypocladium is a fungus found in both soil and water. It has been reported to colonise a wide range of insects, and effectively controls mosquitos in experimental conditions.

Mode of Attack

Penetration: In general, when a spore comes in contact with the insect cuticle, the spore attaches, germinates, penetrates the cuticle (from outside in, or inside out, depending on whether the spore lands on the outer surface, or is ingested). Penetration may involve physical force, or more commonly physical and enzymic action combined. Inside the haemocoel, the form of the fungus may change to yeast-like, though filaments and wall-less protoplasts are also found.

Secondary Metabolites: Many insect pathogens release metabolites that increase the chance of insect death. A wide range of pathogenicity factors are known and they come from different metabolic pathways. The factors may be toxins, immunosuppressants or antibiotics. Destruxens from Metarrhizium are polypeptides that appear to be directly toxic to the insect. Cyclosporin-like immunosuppressants from Tolypocladium increase the rate at which the fungus colonises the haemocoel. Better understanding of the formation and mode of action of secondary metabolites will be important if rapid knock-down of insects remains the goal of inundative biocontrol.

Other Approaches to Insect Control

Classical

The control of pasture cockchafers (Aphodius sp) has been achieved in the Adelaide Hills of South Australia by the release of an isolate of Metarrhizium specific to the host. The larvae of the insect burrow through soil where they feed from roots of pasture grasses and incidentally, contact spores of the fungus. Colonisation of the larvae results in death after several days, with the subsequent release of spores of the pathogen, followed by subsequent cycle of contact, colonisation and insect death.

In principle, the density of the fungus increases when high levels of insect are present in the soil. As a result the insect population declines, enough for most of the larvae to miss contact with the fungus. However, the fungus is sustained at a low density, to become active only as the larval densities increase. The classic pattern of population change is shown in the figure.

Variations:

Several insects are of major concern in Australia. One, Heliothis (Helicoverpa) armigera is a serious pest of many different crops and pastures. In cotton, larvae eat foliage and bore into boles. Heliothis has several stages where it is vulnerable to biocontrol agents: as eggs on the plant surface, as larvae on the plant surface and as long-lived pupae in the soil. Fungi may play a role at each stage. Fungi that live on the leaf surface may colonise eggs. Fungi that live in the leaf may deter or poison larvae feeding from the leaf, or be treated using inundative approaches, and fungi may colonise pupae in soil.

Insects and fungi interact in various ways. The interactions where fungi kill the insect may be used in specific cases, but our overriding impression is that we know so little about the interactions, we are probably going to find many associations that can be manipulated resulting in a decline of the pest.

HYGIENE STRATEGY OF ANTS AVOID CHANGES OF THEIR FUNGAL CULTIVAR

Leaf-cutting ants maintain an obligate symbiosis with their fungal garden. The ants and their cultivar have to cope with hundreds of endophytic fungal species which are brought in with the leave cuts.

Endophytes: Endophytes are a bacterium or fungus, that live within a plant for at least part of its life without causing apparent disease. Endophytes may benefit host plants by preventing pathogenic organisms from colonizing them. Fungal endophytes may outcompete the fungal cultivar of the nest.

Van Baels and colleagues 2009 found that ants control endophytes applying strategies like preferring to cut leaves from low endophyte densities, they reduce the amount of endophytic fungi in leaves before storing them in their nest and the original fungal cultivar inhibits the growth of foreign fungi that might out-grow the cultivar.

FIFTY MILLION YEARS OF EVOLUTION MOULDED THE LEAF-CUTTER ANTS

Entomologists Ted Schultz and SeÃin Brady explain that agriculture is a specialized form of symbiosis which evolved in only four animal groups: humans, bark beetles, termites, and ants. Evolution produced the five distinct agricultural systems of the fungus-growing ants, resulting from a single ancestor living 50 million years ago.

In the past 25 million years, four different specialized agricultural systems have evolved, leading to the most recently evolved and best-known fungus-growing ant species-leaf-cutter ants. Leaf-cutter ants originated recently-less than 10 million years ago.

Ants Depend on Fungus but the Fungus does not Depend on the Ants

The fungal garden which serves as food for the ant colony is grown from a few strands of the garden of origin. The queen carries them in her mouth on her mating flight to start a new colony. Mikheyev, Mueller and Abbot 2006 describe evidence of recombination in attine fungal cultivars, contradicting widely held perceptions of obligate clonality.

Mikheyev and Mueller 2006 studied the coevolutionary process of the dependence of fungus-farming ants on fungus gardens for food. The researchers found that the ants were dependent on the fungus, but no dependence of the fungus on the ant was stringent. The fungus-growing ants were seen as a coevolutionary integration.

The fungi occasionally reproduce sexually by spores carried by wind. It exchanges genes with other ant-cultivated fungi. According to genetic analysis made by the authors the fungus of Cuba exchanged genes with mainland fungi populations of Central and South America brought in by winds. The fungus is still using its sexual reproduction and is not completely dependent on asexual, clonal reproduction through their ant farmers. This is an example that coevolution does not imply mutual dependence.

Coevolution: Biological coevolution is the change of a biological object triggered by the change of a related object. Coevolution can occur at multiple levels of biology: it can be as microscopic as correlated mutations between amino acids in a protein, or as macroscopic as covarying traits between different species in an environment. Each party in a coevolutionary relationship exerts selective pressures on the other, thereby affecting each others' evolution. Species-level coevolution includes the evolution of a host species and its parasites, and examples of mutualism evolving through time

A species may evolve in response to a number of other species, each of which is also evolving in response to a set of species. This situation has been referred to as "diffuse coevolution". And, certainly, for many organisms, the biotic (living) environment is the most prominent selective pressure, resulting in evolutionary change.

Yeasts Associated with Leaf-cutting Ants

Carrero and colleagues 1997 isolated 137 yeasts associated with the leaf-cutting ant Atta sexdens rubropilosa Forel, 1908. Candida, Cryptococcus, Rhodotorula, Sporobolomyces, Tremella, Trichosporon, Pichia and black yeasts were isolated by the authors. The genus Candida was widely distributed, with C. homilentoma, C. colliculosa-like, C. famata and C. colliculosa being the most prevalent.

Leave-cutting Ant Control Strategies, Chemical Control

Atta sexdens rubropilosa Forel, 1908 "Lemon fire antâ□•:

Crush the head of a soldier ant between your fingers. A lemon-like odour characterises this ant.

Atta sexdens piriventris Santschi, 1919 "Southern lemon fire antâ□•: Same identification as foregoing. Atta laevigata F. Smith, 1858 "Glas had fire ant ": The largest workers of the colony have a smooth and shining head, similar to glass.

Atta bisphaerica Forel, 1908 "Yellow fire ant":

Its colour is brown-yellow not shining.

Atta capiguara GonÃ§alves, 1944 "grass-cutting ant":

It prefers grass instead of leaves. It differs from the foregoing ants in not producing a lemon smell, it is not brilliant,Its head is not brown yellow

It is recommended to find vertical holes which lead directly to the centre of the colony to apply the liquid poison, powder or fumes. Baits should be left on the trail of the ants. Application of poison must follow the instructions of the producer. Chemical used in form of gases and powder are applied directly to the nest. Baits are carried to the fungal garden by worker ants but are not accepted by all sort of ants.

The strategy varies according to the ant which is to be controlled. It is difficult to eradicate leaf cutter ants with dust, liquid or granular insecticides because the centre of the nests, with the fungal garden is often missed. Bits are often not accepted. Baits with hydramethylnon are in use to control leaf cutting ants.

Phorid Flies

Wuellner and colleagues 2002 suggest biocontrol of leave-cutting ants using phorid flies such as Pseudacteon tricuspis and Pseudacteon obtusus for the control of the red imported fire ant in geographical areas where polygyne colonies dominate, such as Texas. For Florida, where monogyne colonies dominate, Pseudacteon litoralis is being recommended by the authors.

Gyne: Gyne is the primary reproductive female caste of social insects (especially ants, wasps, and bees of order Hymenoptera). Gynes are those destined to become queens, whereas female workers are typically sterile and cannot become queens.

A colony with multiple queens is said to be a polygyne form, where as with only one is a monogyne form. The fire ant Solenopsis invicta is known to have colonies in both polygyne and monogyne forms.

Female phorid flies Pseudacteon obtusus were released in Texas in 2008-2009. The flies inject their eggs into the ants. The resulting larvae migrate to the ant's head where it feeds from the ant's brain and decapitates its host to hatch.

The combination of biocontrol using phorid flies in combination with chemical poisoning of ant colonies is being suggested. The chemical control is very effective, but does not avoid a new infestation from colonies which had been overseen.

Pheromone Studies may Lead to New Biologic Control of Leave-cutting Ants

Insect hormones. More than 90% of insect hormones are neuropeptides. Researchers try to understand the function of insect neuropeptides, which are hormones which have important functions during insect developmental and as adults. With increase knowledge on this matter dependence on poisonous chemicals may be reduced.

Choi, Raina and Vander Meer 2009, studying the fire ant, Solenopsis invicta, found the pyrokinin/pheromone-biosynthesis-activating neuropeptide (PBAN), which is a family of peptides. This pentapeptide stimulates the production of pheromone in female moths, muscle contraction, induction of embryonic diapause, melanization, acceleration of puparium formation, and termination of pupal diapause.

The authors hope to increase knowledge about the regulation of pheromone production their release, and how neuropeptides they influence the development of ants. Researches in this field may create non-insecticide methods for fire ant control by interfering in the normal neuropeptide hormone function.

VARIABILITY OF FUNGI

Many plant disease fungi are known to be complex and variable, but the degree of complexity is imperfectly known for most of them and is not well enough known for any of them. This complexity is not surprising because many of the crop plants in which the fungi live are equally complex. There are more than 14,000 varieties of wheat; there are unnumbered varieties of apples, beans, corn, roses, iris, and other cultivated and wild plants. Moreover, man and Nature are continually making new varieties by mutation and by hybridization and selection.

PLANT DISEASE FUNGI

Plant disease fungi also are plants. They are parasitic on higher plants. Most are microscopic in size and relatively simple in structure, but they are plants nevertheless. They germinate, grow, and fructify; they multiply by the countless billions; they Mutate and hybridize; consequently Nature has made countless kinds in the past and is continually making More kinds through mutation and hybridization. As plant breeders produce better kinds of useful plants, therefore, Nature may produce more destructive fungi.

It is the job of plant pathologists to learn what the parasitic fungi are now doing and what they may do in the future to jeopardize or nullify man's success in plant improvement. To understand present Plant disease situations and to predict Possible future developments, it is essential to learn as much as possible about the variability in fungi pathogenic to plants. Other kinds of causal agents of plant diseases, such as viruses and bacteria, must be studied similarly, but the present discussion is restricted to fungi.

The principles of growth and reproduction in plant pathogenic fungi are similar to those for higher plants, except that fungi have no chlorophyll, which enables green plants to make their food from simple materials in the soil and air. The fungi therefore must live on living plants or animals or their products. The fungus parasites of plants live in or on living plants and get their nourishment from them by means of minute tubes, or hyphae, that grow, branch, and form a network known as the mycelium, which finally produces many tiny spores. Spores serve the same propagative and reproductive function as the seeds of higher plants, but spores are much smaller and simpler in structure than seeds.

Spores usually have one or at most a few cells and do not contain an embryo as do seeds. Nevertheless they germinate and produce new fungus plants with the parental characters. The conditions required for germination are similar to those for seeds, moisture and suitable temperature being the most important. The time needed for germination varies with the kind of spore and the temperature; it varies from about an hour to several days. Many fungi produce millions or even billions of spores in a few days to a few weeks and can therefore produce countless numbers in a few short generations. That is one reason why they are so dangerous.

Some spores are produced asexually and some only as a result of sexual fusions. The same fungus may produce several kinds of spores asexually, but usually produces only one kind sexually. Successive crops of asexual spores are generally produced abundantly and quickly when conditions favour growth of the fungus, and the sexual spores are usually formed when the vegetative period is terminating. Each kind of spore has a special name; thus some of the rusts produce five kinds of spores or sporelike bodies teliospores, sporidia, pycniospores, aeciospores, and uredospores, each with special form and function.

The principles and procedures in classifying and naming fungi are similar to those for higher plants. Species of higher plants are determined by morphologic characters of the vegetative and fruiting structures, including seeds. Similarly species of fungi are determined by the characters of the mycelium, of the spore-producing organs, and of the spores themselves. The unit of measurement for fungus spores is the micron, which is one one-thousandth of a millimetre; 25 millimetres make an inch. The range in size is from about 4 microns to more than 100 microns and, together with shape, colour, and certain other characters, is characteristic for each species.

Visible characters are used for grouping higher plants and plant pathogenic fungi into classes, orders, families, genera, species, and varieties. Thus it is easy to distinguish larger groups of agricultural plants, such as grasses and legumes. It is equally easy to distinguish larger groups of fungi, such as smuts and rusts. But there are many kinds of wild and cultivated grasses, such as timothy, bromegrasses, oats, barley, rye, rice, corn, and wheat; and there are many kinds of legumes, such as beans, peas, soybeans, alfalfa, and clover. Likewise, there are many kinds of smuts, such as the loose smut of wheat, the stinking smuts of wheat, common corn smut, and rye smut, and there are more than 3,000 kinds of plant rusts. It is relatively easy to recognize these major groups of higher plants and of fungi, but it becomes increasingly harder to classify the smaller groups, such as species and varieties.

The agronomist must know not only wheat, a genus scientifically designated by its Latin name Triticum, but he must know the species of wheats, such as the common bread wheat, Triticum vulgare; the durum wheats, T. durum; and several other species or subspecies. He must know also thousands of varieties of bread wheats and hundreds of varieties of durum. And he must learn about the new lines and potential varieties that continually are being produced. Likewise the plant pathologist must know the genera, species, varieties, and "lines" of the fungi that parasitize different kinds of wheat, and he must learn about the new ones that continually appear.

Moreover, the agronomist must know not only what crop plants look like, but also how they behave and what they are good for. He must know whether a variety of wheat is a winter wheat or a spring wheat, whether it is winter hardy, where it grows well, whether it makes good flour or poor flour, and whether it is susceptible or resistant to disease; and the plant pathologist must know not only what species of plant disease fungi look like but must also know their pathogenicity for the many different kinds and varieties of plants. Most species of plant pathogenic fungi are equally as complex in composition as species of crop plants, and it is therefore necessary to find out what is within the species. As an example, the species Puccinia graminis, the stem rust of small grains and grasses, is recognized easily by its general appearance and by certain characters of its spores.

But there are several varieties within the species that are alike in some characters and different in others, including the kinds of crop plants that they can attack. Thus the variety tritici of Puccinia graminis parasitizes wheat, barley, and many wild grasses; the variety avenae parasitizes oats and certain wild grasses but not wheat and barley; within the tritici variety there are races that differ in their ability to attack certain varieties of wheat; within the variety avenae there are races that differ in ability to attack varieties of oats.

Kanred, as an example, is immune to some races of the variety tritici and susceptible to others. These rust races in turn can be subdivided into the ultimate subdivisions, the biotypes. A biotype is a population of individuals that are identical genetically.

The descendants of a single non- Sexually produced stem rust spore constitute a biotype, unless mutation occurs to cause genetic diversity. The test of genetic purity of a population of a pathogenic fungus is the consistency of its behaviour. Most species of plant pathogenic fungi comprise many biotypes, and attempts are made to group the most closely related ones into races and then to group the most closely related races into varieties, which in turn are grouped into species. Actually the procedure usually is in the reverse direction: The larger groups are recognized first, and the successively smaller groups are discovered by successively more refined methods. This can be illustrated by results of investigations of Puccinia graminis, the fungus that causes stem rust of small grains and grasses.

Fig. Puccinia Graminis

(The discussion of stem rust is based on results of cooperative investigations between the United States Department of Agriculture and the

University of Minnesota, which were begun in 1914 and continued by many investigators under the general supervision of the senior author, who regrets that the contributions of the many administrators and investigators cannot be specifically mentioned without interrupting the continuity of the discussion.)

Puccinia graminis is a good species Within which there are clearly recognizable varieties, races, and biotypes.

There are at least six rather distinct varieties that differ in size of spores and in the kinds of plants that they can attack:

- Tritici (wheat), whose host plants are wheat, barley, and many wild grasses;
- Secalis (rye), rye, barley, and many wild grasses like those attacked by tritici;
- Avenae (oats), oats and wild grasses different from those attacked by tritici;
- Aleifiratensis (timothy), timothy and certain wild grasses;
- Agrostidis (redtop), redtop and other species of Agrostis;
- Poae (bluegrass), Kentucky bluegrass (Poa pratensis) and related species.

DIFFERENTIAL VARIETIES OF WHEAT

There probably are more varieties than the six listed, and new ones may appear in future as a result of mutation or hybridization. Races are known within the varieties tritici, avenae, and secalis. Although there sometimes are at least slight morphologic differences between races within a variety, the most important and most easily recognizable differences are in the degree of infection (pathogenicity) on certain varieties within the genera Triticum (wheat), Avena (oats), and Secale (rye), respectively. The crop varieties, designated as differential varieties, now used to distinguish races within the tritici and avenae varieties, were selected from hundreds that were tested, and, at the time of selection, appeared to be representative and adequate for distinguishing races.

The differential varieties of wheat are:

- *Triticum compactum (club wheats)*: Little Club, C. I. 4066 (Cereal Investigations accession number, Department of Agriculture). (Certain lines of Jenkin, C. I. 5177, notably Hood, C. I. 11456, may be substituted for Little Club.)
- *Triticum vulgare (bread wheats)*: Marquis, C. I. 3641; Reliance, C. I. 7370 (certain lines of Kanred, C. 1. 5146, can generally be substituted for Reliance); Kota, C. I. 5878.
- *Triticum monococcum (Einkorn)*: Einkorn, C. I. 2433.
- *Triticum durum (durum or macaroni wheats)*: Arnautka, C. I. 1493; Mindum, C. I. 5296; Spelmar, C. I. 6236; Kubanka, C. I. 2094; Acme, C. I. 5284.

When inoculated with different races, these varieties may be susceptible (S), resistant (R), or heterogeneous (mesothetic) (M). Each of these classes is subdivided into infection types, based on size of pustules (small blisterlike eruptions containing from a few to 250,000 rust spores), and the condition of the wheat tissue in which pustules are formed.

The rust stage which is used in identification is the repeating or summer-spore stage, called also the uredial stage; and the pustules of this stage are called also uredia (singular, uredium).

- *Immune*: No rust pustules developed; small flecks of dead host tissue sometimes present and designated by a semicolon, thus, o;.
- *Very Resistant:* Rust pustules extremely small and surrounded by dead areas.
- *Moderately Resistant:* Pustule small to medium; usually in green islands of host tissue surrounded by a band of yellowish (chlorotic) or dead tissue.
- *Moderately Susceptible*: Pustules medium in size; usually separate, no dead areas, but yellowish (chlorotic) areas may be present, especially under unfavourable conditions.
- *Very Susceptible*: Pustules large and often united; no dead tissue, but there may be leaf yellowing under unfavourable growing conditions.
- Heterogeneous Size of pustules variable, sometimes including all of the above types and intergradations between them on the same leaf.

THAT grouping indicates the types of infection produced by rust races on the differential varieties of wheat. As indicators of resistance or susceptibility of the wheat varieties, the infection types are grouped as follows: Types 0, 1, and 2 are resistant (R); types 3 and 4 are susceptible types (S); and type x is mesothetic (variable) (M).

It will be seen, therefore, that varieties are considered resistant when the rust is unable to grow extensively and produce pustules on them or when the pustules are extremely small, as indicated by types 1 and 2. The small pustules are likely to be surrounded by discoloured or dead host tissue, but the areas killed are almost always very small, so that the rust actually does very little damage to the variety. The production of infection types 3 and 4 on varieties indicates that they are susceptible, because the rust can grow well and produce large pustules with large numbers of spores, without killing the host tissues of the wheat plant quickly and thus limiting its own growth, as the rust cannot grow in dead tissue. The infection type x is variable and therefore is an indication that the resistance or susceptibility of the variety is also variable.

To identify rust races, seedling plants of the 12 differential varieties are inoculated in the greenhouse by applying rust spores to them, after which the inoculated plants are kept in a moist chamber for 24 hours to provide moisture for the spores to germinate and cause infection. It takes 1 to 2 weeks for the rust to produce a new crop of spores. For the identification of races, only three reaction classes are used, namely: resistant (R), susceptible (S), and mesothetic (M). The races are designated by numbers; the differences between races 56 and 59 are shown in the table:

	Race 56	Race 59
Little Club	S	S
Marquis	S	R
Reliance	S	R
Kota	S	R
Amautka	R	R
Mindum	R	R
Spelmar	R	R
Kubanka	S	M
Acme	S	S
Einkorn	R	S
Vernal	R	R
Khapli	R	R

The table shows that there are decided differences between races 56 and 59. Both races attack Little Club normally. Marquis, Reliance, and Kota are susceptible to race 56 and resistant to race 59. Arnautka, Mindum, and Spelmar are resistant to both races. They differ again on Kubanka and are the same on Acme. They differ on Einkorn and produce the same effect on Vernal and Khapli. They produce the same effect on seven varieties but differ on five of them. They are therefore distinctly different races, as determined by their ability to attack certain varieties of wheat.

RACE 59 comprises several known biotypes. The collections of rust from the United States and Mexico that were identified by E. C. Stakman and colleagues as race 59 between 1929 and 1944 produced the infection types given in the foregoing. Collections of rust taken from the field may contain several races, which are separated from each other by appropriate inoculations on differential varieties and the different races can then be isolated and cultured separately. Each purified culture is called an isolate. During a season when many collections of rust are identified, many isolates of the same race are made, and a number of them usually are kept for comparative tests at the same time and under the same conditions.

In 1944 an isolate from a collection that was obtained directly from barberry bushes in Massachusetts produced infection type 2 on Reliance wheat instead of the type o, which was characteristic of the race; soon thereafter a similar isolate was obtained from a Collection of rusted barberry leaves from Washington State. When comparative tests were made between isolates that produced type o and those that produced type 2, the difference proved to be consistent, indicating that the isolates were genetically different. As the isolates could not be subdivided further, it was concluded that each was pure genetically, and they were considered as two biotypes Of the same race, 59, because they were alike on 11 varieties and the difference on the one variety, Reliance, was slight; Reliance was immune to One and highly resistant to the other. Tile new biotype was therefore designated as race 59A. Later in 1944 two other isolates, also from barberry, differed slightly but consistently from the other two on Kota wheat, and they were designated as 59B, and 59C, respectively.

The infection types produced by the four biotypes on Reliance and Kota are:

	Reliance	Kota
59	0	0;
59A	2	0;
59B	2	2
59C	0	2

(The semicolon after o indicates that there were small flecks of dead wheat tissue, but no pustules.)

These minor differences in infection types on Reliance and Kota seemed scientifically interesting rather than practically important until it was found that 59A is much more virulent than 59 on certain previously resistant varieties of barley. Clearly, then, the differences between at least two biotypes of race 59 on the differential varieties of wheat are not the only differences between them, and, if they could be tested on all known varieties of wheat and barley, additional differences might be found which would justify considering them as separate races. As new varieties of wheat are produced, rust isolates that now appear identical may prove to be different. Race 15 is a good example. Race 15 of wheat stem rust was discovered in 1918.

The race was easy to identify; all of the 12 differential varieties of wheat except Khapli are susceptible to it. The identification of rust collections from different places in the United States disclosed that some isolates of race 15 were slightly more virulent than others, but the differences were neither great nor consistent enough to justify the conclusion that they were genetically different that is, different biotypes. The first conclusive evidence that some isolates were less virulent than others was obtained by studying a collection from Japan, which produced smaller pustules on some differential wheat varieties than any of the United States collections. All of the differential varieties except Khapli were susceptible, but some were distinctly less susceptible than to isolates from the collections previously studied. The Japanese rust was therefore designated as 15A, a distinct biotype. Race 15 therefore included at least two biotypes, and it was suspected but not proved that there were still others. In 1937, however, an isolate was obtained that was distinctly more virulent on some differential wheat varieties than any previously studied, and it was therefore designated as 15B, a third biotype within race 15.

Although clearly different genetically, the known differences between the biotype originally designated as race 15 and those subsequently found and designated as 15A and 15B were not great enough to justify considering them as separate races. As 15B appeared to be a potential menace to resistant varieties grown in northern United States, attempts were made to learn as much as possible about it. As it occurred only sparingly in nature, however, studies had to be limited to the very few isolates that could be obtained. Attempts were made to find varieties that would distinguish more clearly between 15 and 15B. After extensive investigation, research workers found that a relatively new variety, Rival, and an Australian hybrid, Kenya x Gular, were fairly good differentials. Finally the new variety Lee (Hope x Timstein), released to growers by the Minnesota Agricultural Experiment Station and the Department of Agriculture in 1951, proved resistant to 15 and susceptible to 15B, thus serving as a new indicator or differential variety for distinguishing clearly between 15 and 15B on the basis of resistance and susceptibility instead of on the less satisfactory basis of degrees of susceptibility. That seemed like progress in understanding a complex situation, but the situation itself soon became more complex.

In 1950 race 15B became widespread and prevalent in North America for the first time at least since races of wheat stem rust were recognized. Previously it had been found occasionally near barberry bushes, especially in the Eastern All of the 317 isolates that were identified as 15B in 1950 produced essentially the same infection types on the 12 differential varieties of wheat and on Lee wheat; otherwise they would not have been designated as 15B.

There were some indications, however, that some of the isolates were not completely identical. These results were obtained by the senior author and his colleagues at the Federal Rust Labouratory, University Farm, St. Paul, Minn. Therefore 17 selected isolates were tested on 99 additional varieties of wheat and on 17 varieties of barley.

Those tests which were made by Rosendo Postigo, a graduate student at the University of Minnesota, with the collabouration of Helen Hart and E. C. Stakman disclosed that 14 of the 17 isolates differed clearly in their effects on one or more of these varieties, some of which had been produced recently. It is evident, then, that biotypes of rust that seem to be identical because of their identical behaviour on many varieties may prove to be different if additional biologic indicators can be found. The wheat varieties are, of course, really biologic indicators or testers, and must be depended on for distinguishing between rust biotypes, because the rust grows only on living plants and there is no known method of distinguishing between biotypes except by their effects on varieties of living plants.

The experience with race 15 and 15B shows that recognition of differences between biotypes and between races depends on the availability of host varieties that react differently to them, and varieties may serve as additional differentials. Conclusions regarding rust races therefore are limited by the varieties available at a given time; furthermore, as new. races are continually being produced, it is necessary periodically to revise the classification of races.

Puccinia graminis is a good example of the complexity within the species, within the variety tritici, and even within races. There are a great many biotypes and their recognition and grouping into races for practical purposes are subject to limitations of suitable differential varieties at any given time and place. To recognize the differences between all biotypes n existence would require that all varieties of wheat, barley, and many grasses be inoculated under several combinations of environmental conditions with an adequate sample of rust collections from the world and while that was being done, many new biotypes probably would have been produced by mutation and hybridization.

The problem of identifying races is difficult not only because different biotypes may appear alike on many varieties and under many environmental conditions. It is difficult also because the infection types produced by a single biotype on a single variety may vary widely with environmental conditions. As an example, Australian and American isolates of race 34 of P. graminis tritici differ in temperature relations and in colour, thus suggesting that different biotypes may behave alike at certain temperatures but not at others. Certain races produce a type x infection under some conditions but not under others. Thus the only known difference between races 17 and 37 of P. graminis tritici is that the latter produces type x on Kubanka while race 17 produces type 3 to 4. Likewise, the difference between races 56 and 125 is that the latter produces type x on Kubanka; the difference between the races is genetic, but it is slight and variable enough to be obscured by differences in light and temperature. The same is true of some of the races of oats stem rust, Puccinia graminis avenge.

Despite the difficulties, it has been possible in the past to explore within the species Puccinia graminis sufficiently to explain differences in behaviour of stem rust of wheat and oats in different times and places and to predict what is likely to happen in the future. Some races are relatively stable genetically and therefore are consistent in behaviour historically and geographically; others are unstable and variable. Knowledge regarding the diversity within the species is incomplete. Were it possible to grow all stages of the rust on artificial media, it probably would be possible to recognize many more biotypes than is now possible.

Although most plant scientists are familiar with the life history of P. graminis, its relation to genetic studies deserves emphasis. There are five stages: Telial, sporidial, pycnial, aerial, and uredial. The mature black spores or teliospores are diploid, but reduction division occurs when they germinate, and the sporidia formed on the four-celled promycelia are haploid and of two sex groups. The sporidia can infect only certain species of barberry (species of Berberis and Mahonia) on which they produce pycnia containing pycniospores, which are in reality specialized sex cells or gametes. Although the pycnia and pycniospores all look alike, they are of two kinds sexually, having been produced by sporidia of different sex. All the pycniospores produced within a single pycnium are of the same sex, so that pycniospores from pycnia of different sex must be brought together (in nature this is usually done by insects) in order to initiate the sexual stage. This is therefore roughly analogous to cross-pollination in dioecious higher plants. The nuclei of opposite sex pair at the base of the aecium, divide simultaneously, but do not fuse for a long time.

From the "fertilized" pycnia are formed the cluster cups or aecia on the barberry, and in them are produced the aeciospores, each with paired nuclei of opposite sex. The aeciospores cannot infect the barberry, but only appropriate grains and grasses on which the uredospores are produced in pustules, or uredia. The uredospores contain paired nuclei also, and they can infect only grains and grasses, but this is the repeating stage of the fungus, and successive generations of uredospores can be produced indefinitely, provided susceptible kinds of grains or grasses are available and in vegetative condition. When conditions become unfavourable for the development of the uredial stage, the telial stage is formed. The young teliospores have paired nuclei, which fuse within the spore, and thus the final stage in the long sexual process results in the formation of diploid nuclei. As the fused nuclei may differ in factors for pathogenicity, segregation and recombination can result in the production of new biotypes.

The fact that the uredial stage of the rust is the only one that can be clonally propagated means that races must be identified in this stage, and, since it has paired nuclei of opposite sex that later fuse in the telial stage, a single heterozygous race may yield many different races as a result of segregation and recombination even if it is selfed. In reality so many biotypes often can be isolated from or near barberry that it is extremely difficult to group them into races, because many are so nearly alike and because it often is hard to find varieties that show even the larger differences.

The common corn smut, Ustilago zeae, is suitable for studying biotypes within species. The smut spores (chlamydospores) result from the fusion of haploid nuclei. Accordingly the chlamydospore nucleus has a set of chromosomes from each parental line that is, double the number (diploid) in each sex cell (gamete). On germination, the spore sends out a four-celled tube, the promycelium, like that produced by teliospores of stem rust. Each cell likewise produces a small, sporelike body, known as asporidium, which has a single nucleus with half as many chromosomes (haploid) as there are in the chlamydospore nucleus; this results from reduction division early in the process of germination. The sporidia are therefore sex cells or gametes, like sperms and eggs, except that those of different sex cannot be distinguished by appearance; they look alike, but those of opposite sex will conjugate (fuse) while those of the same sex will not. The test of sex, therefore, is behaviour and not appearance. But these gametic sporidia are unlike most specialized sex cells because they can either multiply by yeastlike budding or germinate like spores and produce an extensive mycelium. This stage of the smut can be grown on or in sterilized nutrient media. Although only about 8 microns long (one three-thousandth of an inch), single sporidia can be picked up with a micromanipulator and placed on solid nutrient media in sterilized flasks; within about 3 weeks colonies have formed, consisting of about 2 billion sporidia or a network of hyphae comprising a billion or more cells, all derived from a single, haploid cell. Such cultures are called monosporidial lines and are a single biotype, unless new biotypes are formed by mutation. There is no chance for segregation, because the original sporidia themselves are the product of segregation. The colonies produced from different sporidia often differ in many characters, such as size, colour, and surface patterns. It is like looking at different kinds of forests from an airplane, with each forest comprising one kind of tree.

E. C. Stakman and L. J. Tyler at the University of Minnesota in 1934 crossed two monosporidial lines with five clearly contrasting characters. By 1952 many investigators there had produced about 10 thousand different biotypes from recombinations and mutations among the descendants of the two original sporidia. These biotypes differ from each other in one or many major or minor characters, including size, colour, and topography of colonies, many other cultural characters, mutability and sex and pathogenicity.

Occasiorial monosporidial lines are diploid instead of haploid, but they are exceptional. Colour of colonies among these biotypes ranges from aniline black through many shades and tints of purple, brown, pink, drab, and white. Innumerable patterns and combinations of topographic patterns are superimposed on the profuse array of colours various kinds of concentric bands and ridges and many kinds of radial lines, folds, furrows, and ridges, some straight, some curving clockwise and some counterclockwise. There are many types of marginal growth. Some biotypes produce sporidia only. Others produce mycelium only. There also are many intermediate kinds, in which the growth type is determined by the amount of sugars in the medium.

Temperature requirements differ and may affect relative survival in mixtures of biotypes. There are multiple sex groups and many gradations in pathogenicity. There are wide differences in mutability, and it is becoming evident that in cultures of some of the most constant monosporidial lines there are mutants whose presence is masked under most conditions by the growth of the original lines. These masked mutants may persist indefinitely without betraying their presence unless special methods or media are used to detect them.

Possibly Ustilago zeae represents the extreme of intraspecific diversity, or possibly it is only better known than many other plant pathogens. Covered kernel smut of sorghum, Sphacelotheca sorghi, and the grass smuts, Sorosporium syntherismae and Ustilago sphaerogena, also comprise many biotypes. Some smuts, however, such as loose smut of barley (Ustilago nuda), loose smut of wheat (U. tritici), loose smut of oats (U. avenge), and covered smut of barley (U. hordei) seem to be somewhat more homogeneous and constant. Investigations on rusts and smuts have been mutually complementary in explaining variability of pathogenic behaviour. In the rusts only the binucleate uredial stage can be propagated clonally. In the smuts only the haplophase can be so propagated.

In the smuts and rusts, three distinct phases correspond with the nuclear condition: The haplophase, in which the cells have nuclei with the reduced number of chromosomes, such as the sporidia of both groups and the pycnial stage in the rusts; the dicaryophase, in which two nuclei of opposite sex are paired in the cells but not fused; and the diplophase, in which the two nuclei of the dicaryophase have fused. The haplophase in the rusts (sporidia and pycnial stage) is restricted almost entirely to a sexual function. In some of the smuts, however, the sporidia (haplophase) can function both as propagative cells and as sex cells, but the ability of the haplophase to grow parasitically is limited. In both the smuts and rusts, the dicaryophase is the important parasitic phase, and it is important to realise that there are dicaryotic hybrids. This simply means that each cell can contain two nuclei of different sex and with different factors for pathogenicity and for other characters. The final hybridizing act is not complete until the two nuclei fuse to form the diplophase, which is restricted largely to the resting spores, the teliospores of the rusts and the chlamydospores of the smuts, except that diploid sporidial lines of smuts are encountered occasionally.

The uredial stage of rusts can be propagated indefinitely on appropriate hosts; hence, repeated experiments can be made on the pathogenicity of dicaryotic hybrids, where minor differences in pathogenicity are perceptible. In the smuts, on the other hand, the gametic haplophase can be propagated saprophytically on the artificial media where conditions can be varied and controlled, but the pathogenic dicaryophase cannot be propagated clonally even on living plants. It comes to an end with the formation of diploid spores. Accordingly the progeny of a single uredospore in the rusts constitute a biotype, barring mutation; in the smuts the progeny of single haploid sporidium constitute a biotype, barring mutation. Likewise the monosporidial diploid lines that occasionally appear in smuts are biotypes, barring mutation and segregation. A physiologic race in the rusts, then, can be purified to a single biotype, while in the smuts the most practicable way of determining the parasitism of so-called physiologic races is to inoculate differential varieties with successive crops of chlamydospores.

The term physiologic race, then, does not have the same genetic connotation in the rusts and smuts. In many of the Fungi Imperfecti, which are imperfectly known genetically, the only criterion of races must be behaviour in culture and on host plants. As commonly used, therefore, the term physiologic race means a biotype or group of biotypes that can be distinguished with reasonable certainty and facility by its physiologic characters, including pathogenicity, and, in some fungi, growth characters on artificial media.

Under that definition, most species of plant pathogenic fungi comprise physiologic races. Some of the downy mildews (including Phytophthora infestans, which causes late blight of potato), some of the powdery mildews, and several of the wood-rotting fungi comprise many races. The species of Helminthosporium that cause leaf spots, seedling blights, and root rots of small grains and corn all are complex in composition.

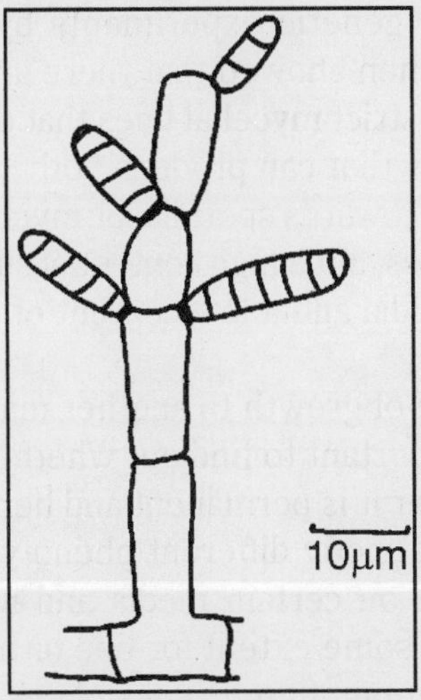

Fig. Helminthosporium

Helminthosporium sativum, H. gramineum, H. Ceres, H. carbonum, and H. victoriae comprise many races, whether judged by behaviour on artificial media or on host plants or both; H. sativum is known to be in the same category as Ustilago zeae with respect to diversity. Various species of Fusarium, such as those that cause flax wilt, melon wilt, tomato wilt, and head blight or scab of cereals, comprise many races. Colletotrichum lini, which causes anthracnose of flax, C. lindemuthianum, the cause of bean anthracnose, Rhizoetonia solani, which causes blights and rots of hundreds of kinds of plants, the apple scab fungus, the fungus causing brown rot of stone fruits, and many others comprise numerous physiologic races.

The determination of phenotypic variability of fungi the temporary changes in appearance or behaviour due to environment is relatively easy when the fungi can be grown on nutrient media, as environmental factors usually affect the fungus and not the medium. Similar determinations are more difficult when the fungus is growing on living plants, because environment affects the fungus, the host plant, and the interactions between the two. The phenotypic variability caused by nutrients, temperature, and other factors is often so great as to obscure genetic differences between biotypes or races when they are determined partly by cultural characters and by Pathogenicity. It is essential therefore to standardize media and other conditions with varying degrees of exactness, depending on the fungus and the objectives of the investigation.

A primary requisite is to determine the limits of phenotypic variability of isolates before making conclusions regarding genetic differences between them. Variability in growth type of corn smut due to nutrition, temperature, and other factors can be wide or narrow, depending on the biotype. An example: Mycelial patches or sectors in colonies of certain sporidial lines seemed to be mutants until it was found that subcultures from dissimilar sectors sometimes were identical. Physiologic and genetic experiments by M. F. Kernkamp, at the University of Minnesota, then showed that there are strict sporidial lines that cannot produce mycelium, strict mycelial lines that do not produce the sporidia, and sporidial-mycelial lines that can produce both sporidia and mycelium.

The relative tendency towards sporidial or mycelial production in the latter group differs in different lines, but a high content of sugar in the nutrient medium favours production of sporidia and a low content of sugar favours formation of mycelium.

Change from one type of growth to another may ma result from mutation also; hence it is always important to find out whether the change is temporary and nonheritable or whether it is permanent and heritable. The same biotypes, then, may have several distinctly different phenotypes. Conversely, different biotypes may appear alike on certain media and not On others. Phenotypic differences may Persist to some extent for one or more Cultural generations; consequently the classification of groups of isolates should be made only after they have been grown for one or more generations under identical conditions. Variability in development of fungus pathogens on living hosts can be studied advantageously in some of the rust fungi because of the distinct infection types produced.

From investigations by W. L. Waterhouse and associates at Sydney University, in Australia, Thorvaldur Johnson and others at the Dominion Labouratory of Plant Pathology, Winnipeg, and by Helen Hart and associates at the University of Minnesota, it is now known that temperature and light can cause fundamental differences in the development of certain races of stem rust on certain varieties of wheat and oats.

Certain Kenya wheats and some hybrids, for example, are almost immune to race 15B of wheat stem rust at 65 F. and completely susceptible at 85. Conversely, race 34 develops best on some varieties of wheat at moderate temperatures and relatively low light intensity, and the infection type is depressed by high temperature and high light.

The oats variety Hajira is resistant to known races of P. graminis avenge below about 80 but it is susceptible to race 6 at 85 and to races 7 and 8 at 90. The implications of phenotypic variability in the identification of races are important, because a given rust race may produce infection type I on a host variety at temperatures as high as 75 and type 4 at 85. This is true of some race-variety combinations but not of others, and must therefore be studied for each combination.

PROBLEMS OF VARIABILITY IN FUNGI

Many plant disease fungi are known to be complex and variable, but the degree of complexity is imperfectly known for most of them and is not well enough known for any of them. This complexity is not surprising because many of the crop plants in which the fungi live are equally complex. There are more than 14,000 varieties of wheat; there are unnumbered varieties of apples, beans, corn, roses, iris, and other cultivated and wild plants. Moreover, man and Nature are continually making new varieties by mutation and by hybridization and selection.

Plant disease fungi also are plants. They are parasitic on higher plants. Most are microscopic in size and relatively simple in structure, but they are plants nevertheless. They germinate, grow, and fructify; they multiply by the countless billions; they Mutate and hybridize; consequently Nature has made countless kinds in the past and is continually making More kinds through mutation and hybridization. As plant breeders produce better kinds of useful plants, therefore, Nature may produce more destructive fungi. It is the job of plant pathologists to learn what the parasitic fungi are now doing and what they may do in the future to jeopardize or nullify man's success in plant improvement. To understand present Plant disease situations and to predict Possible future developments, it is essential to learn as much as possible about the variability in fungi pathogenic to plants. Other kinds of causal agents of plant diseases, such as viruses and bacteria, must be studied similarly, but the present discussion is restricted to fungi.

The principles of growth and reproduction in plant pathogenic fungi are similar to those for higher plants, except that fungi have no chlorophyll, which enables green plants to make their food from simple materials in the soil and air. The fungi therefore must live on living plants or animals or their products. The fungus parasites of plants live in or on living plants and get their nourishment from them by means of minute tubes, or hyphae, that grow, branch, and form a network known as the mycelium, which finally produces many tiny spores.

Spores serve the same propagative and reproductive function as the seeds of higher plants, but spores are much smaller and simpler in structure than seeds. Spores usually have one or at most a few cells and do not contain an embryo as do seeds. Nevertheless they germinate and produce new fungus plants with the parental characters. The conditions required for germination are similar to those for seeds, moisture and suitable temperature being the most important. The time needed for germination varies with the kind of spore and the temperature; it varies from about an hour to several days. Many fungi produce millions or even billions of spores in a few days to a few weeks and can therefore produce countless numbers in a few short generations. That is one reason why they are so dangerous.

Some spores are produced asexually and some only as a result of sexual fusions. The same fungus may produce several kinds of spores asexually, but usually produces only one kind sexually. Successive crops of asexual spores are generally produced abundantly and quickly when conditions favor growth of the fungus, and the sexual spores are usually formed when the vegetative period is terminating. Each kind of spore has a special name; thus some of the rusts produce five kinds of spores or sporelike bodies teliospores, sporidia, pycniospores, aeciospores, and uredospores, each with special form and function.

The principles and procedures in classifying and naming fungi are similar to those for higher plants. Species of higher plants are determined by morphologic characters of the vegetative and fruiting structures, including seeds. Similarly species of fungi are determined by the characters of the mycelium, of the spore-producing organs, and of the spores themselves. The unit of measurement for fungus spores is the micron, which is one one-thousandth of a millimeter; 25 millimeters make an inch. The range in size is from about 4 microns to more than 100 microns and, together with shape, color, and certain other characters, is characteristic for each species.

Visible characters are used for grouping higher plants and plant pathogenic fungi into classes, orders, families, genera, species, and varieties. Thus it is easy to distinguish larger groups of agricultural plants, such as grasses and legumes. It is equally easy to distinguish larger groups of fungi, such as smuts and rusts. But there are many kinds of wild and cultivated grasses, such as timothy, bromegrasses, oats, barley, rye, rice, corn, and wheat; and there are many kinds of legumes, such as beans, peas, soybeans, alfalfa, and clover. Likewise, there are many kinds of smuts, such as the loose smut of wheat, the stinking smuts of wheat, common corn smut, and rye smut, and there are more than 3,000 kinds of plant rusts. It is relatively easy to recognize these major groups of higher plants and of fungi, but it becomes increasingly harder to classify the smaller groups, such as species and varieties.

The agronomist must know not only wheat, a genus scientifically designated by its Latin name Triticum, but he must know the species of wheats, such as the common bread wheat, Triticum vulgare; the durum wheats, T. durum; and several other species or subspecies. He must know also thousands of varieties of bread wheats and hundreds of varieties of durum. And he must learn about the new lines and potential varieties that continually are being produced. Likewise the plant pathologist must know the genera, species, varieties, and "lines" of the fungi that parasitize different kinds of wheat, and he must learn about the new ones that continually appear.

Moreover, the agronomist must know not only what crop plants look like, but also how they behave and what they are good for. He must know whether a variety of wheat is a winter wheat or a spring wheat, whether it is winter hardy, where it grows well, whether it makes good flour or poor flour, and whether it is susceptible or resistant to disease; and the plant pathologist must know not only what species of plant disease fungi look like but must also know their pathogenicity for the many different kinds and varieties of plants.

Most species of plant pathogenic fungi are equally as complex in composition as species of crop plants, and it is therefore necessary to find out what is within the species. As an example, the species Puccinia graminis, the stem rust of small grains and grasses, is recognized easily by its general appearance and by certain characters of its spores. But there are several varieties within the species that are alike in some characters and different in others, including the kinds of crop plants that they can attack. Thus the variety tritici of Puccinia graminis parasitizes wheat, barley, and many wild grasses; the variety avenae parasitizes oats and certain wild grasses but not wheat and barley; within the tritici variety there are races that differ in their ability to attack certain varieties of wheat; within the variety avenae there are races that differ in ability to attack varieties of oats. Kanred, as an example, is immune to some races of the variety tritici and susceptible to others. These rust races in turn can be subdivided into the ultimate subdivisions, the biotypes. A biotype is a population of individuals that are identical genetically.

The descendants of a single non- Sexually produced stem rust spore constitute a biotype, unless mutation occurs to cause genetic diversity. The test of genetic purity of a population of a pathogenic fungus is the consistency of its behavior. Most species of plant pathogenic fungi comprise many biotypes, and attempts are made to group the most closely related ones into races and then to group the most closely related races into varieties, which in turn are grouped into species. Actually the procedure usually is in the reverse direction: The larger groups are recognized first, and the successively smaller groups are discovered by successively more refined methods. This can be illustrated by results of investigations of Puccinia graminis, the fungus that causes stem rust of small grains and grasses. (The discussion of stem rust is based on results of cooperative investigations between the United States Department of Agriculture and the University of Minnesota, which were begun in 1914 and continued by many investigators under the general supervision of the senior author, who regrets that the contributions of the many administrators and investigators cannot be specifically mentioned without interrupting the continuity of the discussion.)

Puccinia graminis is a good species Within which there are clearly recognizable varieties, races, and biotypes. There are at least six rather distinct varieties that differ in size of spores and in the kinds of plants that they can attack: (1) tritici (wheat), whose host plants are wheat, barley, and many wild grasses; (2) secalis (rye), rye, barley, and many wild grasses like those attacked by tritici; (3) avenae (oats), oats and wild grasses different from those attacked by tritici; (4) Aleifiratensis (timothy), timothy and certain wild grasses; (5) agrostidis (redtop), redtop and other species of Agrostis; (6) poae (bluegrass), Kentucky bluegrass (Poa pratensis) and related species.

There probably are more varieties than the six listed, and new ones may appear in future as a result of mutation or hybridization. Races are known within the varieties tritici, avenae, and secalis. Although there sometimes are at least slight morphologic differences between races within a variety, the most important and most easily recognizable differences are in the degree of infection (pathogenicity) on certain varieties within the genera Triticum (wheat), Avena (oats), and Secale (rye), respectively. The crop varieties, designated as differential varieties, now used to distinguish races within the tritici and avenae varieties, were selected from hundreds that were tested, and, at the time of selection, appeared to be representative and adequate for distinguishing races.

TERM ADAPTATION OF FUNGI

The term adaptation, as used here, means the capacity of a single biotype to acquire and transmit the ability to do something that it either could not do originally or could not do well. Thus, race 19 of wheat stem rust can produce only small pustules on Marquis wheat. It would be a case of adaptation if genetically pure race 19 were grown on Marquis several successive rust generations and produced successively larger pustules until it acquired the ability to grow well on this variety, or at least better than it did originally. Likewise it would be adaptation if a single biotype of the corn smut fungus (Ustilago zeae) were grown for several successive generations on a medium containing arsenic and thus acquired the ability to tolerate several times as much arsenic as it did originally.

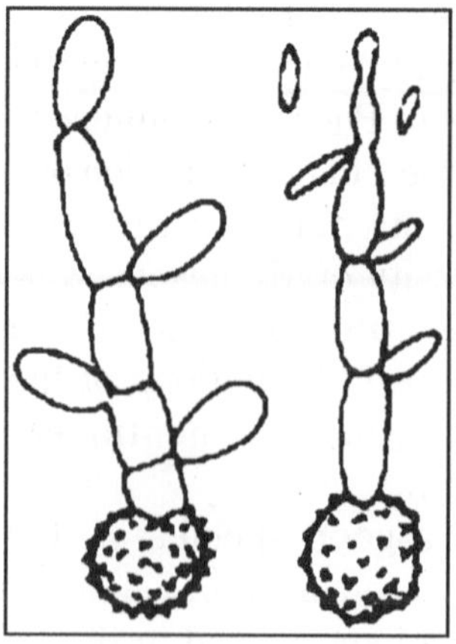

Fig. Ustilago Zeae

Between 1900 and 1910 the question of adaptation was studied by several investigators, who concluded that certain rust fungi and powdery mildews could adapt themselves quickly to resistant varieties. Marshall Ward made many experiments with rust of bromegrasses (Puccinia dispersa) at Cambridge University, England, and published a paper on adaptive parasitism in 1903. He concluded that the brome rust could develop the ability to grow well on a resistant species of Bromus if grown on the resistant species for one or more successive rust generations.

He also concluded that there were "bridging species" of Bromus that enabled the rust to attack very resistant varieties; for example, the rust could grow well on variety A but not on variety C. If, then, a variety B could be found that was intermediate taxonomically between A and C, the rust could be grown on B and there it acquired the ability to infect C. Variety B, therefore, was considered a bridge or "bridging host" between the susceptible A and the resistant C.

Fig. Puccinia Dispersa

In cooperative investigations between the United States Department of Agriculture and the University of Minnesota, E. C. Stakman, F. J. Piemeisel, and M. N. Levine studied the possible adaptation of stem rust of wheat and other cereals and grasses and of powdery mildew of wheat and barley, but obtained no evidence that these fungi could adapt themselves to resistant varieties by any of the methods previously tried nor by any new methods that could be devised.

They also pointed out that differential hosts must be used to separate different biologic forms from mixtures before making experiments on adaptation; otherwise what appears to be adaptation may be merely the result of the selective effect of host plants on a mixed population of the pathogen. At that time, the now obsolete term "biologic form" was used to designate what are now called varieties of stem rust; races were just being discovered within the tritici and other "biologic forms"; and eventually new facts led to new concepts and to changed terminology. These results seemed to prove that stem rust and powdery mildew of cereal grains did not change by adaptation. But the rusts and powdery mildews are obligate parasites; it has not been possible to grow them on anything except living plants.

Accordingly it was desirable to study possible adaptation or changes in virulence in fungi that can grow both in living host plants and on nutrient media that is, both as parasites and as saprophytes. The results of extensive experiments made by J. J. Christensen and C. L. Schneider at the University of Minnesota from 1946 to 1950 with Helm inthosporiun? sativum, which causes leaf spots, foot rots, and root rots of barley, wheat, and other cereals and grasses, support the view that genetically pure lines of fungi do not change their virulence easily.

The isolate of H. sativum they studied had been grown on artificial nutrient media for 28 years and had been purified by making many successive single-spore isolations. Wheat plants were inoculated with single spores; when new spores were produced, single ones were again picked and inoculated into plants with a fine needle. This process was repeated ten times in succession, and there never was any evidence of change in virulence. This extreme refinement of method was necessary because H. sativum mutates rather freely, producing many mutants that are less pathogenic and a few that are more pathogenic than the parental line.

Some of the less pathogenic lines grow very rapidly and therefore tend to overgrow the original parental line. Unless the fungus is grown under conditions that permit the recognition of mutants and their separation from the parental line, an initially pure isolate may soon comprise a diverse mixture of mutant lines, and the original line may even have been lost in the process of making periodic transfers to new tubes of nutrient media.

Obviously, then, an isolate of H. sativum derived from a single spore may change in virulence, but the change is due to the production of new biotypes resulting from mutation and not to a change in virulence of the original biotype. It would be easy to conclude that some of the smut fungi increased in virulence for resistant varieties as a result of successive passages through those varieties if the intraspecific complexity of smut species were not known.

Many investigators, notably W. A. R. Dillon-Weston in England, tested the resistance of varieties of wheat, barley, oats, and other crop plants to various smuts. When first inoculated with a collection of smut spores (chlamydospores), very little smut develops on some varieties.

If the smut from a resistant variety is used to inoculate plants of the same variety, heavy infection may result, because nonpurified collections of smut spores are likely to comprise many parasitic races that can be separated from each other by certain varieties. Assuming that three wheat varieties, A, B, and C, are inoculated with a smut collection containing races 1, 2, and 3 in the ratio of 90:9:1 and assuming that all three races infect variety A normally, that race 2 infects B but not C, and that race 3 infects C and not B, the percentages of infection theoretically should be the following:

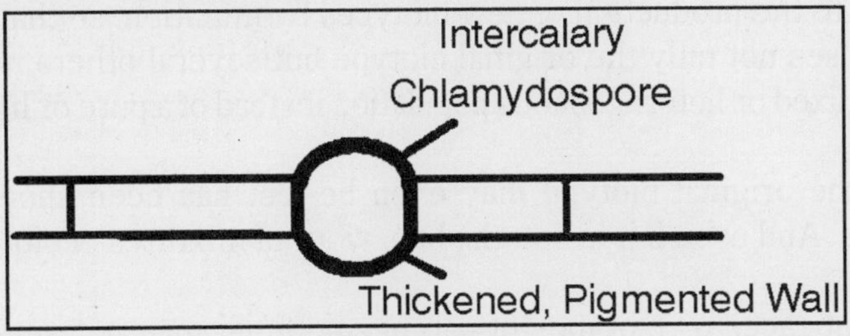

Fig. Chlamydospores

Obviously, if plants of variety A are then inoculated with the smut produced on A, spores of all three races will again be produced. But if variety B is inoculated with spores from B, 100 per cent of them will be of race 2; and if variety C is inoculated with spores from C, 100 per cent of the spores will be of race 3, the only one in the original mixture that can attack C. Thus, varieties B and C are biologic indicators, or differential varieties, that make it possible to find out not only which races were present in the original smut collection but also the relative percentages of each.

Moreover, the apparent adaptation of the smut to varieties B and C is not adaptation at all but is due to selection of the races that can attack those varieties out of a mixture in which they were present in small amounts only.

Even after races of smuts have been isolated and purified as we have described, the degree and permanence of purity are only relative because the spores are not always exactly alike. And, even if they were, the chlamydospores of smuts are the result of sexual fusions. When the spores germinate, segregation and recombination usually occur, so that the progeny of a single spore might be diverse genetically.

For practical purposes, physiologic races of smuts are usually considered to be collections of chlamydospores that behave with relative constancy in successive generations.

Because of sexual recombinations and mutations within the races, however, they often comprise many biotypes, usually closely enough related so that the race behaves fairly consistently.

The degree of refinement commonly practiced in classifying smut races usually suffices for determining the relative resistance of crop varieties, but certainly not for studies on adaptation or other physiologic and genetic phenomena.

It suffices for practical procedures but not for scientific understanding. There are several other reasons why races or biotypes of fungi may appear to have changed, when in reality there has been a change in the kinds of biotypes. It has been pointed out in the discussion of

Helminthosporium sadvum that apparent changes in a monosporous line may be due to the production of new biotypes by mutation, so that the culture soon comprises not only the original biotype but several others in addition; it becomes a mixed or heterogeneous population instead of a pure or homogeneous one.

That the original biotype may even be lost has been shown by many experiments. And even if it is not, the behaviour of mixtures of biotypes cannot be predicted.

Different races or biotypes of pathogenic fungi may not survive equally well in mixtures. This is to be expected if the environmental conditions favour one race more than the others.

But it is sometimes true when conditions, as far as can be determined, are equally favourable to the different races. Extensive studies of the relative survival of races of wheat stem rust have been made at the University of Minnesota and at Sydney University, and typical results were published by I. W. Watson in 1942 and by W. Q. Loegering in 1951.

As the principle is more important than the details, which must be ascertained for each individual mixture of each fungus, a few examples are given to illustrate differential survival of crop plants and of fungi. H. V. Harlan and Mary L. Martini of the

Department of Agriculture published in 1938 the results of experiments on the relative survival of barley varieties in a mixture grown in successive years. H. H. Lande and A. F. Swanson made similar experiments on winter wheats in Kansas and published their results in 1942.

The experiments with barley and winter wheats established the important principle that certain varieties persisted better than others in mixtures in which the relative percentage of each variety was determined in the first year and in each successive year, when seed was taken each year and used for the next year's planting.

The Darwinian principles of competition, natural selection, and survival of the fittest applied to these mixtures of varieties of barley and of wheat. Watson and Loegering tested the validity of the same principle on mixtures of races of wheat stem rust. Mixtures of other fungi may behave like those of wheat stem rust, which have been used as an example because of the relative ease and precision of identification.

Paul E. Hoppe at the University of Wisconsin inoculated corn with equally pathogenic strains Of Diplodia zeae, which causes ear rot and stalk rot, and the antagonism between strains was so great that only One survived. Similarly V. F. Tapke made experiments with different races of covered smut of barley and observed that there was differential survival of races.

Fig. Diplodia Zeae

It appeared for a number of years that Phytophthora infestans, the fungus that causes late blight of potatoes, could adapt itself to resistant varieties, as Donald Reddick and associates, in their attempts to develop blight-resistant varieties in New York, observed that resistant varieties might be slightly infected when first inoculated with certain races of the blight fungus, but became more severely infected when successive inoculations were made with the blight fungus from the same variety or certain others.

Helena L. G. DeBruyn made similar experiments in the Netherlands and confirmed the results of Reddick and others. H. D. Thurston and C. J. Eide, at the University of Minnesota, have obtained results that indicate that the apparent adaptation of the potato blight fungus is due, in some cases at least, to the selective effect of varieties on a mixed population of biotypes. As an example, two races of the blight fungus infect Irish Cobbler equally well, but one attacks Cherokee and the other does not.

When these races were grown in mixture on Irish Cobbler, the Cherokee race decreased rapidly in prevalence but did not disappear entirely; consequently, when the Cherokee variety is inoculated with the blight fungus from the Cobbler variety the infection is very light, because so small a proportion of the racial mixture can attack it.

But the blight that does develop is caused by the Cherokee race, and when this is used to inoculate Cherokee again, there is abundant infection because all of the inoculum is potentially effective.

The present authors and some of their associates began a comprehensive investigation of adaptation in 1935.As they had not been able to obtain evidence for adaptive changes in parasitism, they investigated the possibility of adaptation to chemicals in nutrient media.

Certain protozoa and bacteria were reputed to develop the ability to adapt themselves to deleterious chemicals; therefore studies were made independently by three groups of investigators to find out if this were true also of fungi.

The results were not uniform. In some instances mutation clearly accounted for what could have been considered adaptation. J. J. Christensen and several associates grew the asexual (Fusarium) stage of Gibberella zeae, which parasitizes corn and causes head blight of wheat and barley, on nutrient media containing, respectively, ethyl mercury phosphate, a widely used seed disinfectant; mercury bichloride; and malachite green in concentrations that dwarfed the growth of the fungus.

Fig. Fusarium

The fungus produced numerous clearly visible mutants on each medium. Some grew more poorly than the parent on the special media. Some grew about equally well. Some made from 5 to 15 times as much growth as the parent. Some of the mutants, in fact, grew so much faster on the mercury and malachite green media that they soon overgrew the original line completely. The relative ability of the original line and its mutants to tolerate these chemicals persisted through many generations. In the experiments there was no increase in ability to tolerate the chemicals tried, except as a result of visible mutations, which might easily have been overlooked if special methods had not been used to detect them.

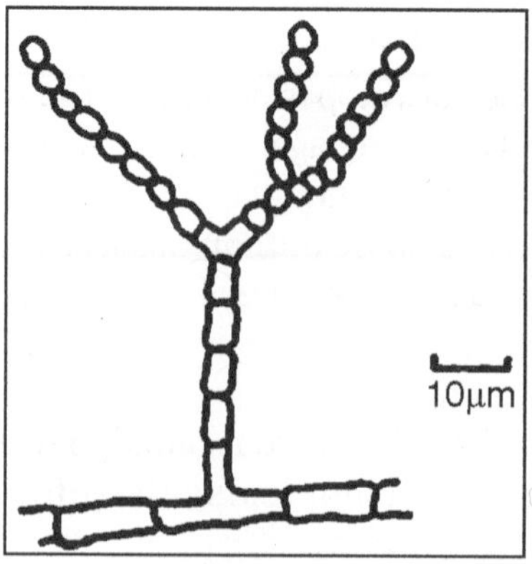

Fig. Monilia

E. O. Mader and C. L. Schneider grew the asexual, or conidial, stage (Monilia) of Sclerotinia fructicola, the cause of brown rot of peaches, plums, and similar fruits, on nutrient media containing injurious amounts of copper sulfate. Some of their results were comparable with those of Christensen: Mutants were produced that grew much better than the parent on the copper media and they differed also in appearance. The increased tolerance for copper was persistent and constant in the mutant lines.

But the fungus also acquired increasing tolerance for copper sulfate, without visible mutation, when grown in successive transfer generations on copper-containing media. When the fungus was grown for successive generations on copper-free media, however, it lost the acquired ability.

The third series of Minnesota experiments were made by Coyt Wilson, Frank Stevenson, Donald Munnecke, J. M. Daly, Elisa Hirschhorn, and E. C. Stakman. This group studied the adaptation of corn smut (Ustilago zeae) to arsenic. They grew mutable and relatively non-mutable monosporidial lines on nutrient media containing 2,400 parts per million of calcium arsenite.

All lines grew poorly at first, but after having been transferred successively for about 10 generations to increasing concentrations of arsenic, all lines eventually grew as well on media containing 12,000 parts per million of calcium arsenite as they had originally grown on 2,400 parts per million, and some lines even grew on 14,000 parts per million.

When grown for five successive generations or more on arsenic-free media, the arsenic-adapted lines lost their acquired ability. Numerous mutants appeared in the mutable line growing on arsenic media.

Some appeared in the relatively stable line, but only two grew slightly better than the parents on arsenic. It was the consensus of this group of investigators, some of whom worked independently at different times with different lines and with different methods, that perceptible mutations did not account for the results.

All smut lines that were tried, including the least mutable ones among thousands that were studied by Stakman and many associates for more than 20 years, increased their ability to grow on arsenic. It is possible that there had been unseen mutation in physiologic characters, but, if so, these mutants escaped detection.

From extensive experiments with several fungi, at Michigan State College about 25 years ago, P. D. Caldis and G. H. Coons concluded that white variants that appeared in single-sporecultures of several fungi were "semi- permanent) variations, which differed from the parent form somatically rather than genetically and were corn parable with the "dauermodifications" described for paramecia by Victor Jollos.

The genetic explanation for this type of semi-permanent change is still lacking. It is commonly observed that there may be what is sometimes called a "hangover" effect when fungi are grown for some time on one kind of nutrient medium and then transferred to another kind. This may be accounted for partly by the production of adaptive enzymes or other chemical substances, but the exact explanation cannot be given for many of these temporary changes.

Although it cannot be asserted categorically that parasitic fungi never adapt themselves to resistant varieties of plants, there is convincing evidence that apparent adaptation often is due to the selective effect of crop varieties on a mixed population of biotypes, either because the isolate or line of the fungus was genetically diverse at the beginning, even though it appeared pure by all tests then available, or because it became mixed as a result of inconspicuous mutations.

The apparent loss of virulence in a single biotype frequently unquestionably is due to the production of mutants that look like the original biotype but have lost genetic factors for virulence. The apparent acquisition of ability to tolerate injurious chemicals certainly is due to mutation in some instances but appears to occur independently of mutation in others.

This is still an unsolved problem and one that requires intensive research. The difficulty of obtaining and maintaining genetically pure cultures of pathogenic fungi cannot be overemphasized in research of this kind.

7

Damping Off Fungal Pathogens

Damping off fungi, white rusts and downy mildews all belong to the order Peronosporales. Many fungi in this group are parasites and may cause diseases which cause great economic loss, such as late blight of potato, downy mildew of grapevine and white rusts of crucifers. Some of them, particularly the members of the Peronosporaceae and Albuginaceae appear to be obligate parasites and do not survive saprophytically in soil or decaying vegetable matter.

Peronosporales are coenocytic mycelial fungi which are parasitic in plants or saprobic in water or soil. The cell walls are mainly composed of a glucan-cellulose complex in addition to protein containing hydroxyproline and lipid. The parasites form haustoria.

Asexual reproduction occurs by zoosporangia or conidia which vary greatly in shape. The sporangia possess one or few pores with or without a papilla and produce biflagellate zoospores or a plasma, either in a vesicle or naked, or a germ tube. Sexual reproduction is oogamous.

FUNGAL FACTORS

Most of the fungi that infect humans and cause disease are classified by tissue or organ levels that are primary sites of colonization.

SUPERFICIAL FUNGAL INFECTIONS

Superficial fungal infections involve only the outermost layers of the stratum corneum of the skin (*Phaeoannellomyces werneckii* [syn. *Exophiala werneckii*] and *M furfur*) or the cuticle of the hair shaft (*Trichosporon beigelii* and *Piedraia hortae*). These infections usually constitute cosmetic problems and rarely elicit an immune response from the host (except occasionally *M furfur* infections).

Recently *T beigelii* and *M furfur* were implicated as opportunistic agents of disease, particularly in immunosuppressed or otherwise debilitated patients. Patients are accidentally infected with these common organisms via indwelling catheters or intravenous lines. Virtually nothing is known concerning the pathogenic mechanisms of these fungi.

DERMATOPHYTE INFECTIONS

The dermatophytes are fungi that colonize skin, hair, and nails on the living host. These fungi possess greater invasive properties than those causing superficial infections, but they are limited to the keratinized tissues. They cause a wide spectrum of diseases that range from a mild scaling disorder to one that is generalized and highly inflammatory. Studies have shown that the disease-producing potential of these agents depends on various parasite and host factors, such as the species of organism, immunologic status of the host, type of clothing worn, and type of footwear used. Trauma plays an important role in infection.

These organisms gain entry and establish themselves in the cornified layers of traumatized or macerated skin and its integument and multiply by producing keratinase to metabolize the insoluble, tough fibrous protein. The reason why these agents spread no deeper is not known, but it has been speculated that factors such as cell-mediated immunity and the presence of transferrin in serum inhibit fungal propagation to the deeper tissue layers and systemic disease does not occur. Some dermatophytes have evolved a commensal relationship with the host and are isolated from skin in the absence of disease. Little is known about specific pathogenic mechanisms of the dermatophytes, but they do not cause systemic disease.

SUBCUTANEOUS MYCOSES

The fungi that have been implicated in the subcutaneous mycoses are abundant in the environment and have a low degree of infectivity. These organisms gain access to the subcutaneous tissues through traumatic implantation. Again, little is known about mechanisms of pathogenesis. Histopathologic evidence indicates that these organisms survive in the subcutaneous tissue layers by producing proteolytic enzymes and maintaining a facultative microaerophilic existence because of the lowered redox potential of the damaged tissue.

In eumycotic mycetoma there is extensive tissue damage and production of purulent fluid, which exudes through numerous intercommunicating sinus tracts. Microabscesses are common in chromoblastomycosis, but the clinical manifestation of disease indicates a vigorous host response to the organism, as seen by the intense tissue reaction that characterizes the disease (pseudoepitheliomatous hyperplasia).

Although most of the fungi implicated in this category of disease exist in a hyphal morphology, the agents of chromoblastomycosis and sporotrichosis are exceptions. Chromoblastomycosis is caused by a group of fungi that have several features in common. They are all darkly pigmented (dematiaceous) and exhibit a pleomorphism consisting of two distinct morphologies: the organism may exist in a mycelial state or as a thick-walled spherical cell that divides by cleavage. The latter cell morphology, called a muriform cell, sclerotic cell or Medlar body, is the pathologic morphology seen in tissue sections. However, transition to the sclerotic morphology may not be a crucial requirement for pathogenesis. Several dematiaceous fungi cause a disease called phaeohyphomycosis, which clinically consists of a broad group of diseases characterized by the presence of various darkly pigmented yeastlike to hyphal elements, but not sclerotic cells, in pathologic specimens.

Alternatively, the immune reaction of the host may dictate the morphology that the organism assumes. Again, there is no information about mechanisms or the role of morphogenesis in the pathogenesis of this group of fungi.

Sporotrichosis is caused by *Sporothrix schenckii*, which grows as a mold in nature or when cultured at 25°C, but as yeastlike cells when found in tissues. The clinical manifestations of disease caused by *S schenckii* vary, depending on the immune status of the patient. The classic condition, subcutaneous lymphanigitic sporotrichosis, is characterized by numerous nodules, abscesses, and ulcerative lesions that develop along the lymphatics that drain the primary site of inoculation.

The disease does not extend beyond the regional lymph nodes that drain the site of the original infection. Alternatively, infection may result in solitary lesions or pulmonary disease. Clinical manifestations of pulmonary infections vary depending on the immune status of the patient. The immunocompetent individual has a high degree of innate resistance to disease, and when infection occurs the organism is often a secondary colonizer of old infarcted or healed cavities of the lungs. If the patient is immunocompromised, dissemination can occur. There is no information about mechanisms of pathogenesis of this dimorphic fungus.

SYSTEMIC MYCOSES

Of all the fungi that have been implicated in human disease, only the six agents that cause the systemic mycoses have the innate ability to cause infection and disease in humans and other animals. The primary site of infection is the respiratory tract. Conidia and other infectious particles are inhaled and lodge on the mucous membrane of the respiratory tree or in the alveoli, where they encounter macrophages and are phagocytosed. To successfully colonize the host these organisms must be able to survive at the elevated temperature of the body and either elude phagocytosis, neutralize the hostility they encounter, or adapt in a manner that will allow them to multiply.

Several factors contribute to infection and pathogenesis of these organisms. Of the six systemic agents, five, *Histoplasma capsulatum*, *Blastomyces dermatitidis*, *Paracoccidioides brasiliensis*, *Coccidioides immitis*, and *Penicillium marneffei* are dimorphic, changing from a mycelial to a unicellular morphology when they invade tissues, except *C immitis* that forms spherules. The change from mycelial to yeast morphology in *H. capsulatum* appears critical for pathogenicity. Several physiologic changes occur in the fungus during the transition, which is induced by the temperature shift to 37°C.

The triggering event is a heat-related insult: the temperature rise causes a partial uncoupling of oxidative phosphorylation and a consequent decline in the cellular ATP level, respiration rate, and concentrations of electron transport components. The cells enter a period of dormancy, during which spontaneous respiration is maintained at a decreased level. Then there is a shift into a recovery phase, during which transformation to yeast morphology is completed.

Mycelial cells of *H capsulatum* that are unable to undergo this morphologic transition are avirulent. Similar observations have been made when mycelia of *B dermatitidis* and *P brasiliensis* are shifted from 25°C to 37°C, and it has been implied that transformation to the yeast morphology is critical for infection.

Coccidioides immitis is also dimorphic, but its parasitic phase is a spherule. Little is known about the role of morphologic transformation in infection and disease of this organism. Dimorphism does not appear to play a role in *C neoformans* pathogenesis since the organism is an encapsulated yeast both at 25°C and in host tissues. The sexual phase of *C neoformans*, *Filobasidiella neoformans*, is known, and the organism assumes a filamentous morphology, producing small basidiospores. It has been suggested that these propagules are relevant in infection.

In addition to adjustment to the elevated temperature of the host, the infectious propagules must deal with the hostile cellular environment of the lungs. Studies with mutants of *C neoformans* have shown that the acidic mucopolysaccharide capsule is important in pathogenesis. Acapsular variants of the yeast are either avirulent or markedly deficient in pathogenicity. Since these mutants were obtained by mutagenesis, it is difficult to rule out the contribution of other genetic defects to their decreased pathogenicity.

However, at the cellular level, the capsular polysaccharide inhibits phagocytosis of the yeast. Encapsulated *C neoformans* cells are highly resistant to phagocytosis by human neutrophils, whereas acapsular variants are effectively phagocytosed. The active component of the capsular polysaccharide has been identified as glucoronoxylomannan. In addition, the capsular polysaccharide is poorly immunogenic in humans and laboratory animals, and the glucoronoxylomannan component persists for extended periods in the host.

In addition to the capsular polysaccharide, elaboration of phenyl oxidase (an enzyme that catalyzes the oxidation of various phenols to dopachrome) by *C neoformans* appears to be a determinant of virulence, although the role of this enzyme in virulence is unknown. The infectious propagules of *H capsulatum*, *B dermatitidis*, *P brasiliensis*, and *C immitis* are readily phagocytosed by alveolar macrophages. To survive phagocytosis and to multiply, these fungi must neutralize the effects of the phagocytes. The production of reactive oxygen metabolites by phagocytic cells is an important host defense against microorganisms. Studies have shown that the yeast phase of *H capsulatum* fails to trigger release of reaction oxygen metabolites in unprimed murine macrophages despite extensive phagocytosis.

How they avoid destruction by the fungicidal mechanisms within lysosomes is unclear. Arthroconidia of *C immitis* inhibit phagosome-lysosome fusion and survive within normal murine peritoneal macrophages. Phagosome-lysosome fusion takes place after *H capsulatum* infection, but the yeast cells survive in the phagolysosome. It has been speculated that the fungus neutralizes the fungicidal components of the lysosome by a mechanism not yet elucidated.

There is very little information about mechanisms of fungal pathogenicity, in contrast to what is known about molecular mechanisms of bacterial pathogenesis. Fungal pathogenesis is complex and involves the interplay of many factors. Studies to elucidate these mechanisms are needed because of the increasing incidence of opportunistic infections.

DAMPING OFF DISEASE OF SEEDLINGS

Damping off of seedling is very common all over the world. It occurs every where, in agricultural and forest soils, in tropical and temperate climates and in almost every greenhouse. Damping off can be of two types, that is, pre-emergence damping off in which the seed and radicle rot before the seedlings grows out of the soil and the post-emergence damping off in which the newly emerged seedlings are killed at ground level after they emerge from the soil, causing them to collapse or topple over.

This is a common symptom of post-emergence damping off. This disease is most noticeable in nursery beds, greenhouse flats and row crops because symptoms develop suddenly. killing large number of seedlings in local areas. Among the pathogens in this class are 1 he species of *Pythium, Phytophthora, Corticium* and *Fusarium;* the former two will form the subject of our present discussion. Most species of *pythium* mainly affect the juvenile or succulent tissues. This restricts their parasitism to seedlings, the feeder roots or root tips of older plants, and to watery fruits or stem tissues.

They are also cause of fruit rots of crops such as beans, squash and watermelons. At a later stage, when the cells of stems and main roots have developed secondary thickenings, infection is restricted to feeder roots. This causes seedlings to become stunted and chlorotic. Peach decline, which kills thousands of trees each year, has been shown to be a feeder root problem caused by a combination of factors, including *Pythium* spp. Pectolytic and celiulolytic enzymes produced by *pythium* and *Phytophthora* spp. play an important role in tissue disintegration and maceration.

Symptoms

As has already been said, damping off of vegetable seedling occurs in two stages, that is, the pre-emergence and the post-emergence phase. In the pre-emergence phase the young seedlings are killed before they reach the soil surface.

The young radicle and plumule are killed and there is complete rotting of the seedling. The post-emergence phase is characterized by infection of the young tissues of the collar at the ground level. The infected tissues become soft and water soaked. The seedlings collapse. Such symptoms are very common in seed beds in nurseries.

Morphology of the Causal Organism

The mycelium of *P. debaryanum* is branched, hyaline, and coenocytic hyphae, the main hyphae being up to 5 μ. in diameter. The ends of the blunt terminal hyphae penetrate the cell walls of the hypocotyl and ramify within and between the tissues of the cortical parenchyma; sporangia and oospores are formed in the parenchymatous tissues of the host.

Sporangia are terminal or intercalary and are 15-26 μ in size. They are spherical, oval or barrel shaped depending upon their position.

Just before germination, a prominent beak is formed on the sporagium, followed by a thin-walled vesicle into which the sporangial contents are emptied. Zoospores are differentiated in the vesicle. The zoospores are reniform, with two lateral flagella. They measure up to 8 μ in diameter.

Sexual reproduction is oogamous, oogonia spherical and terminal on the short side branches with a smooth and thin wall, 15 to 28 μ antheridia monoclinous or diclinous, one to six per oogonium; oospores smooth. aplerotic 12 to 20 μ Among other common species of *Pythium* are *P. arrhenomanes, P. aphanidermalum* and *P. ultimum* These are characterized below:

P. arrhenomanes-Sporangia inflated, filamentous, lobulate, zoospores about 12 μ in diameter. Oogonia subspherical to spherical, usually acrogenous, smooth, thin-walled, 17-56 μ, antheridia diclinous, 15 to 25 per oogonium, oospores plerotic, 15 to 54 μ. This species is responsible for seedling blight pre-emergence rotting and root rot of maize, wheat, sugar cane and other Gramineae.

P. aphanidermatum-Sporangia inflated. lobulate branched or unbranched, zoospores 7 by 12 *u*, oogonia spherical, terminal, 22 to 27 μ, antheridia monoclinous or diclinous, intercalary or terminal, one or two per oogonium, dome-shaped, 9 to 11 by 10 to 14 μ.

oospores aplerotic, single, 17 to 19 μ. Associated with soft rot of storage organs and damping off of many host species. It is responsible for cottony leak of cucurbit fruits in storage, cottony blight of turf grasses, root and stalk rot of maize, stem or foot rot of papaya, rhizome rot of ginger, rhizome rot and root rot of turmeric. Other hosts are sugarbeet, tobacco, sugarcane, pineapple, bean and cotton.

P. ultimum-Sporangia mostly terminal and spherical, with a diameter varying from 28 μ and more to 12 *u.* and less, but occasionally intercalary and barrel-shaped, with the dimensions varying from 27.8 μ × 22.9 μ to 17 μ × 14 μ antheridia monoclinous, arising just below the oogonium, usually one per oogonium; oospores aplerotic. single, spherical, 14 to 18 μ. It is a common causal organism of damping off, root rot and soft rot of many species.

Disease Cycle and Predisposing Factors

Pythium spp. are natural inhabitants of the soil, where they exist as a low grade parasites on fibrous roots. The oospores serve as the overwintering or oversummering organs.

They are not vigorous competitor and their saprophytic activities are greatly restricted. Soil moisture is important in saprophytic growth of *Pythium* spp. and survival by resistant structures is more important than saprophytic persistencc.

Factors affecting infection include inoculum density, soil moisture, soil temperature, *pH,* cation composition, light intensity and the presence of members of other microorganisms. Soil temperature and moisture are the most important factors.

Control

Most *Pythium* spp. produce oospores and chlamydospores and persist for many years. Direct control of *Pythium* spp. on field scale is difficult and expensive. On a small scale, *Pythium* spp. can be eliminated from the soil for greenhouse use by heating. The common method is by steam treatment or pasteurization. The latter method is preferred as it does not change the microflora drastically.

Fumigation with chloropicrin or methyl bromide is a standard practice in many nursery and horticultural operations.

Very often the best measure against damping off of fungi involves the use of seed protectants to keep away the pathogens in the pre-emergence phase and to adopt sanitary precautions in the nursery to check post-emergence damping off.

The most commonly used seed protectants are organomercurial fungicides, such as Ceresan or Arasan, Phygon and Agrosan GN, Captan, Ferbam, Thiram and Zineb have also been successfully used for the control of damping off fungi. Drenching the soil with suitable fungicides, such as Cheshunt compound, 1% Bordeaux mixture, 0.1% Ceresan, 0.3% Blitox -50 or 0.2% Esso Fungicide 406 has been found useful in eliminating soil-borne infection.

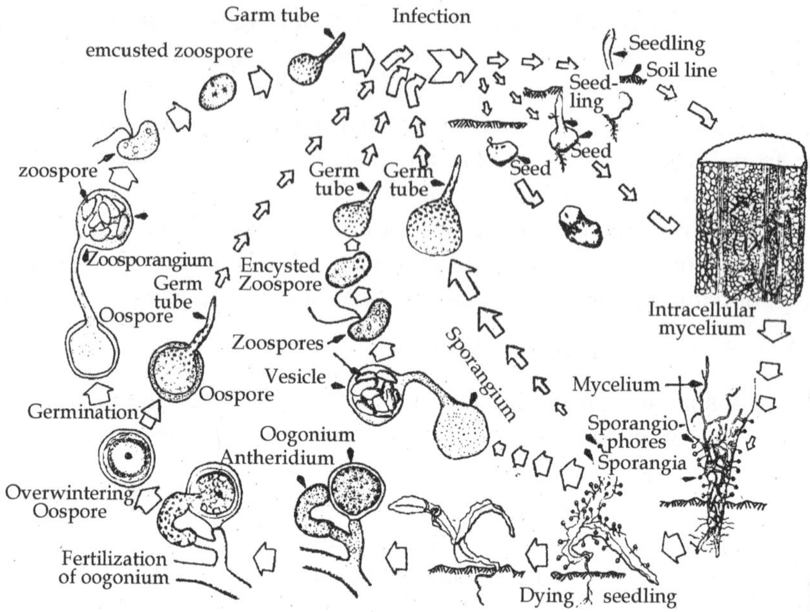

Figure. Disease cycle of damping-off and seed decay caused by *Pythium* sp.

Dexon and other soil fungicides, applied at low rates, have been successfully used to retard the rate of reinfestation of fumigated soils by *Pythium.* The use of healthy seeds in crops (ginger and turmeric) which are vegetatively propagated by rhizomes, etc. forms an important control measure in such diseases. Crop rotation is another method of reducing the populations of soil pathogens, but because the more common and destructive *Pythium* spp. have a wide host range, this practice is not of much help.

SUPERFICIAL FUNGAL INFECTIONS

Superficial fungal infections involve only the outermost layers of the stratum corneum of the skin (*Phaeoannellomyces werneckii* [syn. *Exophiala werneckii*] and *M furfur*) or the cuticle of the hair shaft (*Trichosporon beigelii* and *Piedraia hortae*). These infections usually constitute cosmetic problems and rarely elicit an immune response from the host (except occasionally *M furfur* infections). Recently *T beigelii* and *M furfur* were implicated as opportunistic agents of disease, particularly in immunosuppressed or otherwise debilitated patients. Patients are accidentally infected with these common organisms via indwelling catheters or intravenous lines. Virtually nothing is known concerning the pathogenic mechanisms of these fungi.

DERMATOPHYTE INFECTIONS

The dermatophytes are fungi that colonize skin, hair, and nails on the living host. These fungi possess greater invasive properties than those causing superficial infections, but they are limited to the keratinized tissues. They cause a wide spectrum of diseases that range from a mild scaling disorder to one that is generalized and highly inflammatory. Studies have shown that the disease-producing potential of these agents depends on various parasite and host factors, such as the species of organism, immunologic status of the host, type of clothing worn, and type of footwear used. Trauma plays an important role in infection.

These organisms gain entry and establish themselves in the cornified layers of traumatized or macerated skin and its integument and multiply by producing keratinase to metabolize the insoluble, tough fibrous protein. The reason why these agents spread no deeper is not known, but it has been speculated that factors such as cell-mediated immunity and the presence of transferrin in serum inhibit fungal propagation to the deeper tissue layers and systemic disease does not occur.

Some dermatophytes have evolved a commensal relationship with the host and are isolated from skin in the absence of disease. Little is known about specific pathogenic mechanisms of the dermatophytes, but they do not cause systemic disease.

THE DISSEMINATION OF FUNGI

A few fungi are disseminated principally in the mycelial state as, for instance, the tropical thread blights which blow about with the leaves to which they adhere; the 'Corticium' disease or 'red thread', a common trouble of lawns and sports greens in Britain, produces on the leaves a mycelium of coral-like threads which adhere to the tips of the dying leaves. Others are distributed as vegetative propagants such as the sclerotia and stromatic crusts of Rhizoctonia solani, of which the sporing form, *Corticium solani,* plays little part in spreading the fungus. The long-distance dissemination of fungi is, no doubt, often to be traced to the transference of viable mycelium in plant produce from one country to another; this has happened in elm disease and very probably in potato blight and with many other fungi.

Usually, however, the spores are the organs of dissemination. As they are very small and so light as easily to be caught up into the air as an impalable dust, the wind is the chief agent in their dispersal; there are, however, other important methods of spread.

The spores of fungi may retain their vitality for considerable periods and they show amazing powers of resistance to extremes of heat and cold: sealed in vacuum tubes, slowly dried spores of Mucor and Aspergillus have withstood a temperature of liquid air for nearly 500 hours, followed by immersion of the tubes in liquid hydrogen for 77 hours. Susceptibility to heat varies greatly, for some spores are killed in 10 to 15 minutes at temperatures far below the boiling point, whereas occasionally some can stand the direct action of boiling water or live steam for the same length of time.

So also the spores of some species are killed by drying for 24 hours or even less, while others can be best preserved in a dry condition. As a rule spores are less sensitive than hyphae to adverse surrounding conditions, and vegetative growth is often inhibited at, or near, the freezing point, as is evident from the action of refrigeration in checking mould growth; slow growth on frozen foodstuffs, however, has been recorded, and very slight superficial moldering of meat will occur at a temperature of – 6° C. Unless special percutaneous are taken, moldering as a result of spore germination quickly supervenes on removal from cold storage. These resistant properties of spores are of value in facilitating the aerial dissemination of many fungi, for desiccation in the air and extremes of beat and cold on the ground may have to be faced.

The vertical elongation of the sporophore is often a powerful aid to exposing the spores freely to air currents. Amongst other adaptations which seem to have the same object, the forcible projection of spores from their stalks is often important. When enclosed in a receptacle there are sometimes arrangements for expelling the spores actively into the air.

These are often brought into action by changes in temperature or dryness, as in some Ascomycetes where' puffing' can be induced by concentrating the sun's rays through a pocket~ lens on the.receptacle. Some of the Discomycetes respond similarly if breathed upon when ripe. In the Basidiomycetes it is usual to find a mechanism for the violent release of the basidiospores based on the secretion of a drop of fluid at the junction of the spore and its sterigma. For full advantage to be gained from these various adaptations it is necessary that the spores should be dry and free from mucilaginous envelopment.

In a considerable number of species, including many fungi whose spores are developed in acervuli, the spore beds are surrounded by a gummy secretion. A great many of the bacterial plant pathogens also are immersed in mucilage as they are extruded to the surface of the host plant. When dry, the secretion tends to harden into flakes or sheets, but these are readily soluble in water, as in 'halo blight' of French beans, due to a bacterium.

Dissemination may be effected by the blowing about of the dried gummy shreds, followed by the setting free of the individual cells in rain drops, as sometimes occurs in the American fire blight of fruit trees, due to Bacterium amylovorum, or in *Xanthomonas malvacearum* in cotton, though the latter is often also carried as dust from adjacent cotton fields infected the previous season. The spore bed may be exposed to rain splashing which carries the spores to leaves or stems in the vicinity; species of *Colletotrichum* are often disseminated in this way.

Amongst the Hyphomycetes it is possible to distinguish two main groups according to the slimy or dry character of the spores, and the method of their dispersal depends largely on this character, the dry spores being predominantly suited for distribution by the wind, while the slimy ones are carried by water, by insects, or by adhering to adjacent plants and the like. In a recent survey of the British Hyphomycetes three-fourths of the species belonged to the dry-spored group.

The Coelomycetes, on the other hand, are generally representative of moist sporing fungi. In some of them and in many of the moist-spored Hyphomycetes, the spores are liberated in a mass held together by water or mucilage, and in this form may be subject to aerial dispersal before they are broken up by rain or dew.

Knowledge of the long-distance air-borne dissemination of fungal spores has greatly increased of recent years. Observations on the spread of white pine blister rust in the west of the United States and Canada indicate that it is due to wind-blown infection of currant and gooseberry bushes by aecidiospores from pines that may be hundreds of miles away. Similarly, cereals are infected over great areas by wind-borne uredospores of their rusts. Numerous examinations of the spore-content of the upper air have been made by aeroplane. In early flights, carried out in 1921, spores and pollen grains were found to be fairly abundant up to 11,000 feet, above which they became scarce.

U redospores of *Puccinia graminis* caught at 7,000 feet were germinated successfully. Above 10,000 feet, mould spores, such as those of Aspergillus and *Penicillium,* and bacteria, are most frequently found, occurring at times up to nearly 20,000 feet.

There is circumstantial evidence that the distance to which such spores may be borne in a viable condition during storms is limited mainly by their ability to survive atmospheric conditions. These naturally vary greatly from year to year during the period of spread, but it is reported from Canada that there is rather conclusive evidence that under certain conditions the black rust of wheat has been carried for four or five hundred miles, or even farther, through the air. These spores tend to be deposited in definite showers rather than in a continuous steady fall; in western Canada rust spore-showers occur after a day or two of strong south wind.

The long-distance disstfmination of cereal rusts has been intensively studied. Black rust due to the well-known parasite Puccinia graminis occupies a practically continuous wheat belt in North America from the Gulf of Mexico to the Prairie Provinces in Canada; as the crop is maturing in the south at the same time that it is being sown in Canada, and the intervening areas are more or less intermediate, a regular succession of wheat ready for infection is available from south to north in the spring and summer. The spores seem to follow a main northward drift path from an area in Texas, where the rust over-winters in the uredo stage and the wheat ripens in the late spring, through Kansas in May and June, to the upper Mississippi States.

In Texas itself the summer is intensely hot and the rust is unable to survive until the new crop is sown in the autumn. Reinfection then takes place, apparently chiefly by a reverse flow of spores from the north. So there is an ebb and flow of infection to and from the area in which this fungus over-winters but cannot survive the summer. Confirmatory evidence of this movement is provided by the incidence in particular years of certain physiologic races of the fungus in the areas concerned.

No doubt the spread is not ordinarily carried out in great jumps, though spores not of local origin have been caught as far north as Nebraska in May and June: usually there is time for one or more breaks and the renewal of the inoculum by the infection of local crops in the path from Texas to the northern Mississippi and Canada. Some irregularity of incidence in the upper part of this tract may result from the local over-wintering of the fungus in the teleutospore stage and infection of the alternate host, the barberry, in the spring, for this rust is heteroecious and produces its aecidial stage on barberries, whence it can infect neighbouring wheat and possibly start an epidemic.

But many of the barberries have been destroyed in the barberry eradication campaign and their disappearance in due course will, it is hoped, save much of the chief wheat-growing region in the northern States and Canada from all but the northward spore drift, for the region is much too cold to allow the rust to over-winter in the uredo stage. Attention can then be concentrated on the production of rust-resistant wheats in the comparatively restricted over-wintering area in the south. Nearly all the wheat grown in India is around as a winter crop in the plains.

In these areas rust is unable to survive the intense heat of the summer, and the wheat and barley crops are infected anew each year from rusted cereals grown in the Himalayan foot-hills and some of the hills of Peninsular India. Barberries occur in these hills but.not in the plains, and though the aecidial stage of Puccinia graminis has been found on several of the Himalayan species, it seems to play little part in the perpetuation of the rust.

The uredo stage is abundant when the hill crops are harvested in May and June, and then persists on stubble shoots and volunteer plants until the new hill crops are sown in the autumn. Areas have been found in the Central Himalayas where the early crops are sown in August-September and may be heavily rusted by the first week of December. Inoculum is thus available in time to account for the first outbreaks in the Indo-Gangetic Plain, which usually occur in January-February.

The uredospores have been trapped in the air by means of specially devised aeroscope slides we use before the rust has been found on the crops in the plains. In Peninsular India similar foci of early infection of the growing hill crop have been found and seem to account for extensive distribution of the rust in Madras and the Deccan. A similar chain of infection serves to explain the distribution of wheat brown rust, whereas Puccinia glumarum, the cause of yellow rust, is more sensitive to heat and its uredosori are not ordinarily found in the summer below about the 6,000 feet level in the Himalayas, though it may be the only rust seen above 7,000 feet in N ovember- December.

The long-distance dissemination of cereal rusts in several parts of the world is effected on the same broad lines. Thus the wheat in the Amur region of eastern Siberia gets its inoculum from over-wintering areas in Manchuria; North Africa seems to be infected from Europe, the Balkans from Asia Minor, and the Argentine from Brazil. So to a great extent has been cleared up one of the greatest mysteries with which plant pathologists in certain parts of the world were faced before it was known that spores could be carried over long distances in the upper air.

Such long-distance spore flights are evidently limited to certain types of spores and of prevailing meteorological conditions, at least so far as the parasitic fungi are concerned. If they were general, the object of the plant quarantine services maintained by most countries would be defeated. Heavy wind-drifts or storms no doubt play an important part in them, as does the presence in a receptive condition of a susceptible crop in the area reached.

It is known, furthermore, that sporecontaminated air not far from the ground purifies itself by the fall of the spores at rates varying from o. 5 to 5 mm. per second, while their presence over the ocean has been found to diminish with increasing distance from the land and catches taken on shipboard may fall to nil in individual voyages fairly soon after reaching the open sea; various favourable conditions must coincide to permit successful long-range spread, and they do not seem often to occur.

There is still a great weight of evidence from the known facts of the spread of parasitic fungi into new areas in support of the view that' discontinuous' dissemination is usually effected by transport on or in the host plant, and is linked with the movement of nursery stock or other plant produce along recognised trade routes. The discovery of spore-passage in the upper air and its bearing on the dissemination of the rusts shows that wind-borne distribution can be an important factor in certain circumstances and must always be regarded as a possible menace by those responsible for the protection of crops against foreign disease.

It is also ordinarily accepted as responsible for much of the' continuous' dispersal of fungi within any given land area. But sea-barriers and sometimes mountain ranges have proved effective checks to the dissemination of many plant parasites. Either the spores are unable to survive long passages in the air, or the timing of their flight is such that a suitable host in a receptive condition is not available when the new area is reached, or the meteorological conditions are adverse, or other causes intervene to spoil a safe lodgement.

Many favourable circumstances must combine to render the invasion successful, and experience suggests strongly that they seldom do. Supporting evidence can be found in the history of the spread of American gooseberry mildew in the British Isles and on the continent of Europe; of potato wart disease in many parts of the world, though the resting sporangia of this fungus might be expected to reach the air with dust from infected fields; of blister blight on tea in India; of cacao witches' broom in South America and the West Indies; of several parasites of trees and of ornamental plants, and of many others.

In general it may be said that if human agency be excluded, dispersal within a new area is gradual and continuous, each new generation extending the area by a rather limited distance.

This happened when the, vine mildews and that of the oak were introduced into Europe from America, though the latter at least seems to have had some jumps of about 100 miles, and also with several diseases such as chestnut blight introduced into America from abroad.

Some spores are disseminated largely with the seeds of the plants on which they grow. This is the chief means of spreading bunt of wheat and various other diseases against which seed disinfection is practised.

Soil fungi are carried by irrigation water or surface-wash after heavy rain and are also liable to be distributed by farm implements, or plough cattle, or on the feet of farm workers.

The dissemination of the spores of certain fungi by insects has been noticed by botanists for very many years. Basidiomycetes belonging to the Phalloid or stinkhorn group were particularly marked examples, for their strong and penetrating odour attracts carrion and other flies, and the special adaptation of this group for insect dissemination was early suggested and subsequently fully established.

The sweet honey-dew in which the spores of some fungi are extruded has also long been known to attract insects. These visit the 'Sphacelia' stage of Claviceps purpurea, the cause of ergot of rye, and carry spores away not only as an external contamination but also in the alimentary tract, through which they can pass uninjured. The more recently discovered function of insects in securing the fertilizations of the rusts has been mentioned above.

Naturally, however, it is the role of insects in the spreading of parasitic diseases that has attracted most attention in modern times, and its importance is now fully recognised. Insects are the plain agents of dissemination of virus diseases.

They also play a considerable part in the spread of bacterial diseases, such as fireblight in orchards, where they are attracted to the gummy exudate from twig cankers in the spring and carry the bacterium-laden matter to the blossoms and young growth of healthy apple and pear trees; the honey-bee is considered to be one of the chief agents in initiating the blossom blight form of infection, and the discovery that the organism is able at times to persist in the hive was followed by an embargo being placed on honey from infected countries in some places still free from the disease.

Some of the insects concerned are able to inoculate healthy trees by puncturing the young growth. The organism that causes bacterial wilt or Stewart's disease of maize, *Xanthomonas stewarti*, has been found to over-winter in hibernating *Chaetocnema pulicaria* flea beetles, the chief vectors of the disease in the United States, and to be inoculated into the growing crop by these insects in the spring. Cotton stainers of the genus Dysdercus, and some other insects, are responsible for the introduction of the minute fungi *Nematospora* and its allies into young cotton bools, where they induce internal boll disease; the insects live in between successive cotton crops on various plants and trees, of which the baobab is one of the most important in Africa, but it is not yet known how or where they harbour the fungus, though the evidence suggests that the association is close.

In many cases insects that cause wounds, such as those that bore into stems or gnaw the softer parts, carry parasitic bacteria or fungi on their bodies and serve to spread infection. The Dutch elm disease is mainly disseminated by bark-boring beetles of the genus *Scolytus,* and the watermark disease of willows seems to be spread by several different kinds of insects.

The black rot, and soft rot of cruciferous plants, due to Xanthomonas campestris and Bacterium carotovorum respectively, are also carried passively by insects and other small animals such as slugs. It is easy to demonstrate how readily some of these organisms are picked up by allowing an insect from a heavily diseased plant to pass across the solid surface of some suitable nutrient preparation and observing the colonies of micro-organisms that subsequently develop along the insect's track.

FUNGICIDES AND GERMICIDES FOR PLANT DISEASES

Plant diseases are caused by parasitic fungi or by bacteria, or other vegetable parasites; or by forms of physiological disturbance. Each disease calls for special treatment. Most plant diseases must be prevented, not cured. It should be understood that spraying is only one of the control measures effective against plant diseases. Many diseases are not affected by spraying, though perhaps more are susceptible to this treatment than to any other.

A satisfactory fungicide must be one that does not injure the plants and at the same time is effective against the parasite. For spraying, additional requirements are imposed; it should not dissolve readily in rain water; it should adhere to foliage and fruit; in some cases it should be colorless in order not to make ornamentals more unsightly than when diseased.

The fungicide which has been used most for general purposes is bordeaux mixture. Lately some other preparations, particularly lime-sulfur combinations, have come into use, and in many cases are supplanting bordeaux. There are in addition a large number of other substances which have fungicidal value and are in more limited use for specific cases.

DESTROYING AFFECTED PARTS

It is important that all affected parts should be removed and burned, if possible. In the fall all leaves and fruit that have been attacked by fungi should be raked up and burned. Diseased branches should be severed at some distance below the lowest visible point of attack. Fungous diseases often spread rapidly, and prompt action is usually necessary. Practice clean and tidy culture.

ROTATION OF CROPS

One of the most effective and practical means of heading off fungous diseases. Especially applicable to diseases of roots or root-crops, but also to many other diseases of annual plants.

STERILIZING BY STEAM

An effective fungicidal practice for several soil-inhabiting organisms which attack roots and stems. This includes nematode worms. It is especially applicable in the greenhouse, where it may be applied (a) through sub-irrigation tile or through specially laid perforated steam pipes in the bottom of the bed. Cover the beds with blankets, introduce steam under pressure of 40 to 80 pounds for two hours. Insert thermometers at various places to see that the soil is being uniformly heated. (6) A large galvanized iron tight box may be constructed with finely perforated trays 4 to 6 inches in depth. Soil placed in these trays and steamed for two hours as above will be freed from parasitic organisms. In this case the frames should be sprayed with a solution of formalin, 1 pint in 10 gallons of water.

Steam sterilization of soil may be used on intensively cultivated areas or extensive seed-beds. A portable boiler is necessary. The beds are sterilized after they have been prepared for seed, and just before the seed is sown. A galvanized pan 10 by 6 feet and 6 inches deep is inverted, and the edges are pushed down into the soil one or two inches. The pan is connected with the steam boiler by means of a steam hose and live steam is run into the pan from about forty minutes under a pressure of 100 pounds and up. The higher the pressure the more thoroughly the soil will be sterilized.

The cost of sterilizing is approximately three-fourths of a cent the square foot. It should be noted that soil sterilization has an invigorating effect on the plants, and it will be necessary to run greenhouses at a lower temperature (5°–10°) both night and day. Field sterilization also kills weed seeds, and with the reduction of the cost of weeding makes the process practicable.

BORDEAUX MIXTURE

A bluish-green copper compound that settles out when freshly slaked lime and a solution of copper sulfate are mixed. Many formulas have been recommended and used. The 5-5-50 formula may be regarded as standard. In such a formula the first figure refers to the number of pounds of copper sulfate, the second to the stone or hydrated lime, and the third to the number of gallons of water. Bordeaux must often be used as weak as 2–2–50, on account of injury to some plants.

To make 50 gallons of bordeaux mixture, proceed as follows:

- Pulverize 5 pounds of copper sulfate (blue vitriol), place in a glass, wooden, or brass vessel, and add two or three gallons of hot water. In another vessel slake 5 pounds of quicklime in a small amount of water. When the copper sulfate is all dissolved, pour into a barrel and add water to make 40 or 45 gallons. Now strain the lime into this, using a sieve 50 meshes to the inch or a piece of cheese-cloth supported by ordinary screening. Stir thoroughly, and add water to the 50-gallon mark. The flocculent substance which settles is the effective fungicide. Always stir vigorously before filling the sprayer. Never add the strong lime to strong vitriol. Always add a large amount of water to one or the other first. Blue vitriol used alone would not only wash off quickly in a rain, but cause a severe burning of fruit and foliage. Lime is added to neutralize this burning effect of the copper. If the lime were absolutely pure only slightly more than one pound would be required to neutralize this burning effect. For many purposes an excess of lime is not objectionable and may be desirable. For nearly ripe fruit and ornamentals an excess of lime augments spotting. In such cases the least amount of lime possible should be used. Determine this by applying the cyanide test.
- Secure from the druggist 10 cents' worth of potassium ferrocyanide (yellow prussiate of potash) and dissolve it in water in an 8-ounce bottle. Cut a V-shaped slit in one side of the cork, so that a few drops of the liquid can be obtained. Now proceed as before. Add lime with constant stirring until a drop of the ferrocyanide ceases to give a reddish-brown color.
- When bordeaux mixture is desired in large quantities, stock solutions should be made. Place 100 pounds of copper sulfate in a bag of coffee-sacking, and suspend in the top of a 50-gallon barrel, and add water to the 50-gallon mark. In twelve to fifteen hours the vitriol will be dissolved and each gallon of solution will contain 2 pounds of copper sulfate. Slake a barrel of lime, and store in a tight barrel, keeping it covered with water. Lime so treated will keep all summer. It is really hydrated lime. This is often dried, pulverized, and offered on the market in paper bags of 40 pounds, each, under such names as ground lime, prepared lime, hydrated lime, etc. If the paper is not broken, the lime does not air-slake for a long time. One and one third pounds of hydrated lime equals in value one pound of quicklime. Air slaked

lime cannot be used in preparing bordeaux mixture. Arsenical poisons can be combined with bordeaux mixture.

AMMONIACAL COPPER CARBONATE

For use on nearly mature fruit and on ornamentals. Does not discolor. Weigh out 3 ounces of copper carbonate, and make a thick paste with water in a wooden pail. Measure 5 pints of strong ammonia (26° Beaume) and dilute with three or four parts of water. Add ammonia to the paste, and stir. This makes a deep blue solution. Add water to make 50 gallons.

COPPER CARBONATE

For use in the above formula, it may be obtained as a green powder, or may be prepared as follows: Dissolve 12 pounds of copper sulfate in 12 gallons of water in a barrel. Dissolve 15 pounds of sal soda in 15 gallons of water (preferably hot).

Allow the solution to cool; then add the sal soda solution to the copper sulfate solution, pouring slowly in order to prevent the mixture from working up and running over. A fine precipitate is formed which will settle to the bottom if allowed to stand over night. Siphon off the clear liquid. Wash the precipitate by adding clear water, stirring, and allowing to settle. Siphon off the clear water, strain the precipitate through muslin, and allow it to dry. This is copper carbonate. The above amounts will make about 6 pounds.

CORROSIVE SUBLIMATE (MERCURIC BICHLORIDE)

Used for disinfecting pruned stubs and cleaned-out cankers, at the rate of one part in 1000 parts of water. Can be secured from the druggist in tablet form in vials of 25 each, and costing 25 cents: One tablet makes a pint of solution. Make and store solution in glass and label poison.

FORMALIN

A pungent, clear liquid, very irritating to eyes and nose. Obtained at any drug store at about 40 cents per pint. Used for potato-scab, oat smut, bunt in wheat, soil disinfection, etc.

LIME

Offered for sale in the following forms:

- Ground rock or ground limestone; air-slaked lime is of the same composition, i.e. a carbonate of calcium.
- Lump, barrel, stone, or quick lime; this is burned limestone, and should preferably test 90 per cent oxide of calcium.
- Prepared, ground, or hydrated lime; this is water or steam-slaked quicklime, dried and pulverized.

Used as an applicant to the soil to correct acidity, for club-root of cabbage, etc., and for preparing spray mixtures.

LIME-SULFUR

In the many possible combinations, lime-sulfur is coming to be equally as important as bordeaux mixture, in the control of many plant diseases.

- A mixture of equal parts of dry lime and powdered sulfur is often dusted on plants for surface mildews.
- A paste of equal parts of lime, sulfur, and water. This is painted on the heating pipes in the greenhouse, and is valuable for keeping off surface mildews.
  - Home-boiled dilute lime-sulfur. This solution has been widely used in the past as a dormant spray, particularly for San Jose scale and peach leaf-curl. It is likely to be supplanted.
  - Home-boiled concentrated lime-sulfur.— When a great deal of spraying is to be done, a concentrated lime-sulfur solution may be boiled at home and stored in barrels to be used as needed.

Test with a Beaume hydrometer, which has a scale reading from 25° to 35°. Dilutions are reckoned from a standard solution testing 32°. If the solution tests only 28°, it is not as strong as standard, and cannot be diluted as much as a solution testing 32°.

Decimals are given in ail cases, but for practical purposes the nearest even gallon or half gallon can be used, unless appliances for more accurate measurement are at hand. It is understood in making all dilutions that water is added to one gallon of the concentrate to make the stated amount. Do not measure out the stated amount of water and add the concentrated solution to it.

	1-10	1-15	1-20	1-25	1-30	1-40	1-50	1-60	1-75	1-100
25°	7.4	11	14.7	18.4	22.1	29.5	36.8	44.2	55	73
26°	7.7	11.6	15.4	19.3	23.2	30.9	38.6	46.3	58	77.2
27°	8.1	12.1	16.1	20.2	24.3	32.4	40.5	48.5	60.6	80.7
28°	8.4	12.7	16.9	21.1	25.4	33.8	42.3	50.7	63.5	84.5
29°	8.8	13.2	17.6	22.1	26.5	35.3	44.2	53	66.3	88.2
30°	9.2	13.9	18.4	23	27.6	36.9	46.1	55.3	69	92
31°	9.6	14.4	19.3	24	28.8	38.4	48	58	72	96
32°	10	15	20	25	30	40	50	60	75	100
33°	10.4	15.6	20.8	26	31.2	41.5	52	62.4	78	104
34°	10.8	16.2	21.6	26.8	32.4	43.2	54	64.7	80.8	108
35°	11.2	16.8	22.4	28	33.4	44.9	56	67.4	84.2	112

- Commercial concentrated lime-sulfur. - As manufactured and placed on the market is a clear amber liquid, and should test 32° to 35° Beaume It costs about 20 cents per gallon retail, and comes ready to pour into the spray tank. For apple and pear diseases. Arsenate of lead can be used with this solution, and increases its fungicidal value.

- Self-boiled lime-sulfur. This is a mechanical mixture of the two substances, and is really not boiled, the heat being supplied by the slaking lime. In a small barrel or keg place 8 pounds of good quicklime. Add water from time to time in just sufficient amounts to prevent burning. As soon as the lime begins to slake well, add slowly (preferably through a sieve) 8 pounds of sulfur flour. Stir constantly, and add water as needed. As soon as all bubbling has ceased, check further action by adding a quantity of cold water, or pour into a barrel or tank and make up to 50 gallons. Keep well agitated. Very effective against peach scab and brown rot. Several other formulas have been used: 10–10–50 and 5–5–50. Arsenate of lead can be used with this mixture.

By using boiling water and allowing the hot mixture to stand for half an hour, a stronger spray mixture of the above can be secured. It cannot be used safely on peaches, but has been used successfully on grapes for surface mildew. The addition of sulfate of iron or sulfate of copper, one or two pounds to 50 gallons, has been used for apple rust.

POTASSIUM SULFID (LIVER OF SULFUR)

Simple solution 3 ounces in 10 gallons of water. For mildew in greenhouses, on rose-bushes and other ornamentals.

RESIN-SAL-SODA STICKER

Resin, 2 pounds; sal soda (crystals), 1 pound, water, 1 gallon. Boil until of a clear brown color, i.e. from one to one and a half hours.

Cook in an iron kettle in the open. Add this amount to 50 gallons of bordeaux. Useful for onions, cabbage, and other plants to which spray does not adhere well.

SULFATE OF COPPER (BLUE VITRIOL)

Dissolve 1 pound of pure sulfate of copper in 25 gallons of water. A specific for peach leaf-curl. Apply once before buds swell in the spring. Cover every bud. For use in preparing bordeaux mixture. Costs from 5 to 7 cents per pound, in quantity.

SULFATE OF IRON (COPPERAS)

A greenish granular crystalline substance. Dissolve 100 pounds in 50 gallons of water. For mustard in oats, wheat, etc., apply at the rate of 50 gallons per acre. Also for anthracnose of grapes as a dormant spray.

SULFUR (GROUND BRIMSTONE, SULFUR FLOUR, FLOWERS OF SULFUR)

Should be 99 per cent pure. Valuable for surface mildews. Dust on dry or in the greenhouse used in fumes. Evaporate it over a steady heat, as an oil stove, until the house is filled with vapor. Do not heat to the burning point, as burning sulfur destroys most plants. To prevent burning, place the sulfur and pan in a larger pan of sand and set the whole upon the oil stove.

METABOLITES AND ENZYMES FOR FILAMENTOUS FUNGI

Filamentous fungi have been used as sources of metabolites and enzymes for centuries. For about two decades, molecular genetic tools have enabled us to use these organisms to express extra copies of both endogenous and exogenous genes. This review of current practice reveals that molecular tools have enabled several new developments.

But it has been process development that has driven the final breakthrough to achieving commercially relevant quantities of protein. Recent research into gene expression in filamentous fungi has explored their wealth of genetic diversity with a view to exploiting them as expression hosts and as a source of new genes.

Inevitably, the progress in the 'genomics' technology will further develop high-throughput technologies for these organisms.

CHROMOBLASTOMYCOSIS

A chronic, localised infection infection of subcutaneous tissues caused by several species of dematiaceous fungi.

The 3 most common agents are:

1. Fonsecaea pedrosoi
2. Cladosporium carrionii and
3. Phialophora verrucosa

These fungi, recognised by a variety of names, are saprobes located in soil and decaying vegetation. The route of entry is usually by trauma.

The lesions are sub-cutaneous and the surface can be flat or verrucous. The lesions take several years to develop. These organisms are called dematiaceous fungi, because they have a black colour in the mycelium cell wall (in culture and in tissue). In tissue these fungi form sclerotic bodies which are the reproductive forms dividing by fission.

These organisms induce a granulomatous reaction. The etiologic agents of chromoblastomycosis are septate, mold-like, branching, darkly pigmented which produce asexual fruits called conidia.

We identify these fungi in culture by the shape and formation of the conidia. The fungi have a world-wide distribution especially in warmer climates like the tropics or the southern U.S. The melanin in the pigment may be a virulence factor. These organisms are distributed world-wide. There is no really successful therapy. Excision and local heat have been used with some success. Flucytosine (5-FC), thiabendazole and itraconazole have also been used to treat (or control) this disease. There are no serological tests to aid in the diagnosis.

MYCETOMA (MADUROMYCOSIS)

Mycetomas (fungous tumors) are also chronic, subcutaneous infections. These are called eumycotic mycetoma (tumors caused by the TRUE fungi as opposed to those caused by actinomycetes). These tumors frequently invade contiguous tissue, particularly the bone. A diagnosis of the etiologic agent is essential for patient management because the prognosis and therapy differs.

Mycetoma characteristics:

1. *Tumefaction*: Swelling
2. *Granules*: A variety of colours (white, brown, yellow, black)
3. Draining sinus tracts

The three most common etiologic agents are:

1. Madurella mycetomatis
2. Exophiala jeanselmei
3. Pseudallescheria boydii

Clinical specimens for diagnosis:

1. *Pus*: With granules.
2. *Tissue*: For histological examination.

The colour, size and texture of the granules are an aid in the diagnosis of mycetomas. The agents of mycetoma are all filamentous fungi which require 7-10 days for visible growth on the culture media and then another several days for specific identification. These fungi are identified by the colonial morphology, conidia formation and biochemical reactions. The species of fungi cannot be distinguished in histopathological tissue sections. Treatment is very difficult, but ketoconazole and itraconazole have been used with some success.

ASPERGILLOSIS

Aspergilli produce a wide variety of diseases. Like the zygomycetes, they are ubiquitous in nature and play a significant role in the degradation of plant material as in composting. Similar to Candida and the Zygomycetes, they rarely infect a normal host. The organism is distributed world-wide and is commonly found in soil, food, paint, air vents.

They can even grow in disinfectant. There are more than one hundred species of aspergilli.

The most common etiologic agents of aspergillosis in the United States:

1. Aspergillus fumigatus
2. Aspergillus niger
3. Aspergillus flavus

There are three clinical types of pulmonary aspergillosis:

1. *Allergic*: Hypersensitivity to the organism. Symptoms may vary from mild respiratory distress to alveolar fibrosis.
2. *Aggressive tissue invasion*: Primarily a pulmonary disease, but the aspergilli may disseminate to any organ. They may cause endocarditis, osteomyelitis, otomycosis and cutaneous lesions.
3. Fungus ball which is characteristically seen in the old cavities of TB patients. This is easily recognised by x-ray, because the lesion (actually a colony of mold growing in the cavity) is shaped like a half-moon (semi-lunar growth). The patients may cough up the fungus elements because the organism frequently invades the bronchus. Chains of conidia can sometimes be seen in the sputum.
   - *Culture:* Aspergilli require 1-3 weeks for growth. the colony begins as a dense white mycelium which later assumes a variety of colours, according to species, based on the colour of the conidia. The hyphae are branching and septate. Species differentiation is based on the formation of spores as well as their colour, shape and texture.
   - *Histopathology:* The septate hyphae are wide and form dichotomous branching, *i.e.*, a single hypha branches into two even hyphae, and then the mycelium continues branching in this fashion.
   - *Serology:* There is an excellent serological test for aspergillosis which is an Immunodiffusion test. There may be 1 to 5 precipitin bands. Three or more bands usually indicate increasingly severity of the disease. *i.e.*, tissue invasion.
   - *Treatment:* Itraconazole and Amphotericin B.

ZYGOMYCOSIS

Also known as mucormycosis and phycomycosis. Zygomycosis is an acute inflammation of soft tissue, usually with fungal invasion of the blood vessels.

This rapidly fatal disease is caused by several different species in this class. The zygomycetes, like the Candida species, are ubiquitous and rarely cause disease in an immunocompetent host. Some characteristic underlying conditions which cause susceptibility are: diabetes, severe burns, immunosuppression or intravenous drug use.

The three most common genera causing this clinical entity are:

1. Rhizopus species
2. Mucor species
3. Absidia species

Characteristics: world-wide distribution, commonly in soil, food, organic debris, seen on decaying vegetables in the refrigerator and on moldy bread.

Rhinocerebral infections are common. This disease is frequently seen in the uncontrolled diabetic. Typical case: An uncontrolled diabetic patient comes to ER (may be comatose depending on the state of diabetes) and a cotton-like growth is observed on the roof of the mouth or in the nose.

These are the hyphae of the organism. If untreated, the patient will die within a few hours or days. What do you do to help this patient first? Controlling the diabetic state is most important before administering amphotericin. These fungi have a tendency to invade blood vessels (particularly arteries) and enter the brain via the blood vessels and by direct extension through the cribiform plate. This is why they cause death so quickly.

- *Culture:* A rapid growing, loose, white mold which is visible in 24 to 48 hours. With age, and the formation of sporangia, the colony becomes dark gray. The sporangia contain the dark spores. The mycelium is, wide (10-15 microns), ribbon-like and non-septate (coenocytic). This same appearance is clear in tissue sections. The species are identified by the morphology in culture.
- *Treatment consists of debridement and amphotericin*: There is an immunodiffusion test available, but the physician cannot wait for these results before instituting rapid, vigourous intervention. The diagnosis and treatment must be immediate and based primarily on clinical observations.

FUNGAL DISEASES: DAMPING OFF–CROWN AND ROOT ROTS

Plants in seedbeds may be diseased, either before or after their emergence from the soil, and the disease is called pre- or post-emergence damping off, respectively. In the first case, seedlings do not emerge in patches of the seedbeds. In the second case, plants rot quickly and drop down on the soil. Low temperatures and very wet soils, which delay the growth of the plants, favour infection. A large number of fungi may cause damping off, but *Pythium* spp., *Phytophthora* spp., *Fusarium* spp. and *Rhizoctonia solani* Kühn are the most common. Nowadays, due to the use of improved technology, damping off is no longer a severe disease in greenhouses. However, root rots and crown rots are still destructive in soil, though not in soilless cultures. The most widespread diseases are as follows.

Pythium and Phytophthora Rots

Various *Pythium* spp. and *Phytophthora* spp. may damage the lower part of tomato, pepper, cucumber, carnation, poinsettia, gerbera, etc. both in soil and soilless cultures. In tomato a root and crown rot extending a considerable height above the soil level may occur. The infected area has a dark discoloration and the pith is usually destroyed. *Phytophthora nicotianae* Breda de Haan var. *parasitica* (Dastur) G.M. Waterhouse is the most common pathogen. In pepper a similar disease caused by *Phytophthora capsici* Leonian is very common. Collar, stem and fruit rot as well as leaf spots may occur. In cucumber a soft rot of the young plants at the soil level may occur soon after transplanting. Infected tissues shrink and in wet weather a white mycelium develops. Infected plants wilt and die quickly. Poinsettia grown in pots also suffers from Pythium rot. Severe root rot, extending above ground in succulent plants, and quick death are the main symptoms. In cucumber and poinsettia, *Pythium ultimum* Trow, *Pythium irregulare* Buisman, *Pythium debaryanum* Auct. Non R. Hesse and *Pythium aphanidermatum* (Edson) Fitzp. are mostly involved. Carnations infected by *Pythium* and *Phytophthora* species develop soft rot at the collar and in the root system, resembling Rhizoctonia stem rot

Corky Root Rot of Tomato

The pathogen damages mostly tomato, but also eggplant, melon, etc. Initially tomato leaves turn dull green and growth is stunted. Later, leaves take on a bronze colour and curl downwards. Necrosis of the leaflets follows. Young roots are brown and poorly developed. Scattered lesions appear on the surface of the larger roots which become corky with cracks of different sizes. Yield may be severely reduced. The pathogen survives on the infected root debris due to the presence of minute sclerotia. It is a cool weather disease. In subtropical countries it progresses during the winter and plants start to recover by early spring.

Rhizoctonia Stem Rot (R. solani)

This infects a large number of plants, such as tomato, carnation, poinsettia, etc. causing symptoms resembling Pythium or Phytophthora rots. Rhizoctonia stem rot is mainly confined to the collar. Carnation is very susceptible. Infected plants show pale brown dry lesions, with circular rings, at soil level. Growth is stunted and leaves become dull green. Complete wilting soon follows. Strands of the pathogen develop on the lesions and stems break easily at the infected area. All *Pythium* and *Phytophthora* species as well as *R. solani* are common soil inhabitants. They survive in the soil. Infection usually takes place at the time of planting and symptoms appear very soon in Pythium and Phytophthora rot or several weeks later in Rhizoctonia stem rot *Rhizoctonia solani* may infect at moderate soil moisture levels, but *Pythium* spp. and *Phytophthora* spp. infect only in water-saturated soils.

Crown and Root Rot of Tomato

In plastic greenhouses, a yellowing of the lower leaves appears in infected plants during late winter, when many fruits have already set. In severe infections the whole plant becomes chlorotic and wilts. A dry lesion up to 10 cm long appears on part of or all around the collar. There is a brown discoloration on the root system, predominantly at the end of the main root, the base of the stem and the vascular region of the central root A large number of microconidia, which disseminate the pathogen, appear on the infected stem. The fungus survives by chlamydospores which develop in the soil. The disease is favoured by cool weather.

Black Root Rot of Cucurbits

The disease has been recorded in several countries of northwestern Europe and elsewhere. It infects cucumber and melon, causing a brown rot in the cortical tissue of the root system. Soon a large number of sclerotia develop and the infected tissues turn black. Severely infected plants wither and die. Infection is favoured by cool weather. The pathogen survives in the soil for several years by means of sclerotia.

Control

Effective control of the above-mentioned diseases may be obtained selectively by the following means:

- Use of naturally or artificially suppressive substrates;
- Early drenching by effective fungicides;
- Soil disinfestation;
- Use of resistant cultivars (cvs);
- Grafting on resistant rootstocks;
- Biological control

WILTS: GREENHOUSE CROPS

All major greenhouse crops suffer from one or more wilts. In several crops wilts are the main diseases due to the damage they cause and the difficulty of controlling them.

Fusarium Wilt

The most common Fusarium wilts in greenhouses appear on: tomato [*Fusarium oxysporum* Schlechtend.:Fr. f. sp. *lycopersici* (Sacc.) W.C. Snyder & H.N. Hans.], cucumber (*Fusarium oxysporum* Schlechtcnd.:Fr. f. sp. *cucumerinum* J.H. Owen), melon (*Fusarium oxysporum* Schlechtend.:Fr. f. sp. *melonis* W.C. Snyder & H.N. Hans.), carnation [*Fusarium oxysporum* Schlechtend.:Fr. f. sp. *dianthi* (Prill. and Delacr.) W.C. Snyder & H.N. Hans.], gladiolus [*Fusarium oxysporum* Schlechtend.:Fr. f. sp.*gladioli* (L. Massey) W.C. Snyder & H.N. Hans.], cyclamen (*Fusarium oxysporum* Schlechtend.:Fr. f. sp. *cyclaminis* Gerlach) and chrysanthemum (*Fusarium oxysporum* Schlechtend.:Fr. f. sp. *chrysanthemi* G.M. Armstrong, J.K. Armstrong & R.H. Littrell). Wilt, yellowing, chlorosis, drooping (mostly of the lower leaves), stunting and brown discoloration of the vascular bands up to the top of the stem are the dominant symptoms. Wilting of the lateral shoots and large lesions on the lower part of the stem are also common in Fusarium wilt of carnation and melon. All the above *F. oxysporum* formae have more than one race. Each of them infects cvs of one host but may colonize the root system of other plants as well. They survive in the soil for several years, due to the production of thick-walled chlamydospores, but inoculum is reduced over the years. Fusarium wilt in tomato, watermelon, carnation, cyclamen, chrysanthemum and gladiolus is favoured by higher temperatures than Fusarium wilt in melon.

Verticillium – Phialophora Wilt

This infects a huge number of plants and among them the majority of the plants grown in greenhouses. It is more severe in Solanaceae such as tomato, eggplant and pepper. Of the floral crops, chrysanthemum seems to be more susceptible. Symptoms are very similar to those of Fusarium wilt. Verticillium wilt is favoured by moderate temperatures. *Verticillium dahliae,* which is more common, survives in the soil for many years due to the abundant production of black resistant microsclerotia, while *V. albo-atrum* survives by producing dark dormant mycelium. A similar wilt caused by *P. cinerescens* damages carnations in several areas.

DESTRUCTIVE OF SEVERAL GREENHOUSE CROPS

Powdery mildews are very destructive of several greenhouse crops.

The following are some of the powdery mildew fungi which most attack greenhouse-grown plants:

- On cucurbits *Sphaerotheca fusca* (Fr.) Blumer. [= *Sphaerotheca fuliginea* (Schlechtend.:Fr.) Pollacci], *Erysiphe cichoracearum* DC. and *Leveillula taurica* (Lév.) G. Arnaud (only on cucumber);
- On solanaceous plants L. *taurica* and *Oidium lycopersicum* Cook and Massee (only on tomato);
- On roses *Sphaerotheca pannosa* (Wallr.:Fr.) Lév.;
- On begonia *Microsphaera begoniae* Sivan.;
- On gerbera *E. cichoracearum.*

Powdery mildew fungi, except for *L. taurica,* may attack all green tissues. Initially, white powdery spots, which enlarge and coalesce to cover large areas, are the dominant symptoms. *Leveillula taurica* infects only leaves. Light yellow or yellow-green spots on the upper leaf surface, which later become brown, and scarce white mould on the lower surface are the main characteristics. Powdery mildew-infected plant parts may be chlorotic and distorted.Premature defoliation and poor growth are common features of severely infected plants. Infections take place by conidia. Under favourable conditions powdery mildew progresses rapidly. By the end of the season some powdery mildew fungi, such as *S. fuliginea, E. cichoracearum,* etc. may develop cleistothecia with ascospores, but these do not play an important role in the epidemiology of the disease. Conidia are mostly discharged and transferred by wind currents. Animal pests may also disseminate conidia in greenhouse crops. Young conidia readily germinate on plant surfaces depleted of nutrients.

The relative humidity (RH) favouring infection by powdery mildew fungi and development of the disease differs from species to species. For instance, high RH is more favourable for *S. fuliginea* than for E. *cichoracearum.* Therefore, the first fungus is more frequent on greenhouse cucurbits than the second. High RH may favour spore germination of powdery mildew fungi, but free water may be deleterious. RH at 97–99% is optimal for spore germination of *S. pannosa* and *S. fuliginea.*

At RH below 75% spores of *S. pannosa* do not germinate, but mycelium development and sporulation may occur at RH as low as 21–22%. Powdery mildew fungi overwinter on cultivated plants or weeds, which survive in or outside the greenhouse.

Chemicals such as demethylation inhibitors (DMIs) (triadimefon, fenarimol, etc.), pyrimidines (ethirimol, bupirimate, etc.), pyrazophos and dinocap remain the main means of controlling powdery mildews in greenhouses. Biological control agents have also been effectively tested against *S. fuliginea* and *S. pannosa.* Finally, fully resistant cvs of melon and partially resistant cvs of long-type cucumber are available.

Botrytis Diseases

Lind and Botrytis Gladiolorum

Timmermans are *Botrytis* spp. that most damage greenhouse crops. *Botrytis cinerea* causes grey mould on a large range of hosts, including nearly all the major greenhouse plants. All plant parts at different growth stages may be damaged. Due to the diversity of the infected plant parts, several types of symptoms appear on one or on various hosts. On young stems, leaves, flowers and fruits, initially water-soaked spots occur, which rapidly enlarge under favourable weather conditions.

In tomato fruits green-white circular spots called "ghost spots" also appear. On hard plant parts, such as stems and collars, *B. cinerea* causes cankers and parts above them may die. These symptoms are very common on vegetables such as tomato, eggplant, pepper and cucumber. Infected tissues die soon and a grey mould which consists of conidiophores with clusters of spores develops on their surface.

In plants, like tomato, black sclerotia develop inside the infected stems. *Botrytis cinerea* also causes very characteristic collar rot in lettuce. The infected plants usually develop large brown necrotic lesions on the stem near the soil surface and the lower leaves. The infection gradually progresses upwards. Infected plants may wither and die in a short time.

***Botrytis Tulipae* Causes Tulip Fire Blight**

Spots of various types on leaves and flowers, lesions on the stem, blossom blight and bulb rot are the dominant characteristics. *Botrytis gladiolorum* damages gladiolus and some other Iridaceae. Large spots on leaves and the stem, pinpoint spots on the flowers, neck rot and soft rot of corms are the most common symptoms. *Botrytis* spp. also infect all types of propagating material, which are either destroyed before planting out or become weak plants which may die before or after emergence. Finally, *Botrytis* spp. may cause severe post-harvest losses in plant products during storage or transportation.

Botrytis cinerea develops and sporulates profusely on any organic material. Spores are disseminated by wind over long distances or by water splashes. Healthy plants are infected through wounds, senescent tissues, directly through the epidermis and rarely through stomata. Symptoms may appear very quickly or infection may remain quiescent and symptoms appear later when tissues age or during storage. In greenhouses, initial infection depends on spores transferred from outdoors. Later, the inoculum established in the greenhouse is the main source of infection. In plants grown in non-heated greenhouses, low temperature, high RH and low light intensity, prevalent from late November till late March, create good conditions for infection by *B. cinerea*. Botrytis-incited diseases are prevented by ventilation and heating of greenhouses. Fungicides, mostly benzimidazoles and dicarboxymides, are also used extensively. Nowadays, due to the predominance of resistant strains of the pathogen, they are only marginally effective and growers are advised to combine dicarboxymides with other means of control such as biocontrol preparations. New fungicides have recently been released, but in greenhouses they are used on a limited scale. Formulations of biological control agents such as Trichodex (*Trichoderma harzianum* Rifai T39) are also available.

Downy Mildews

Downy mildews of tomato [*Phytophthora infestans* (Mont.) de Bary], cucurbit [*Pseudoperonospora cubensis* (Berk. & M.A. Curtis) Rostovzev], lettuce (*Bremia lactucae* Regel), rose (*Peronospora sparsa* Berk.) and snapdragon (*Peronospora antirrhini* J. Schröt.) are the most destructive in greenhouse-grown plants. In tomato, leaves and young shoots are infected first. Fruit infection starts mostly near the stalk and spreads very quickly to the whole fruit. Infected tissues of fruits and shoots are firm and brown. In cucurbits downy mildew appears as yellow, angular or circular spots on the upper surface of the mature leaves of the plant.

Soon the tissues at the centre of the spots die and become light brown. Cucumber and melon are more susceptible than watermelon. Downy mildew of lettuce causes scattered light-green to yellow spots on the upper leaf surface. Old spots become brown and dry up. Downy mildew of rose damages all green plant parts, but leaves are more susceptible. Leaf infection resembles the effect of toxins. Infected leaves have purplish red to dark-brown irregular spots and shed readily. Snapdragon plants infected by *P. antirrhini* are stunted and the top intemodes of the young plants are short. The borders of the lower leaves curl down and then dry. Eventually the entire plant dies.

A white fungal growth (brown for cucurbit downy mildew) on the infected tissues under moist conditions is typical of all downy mildews. Plant infection takes place through stomata and mycelium develops intercellularly. Soon branched conidiophores are produced and protrude through the stomata. Infection progresses in the periphery of the spot which gradually enlarges. Conidiospores of downy mildews are ovoid and hyaline, except for *P. cubensis* which are brown. They are discharged by hygroscopic changes and disseminate in greenhouses by wind currents and water splashes. Initial infection may take place by spores transferred long distances on the wind. Abundant oospores of *P. antirrhini* develop on dead plant stems. Oospores of *P. sparsa* also very often develop on infected roses, whereas *P. cubensis* and *P. infestans* oospores are rare.

Phytophthora infestans survives on seed potato tubers and spreads to young potato plants after they have been planted. Inoculum is disseminated from potatoes to neighbouring tomato crops. Cucurbit downy mildew can infect all year round several species of cucurbits, grown either in greenhouses or open fields.

There is evidence that *P. sparsa* survives as a dormant mycelium on the infected stems of roses. *Peronospora antirrhini* perennates as dormant thick-walled oospores in dead plant parts and soil. Free water on plant tissues is necessary for downy mildew fungi to cause infection. High RH is also required for good sporulation. *Peronospora antirrhini* is favoured by low temperature and high RH. Free water or high relative humidity is not often a factor limiting downy mildew development in plastic greenhouses. It seems that temperature is more critical.

For instance, *P. cubensis,* with a high maximum temperature for development and infection, may, under certain conditions, infect all year round, whereas *P. infestans* and *P. sparsa* do not infect during the hot period of the year. Downy mildews complete a cycle within about 6–8 days. Thus, under favourable weather conditions they may have several cycles and spread rapidly.

Chemical fungicides remain the major means of control of downy mildews. Dithiocarbamates, chlorothalonil and the systemic phenylamides (metalaxyl, etc.) are the most commonly used in greenhouses. There are some tomato cvs fully resistant to downy mildew and some partially resistant cucumber cvs suitable for greenhouses, but all rose cvs grown for cut flowers are susceptible to downy mildew. Ventilation of the greenhouses may also effectively prevent infection.

Alternaria Diseases

The following diseases, caused by *Altemaria* spp., seriously affect vegetable and floral crops in greenhouses.

Tomato Early Blight (Alternaria Solani Sorauer)

A collar rot of the young plants before or after transplanting may be the first symptom. In mature plants small irregular brown spots, with or without a yellow halo and concentric rings, appear mainly on leaves. Severely infected leaves are ragged and senescent Similar spots without a yellow ring appear along the stem, leaf stalks, pentucles and the calyx. On fruits, brown to black spots with a leathery surface appear at the stem end. Severely infected plants may be defoliated.

Alternaria Branch Rot and Leaf Spot of Carnation (Alternaria dianthi *Stev. and Hall.*) This mostly infects carnation cuttings during mist propagation and in wet parts of greenhouses. Small purple spots on the leaves are the first symptoms. Soon they enlarge, and their centre turns brown and then black due to the masses of spores which develop. Stem infection usually appears on the knots.

Alternaria Diseases of Minor Importance for Greenhouses

As well as the Alternaria diseases described above, strains of Alternaria alternata (Fr.:Fr.) Keissl. have been recorded:

- Causing cankers in tomato crops;
- Causing leaf spotting in cucumber;
- Causing mostly post-harvest rotting on tomato fruits.

Also *Alternaria cucumerina* J.A. Elliot may on occasion infect cucumber, melon, watermelon and squash. At present, none of them has any economic impact on greenhouse crops. All *Alternaria* species are facultative parasites mostly infecting weak plants.

They survive in the soil on plant debris, but their black spores may also survive on several surfaces in greenhouses. *Alternaria solani* may survive on potato, which is an alternative host.

Spores growing on dead material or on host plants are easily disseminated by wind or by splashed water.

Plant infection takes place through stomata or directly through leaf surface. Spore germination and subsequent infection take place under a wide range of temperature. RH needs to be higher than 97% for rapid germination, but germination may take place in some cases at RH >75%. Senescent tissues are preferentially infected.

The optimal temperature reported for *A. solani* is 18–25 °C and for *A. cucumerina* 20–32 °C. However, temperatures prevailing during the growing period of the respective hosts are not a factor limiting infection.

Control

Alternaria diseases can be prevented by dithiocarbamates, chlorotholonil, iprodione, etc. Hygienic measures and use of healthy propagating material are very important, especially when crops are grown in the soil. Inoculum surviving on plant debris in the soil and spores remaining on the greenhouse frames should be eradicated.

Sclerotinia Rot

This is a common greenhouse disease that damages lettuce, eggplant, tomato, cucumber, pepper, etc. Infection on lettuce begins close to the soil, where a water-soaked area appears. Infection may spread downwards to the roots or upwards to the heart of the plants. Infected leaves fall onto the soil and dry up. The other plants are infected along the stem, leaves, flowers and fruits. Infected areas become water-soaked. Stem infection is more severe. Leaves above the infection area become yellow, wither and die. In wet weather a white mass of mycelia appears on the infected areas, which gradually develops into black sclerotia. *Sclerotinia sclerotiorum,* which is the most common pathogen, produces sclerotia up to the size of bean seeds, whereas *S. minor* produces smaller sclerotia. Sclerotia fall onto the soil where they can survive for several years. When weather conditions are favourable they germinate to produce apothecia which release ascospores and cause new infection. High RH and moderate temperature is required for infection. The elimination of sclerotia and the control measures recommended against grey mould are effective against Sclerotinia rot as well.

Didymella Diseases

Two very severe diseases of greenhouse crops are caused by *Didymella* spp.: Didymella stem rot or canker in tomato and eggplant and gummy stem blight in cucurbits. Both diseases damage all aerial plant parts of their hosts in greenhouses when weather is cool and RH high. They may infect the collar and root system causing yellowing and withering of the plants, which may later die. Cankers along the stem and the petioles are also very common. Plant parts above cankers may die.

Both diseases cause large spots on the leaves which may cover the entire leaf surface. Tomato fruits are infected at the stem end. Initially, the infected area is light brown but it soon turns pink due to the large amount of pycnidio-spores released. Infected parts may cover one third of the fruit surface.

Infection of cucumber and melon fruits by *D. bryoniae* appears mostly at the blossom end. Infection may occur only inside the fruit without being visible on the surface. Soon after infection, a lot of pycnidia appear on the infected areas and their colour turns dark brown. Dark perithecia also appear a little later than pycnidia produced by *D. bryoniae,* while those of *D. lycopersici* are rare. The inoculum remains in plant residues inside and outside greenhouses. In the first case infection starts through the collar. There is good evidence that infection of the aerial parts by *D. bryoniae* is initiated by ascospores released from infected plant material left outside greenhouses.

In greenhouses the two diseases are rapidly spread by water splashes and cultural practices. Soil disinfestation, destruction of plant residue and strict hygienic conditions delay the outbreak of the diseases. However, disinfested soil is readily reinfested. The fungicides commonly used in greenhouse against other fungal diseases are also effective. Moreover, the reduction of the RH and of free water on the leaf surfaces is very effective.

Rust Diseases

These are a very important group, with many common characteristics. The following are the main rusts affecting greenhouse crops.

Carnation Rust

The disease is more severe on leaves, but other green plant parts are infected as well. Initially, small light green spots appear. They gradually turn to powdery brown blisters due to the urediospores developed. Severely infected plant parts are twisted.

Healthy crops are infected by urediospores transferred from neighbouring crops. They are wind or water-splash disseminated and germinate readily on free water. The cycle of the pathogen lasts about two weeks. In greenhouses, where leaves may remain wet for several hours, there may be many disease cycles per crop season.

Rose Rust

The disease is easily identified by the yellow orange rust pustules which develop profusely on the lower surface of older leaves. In greenhouses it is not very destructive.

Several species of *Phragmidium* have been reported to infect rose, but *P. mucronatum* is the most common. It is an autoecious, macrocyclic fungus producing telia by the end of the crop season in the same place as uredospores. They serve as overwintering structures and initiate infection during spring. Free water and temperature 9–27 °C are necessary for the uredospores to germinate.

Chrysanthemum Rust and White Rust of Chrysanthemum

Pale yellow flecks on the leaves followed by dark brown pustules with urediospores are the dominant symptom. Leaves with several pustules may wither and die. No stem infection has been reported. It is a low to moderate temperature disease requiring free water for infection. It survives on infected leaves and is disseminated by wind.

Chrysanthemum white rust is a new and destructive disease of chrysanthemum in Europe and the Mediterranean. Initially, circular white or yellow cushions develop on the lower leaf surface and then soon turn brown. The disease is favoured by high RH and moderate temperatures. Snapdragon rust, geranium rust (*Puccinia pelargoniizonalis* Doidge), etc. are also destructive diseases, but the respective crops are not grown in large acreage. Regular applications of protective fungicides, such as dithiocarbamates and chlorothalonil, or systemic fungicides, such as oxycarboxin and members of the DMIs, are mostly recommended for rust control. Prevention of water condensation is also very effective.

Tomato Leaf Mould

This causes light green to yellow spots on the upper surface of mature leaves. Soon the sporulating fungus growth appears as an olive-green velvety growth on the underside of the yellow spots. The pathogen survives for several months on the greenhouse frame, on the materials used for cropping and in plant debris.

It is disseminated by wind or splashed by water drops. The optimal temperature for infection is 20 to 25 °C. If weather conditions are favourable, leaf mould has several cycles in a season and can destroy the crop completely. There are several races of the pathogen.

Cucurbit Scab

This mostly attacks cucumber, but also squash, melon, etc. It causes nearly circular or angular leaf spots on the leaves, which look water-soaked. Fruit infection is more serious. Initially, water-soaked lesions about 1 cm long, with gummy exudations, develop. A corky tissue usually develops around the lesions, which finally develop a scabby appearance. The pathogen survives on plant debris and spores are air-disseminated. Temperatures of about 15 to 25 °C and RH over 86% favour the disease.

Control

For both diseases, greenhouse ventilation is the best control measure. The disinfestation of greenhouse soil and frames is also very important. Regular application of dithiocarbamates, iprodione, benzimidazoles, etc. are recommended as well. There are several resistant cvs against some races of the pathogens.

8

Beneficial Insect Importance of Plants

ANT

Ants are social insects of the family Formicidae, and along with the related wasps and bees, they belong to the order Hymenoptera. Ants evolved from wasp-like ancestors in the mid-Cretaceous period between 110 and 130 million years ago and diversified after the rise of flowering plants. Today, more than 12,000 species are classified with upper estimates of about 14,000 species. They are easily identified by their elbowed antennae and a distinctive node-like structure that forms a slender waist.

Ants form colonies that range in size from a few dozen predatory individuals living in small natural cavities to highly organised colonies which may occupy large territories and consist of millions of individuals. These larger colonies consist mostly of sterile wingless females forming castes of "workers", "soldiers", or other specialised groups. Ant colonies also have some fertile males called "drones" and one or more fertile females called "queens". The colonies are sometimes described as superorganisms because ants appear to operate as a unified entity, collectively working together to support the colony.

Ants have colonised almost every landmass on Earth. The only places lacking indigenous ants are Antarctica and certain remote or inhospitable islands. Ants thrive in most ecosystems, and may form 15–25% of the terrestrial animal biomass. Their success has been attributed to their social organisation and their ability to modify habitats, tap resources, and defend themselves. Their long co-evolution with other species has led to mimetic, commensal, parasitic, and mutualistic relationships.

Ant societies have division of labour, communication between individuals, and an ability to solve complex problems. These parallels with human societies have long been an inspiration and subject of study.

Many human cultures make use of ants in cuisine, medication and rituals. Some species are valued in their role as biological pest control agents. However, their ability to exploit resources brings ants into conflict with humans, as they can damage crops and invade buildings. Some species, such as the red imported fire ant, are regarded as invasive species, since they have established themselves in new areas where they have been accidentally introduced.

CLASSIFICATION OF ANTS

Little Black Ant

This small, jet-black species is 1.5 mm long. Its petiole has 2 nodes. It is found throughout the United States. In California, it occurs in the southern part of the state and in the San Francisco Bay area. These ants may nest beneath rocks, in rotting wood, in lawns, or in open areas, where the nests are indicated by very small craters of fine soil.

Their colonies are usually moderate to large, and contain many fertile females. Males and winged females may be seen from June to August. The workers are predaceous on other insects, tend honeydew-producing insects, and feed on honeydew as well as nectar and pollen. There have been reports of recently hatched bluebirds and kingfishers being killed by these ants. They may either invade a house from the outside, or establish their nests in the woodwork or masonry of the building. In the house, they feed on both sweets and greasy foods.

Thief Ant

Thief ants are found over most of the United States, and have been found at altitudes up to 5,500 ft (1,700 m). In nature, the nests of *S. molesta* and other tiny thief ants are usually located so that extremely narrow passages from the thief ant nests communicate with the nest passages of a larger ant species. The common name of the thief ant was derived from its habit of steadily pilfering the brood and other food from the nests of larger species. The stinger of the thief ant is rarely used. Although they are in the genus *Solenopsis*, thief ants are not considered to be "fire ants."

The workers are very small, 1.3 to 1.8 mm long, among the smallest ants in the United States. Tle workers are monomorphic (1 form only). The petiole has 2 nodes. The body is largely smooth and shiny, and ranges from yellowish to light or dark brown. Thief ants superficially resemble Pharaoh ants, but are smaller, lighter in colour, and the antenna has a 2-segmented rather than a 3-segmented club. The antennal club is unusually large and elongate, about 1.33 times the length of the remainder of the funiculus. The scape of the antenna extends farther than between the eye and the posterior border of the head.

Biology: The outdoor nesting habits of the thief ant are similar to those of other *Solenopsis* species. There may be several hundred to a few thousand ants in a colony. In nature, they nest in exposed soil or under stones and other objects, and in rotting wood. Copulation takes place during nuptial flights, which have been observed from late July to early fall. A single fertile female is able to establish a nest. *Habits:* Workers are predaceous, and feed on both live and dead insects. They also feed on planted or germinating grains, often causing considerable damage. They seek honeydew and tend honeydew excreting insects. Because they feed on dead rats and mice, they may be carriers of disease-producing organisms to food. They are known to attack young chicks in poultry houses. Workers have been seen carrying gravid segments of the poultry tapeworm, *Raillietina tetragona* (Molin), into their nest, suggesting that the ants may be intermediate hosts for the tapeworm.

When they are found in the home, thief ants usually have their nests in wall voids or elsewhere in the house rather than entering from outside. Their extremely small size enables them to enter containers not accessible to larger ants. They often infest pantries, cupboards, cabinets, and shelves, even though these have been kept closed and scrupulously clean. They are often found around sinks. However, they are so small that they are sometimes not seen, even when crawling on food.

Thief ants will eat practically anything, but appear to prefer foods with a high protein content. They have more of a tendency than the Pharaoh ant to feed on greases, but will eat sweets also.

Pyramid Ant

This small ant is occasionally encountered at lower altitudes throughout California, where it nests in dry and vegetation-free areas. The earth excavated from its nest and gallery system is in a circular crater, usually about 2 to 4 in. (5 to 10 cm) or larger in diameter. It invades the home from the outside, feeds on a wide variety of household foods, but seems to prefer sweets. This species is widely distributed over most of the United States, but is more common in the southern half.

Description and Habits: The pyramid ant is 1.5 to 2.0 mm long, and has a reddish-black head and thorax and black abdomen. Its common name was suggested by the pyramidlike projection on the thorax. The petiole has 1 node. The workers travel in trails, and are fastmoving. They seek honeydew on garden plants, but are also carnivorous and predaceous.

Bicoloured Pyramid Ant

This species is similar to *D. pyramicus*, of which it was once considered to be a variety, but it is slightly larger and has a red head and thorax. It is found in the same geographical areas as the pyramid ant and in similar locations, but is of very little importance as a pest.

Pavement Ant

This species occurs in most of the eastern half of the United States and in California, where it occurs in both the San Joaquin and Sacramento valleys.

Nests occur under stones, along edges of curbing, and in cracks in the pavement. The nests may or may not be surrounded by craters. When nesting in houses, these ants are most likely to be found around or between the lower masonry walls of the foundation. In some areas of the eastern states, the pavement ant is considered to be the most important ant pest in houses, where it may be found throughout the year, but in greater numbers during the summer. It feeds on both greasy and sweet foods.

Description: This species is 3 to 4 mm long, and its body varies from light to very dark brown, with lighter legs and antennae. The head and thorax are furrowed with parallel lines, and the posterior dorsal portion of the thorax bears a pair of small spines. The petiole has 2 nodes. The entire body is covered with stiff hairs.

Velvety Tree Ant

The principal pest in the genus *Liometopum* is the California velvety tree ant, *L. occidentale*. There is only 1 other species in the genus, *L. apiculatum* Mayr, which ranges throughout approximately the same areas as *L. occidentale*, except that it does not occur in California.

Description: This species is 2.5 to 6 mm long, with a glistening, velvety-black abdomen, red thorax, and brownish-black head. Except for the 2 species of *Iridomyrmex*, which because of their economic importance and well-known biologies were used to introduce the ant section of this chapter, the preceding ant species all have petioles with 2 nodes.

Habits: Throughout California, this ant occurs in large colonies in the crotches or hollows of trees, in stumps, and in the ground beneath stones. The insects can be found crawling up trails, up and down oak, alder, elm, sycamore, cottonwood, and other trees, particularly along streams, foraging for insects and honeydew. Areas along streams and under trees also happen to be popular picnic grounds, and the velvety tree ant is therefore the bane of picnickers - biting people, crawling over their food, and producing a disagreeable odor when crushed. The removal of decayed portions of ornamental or shade trees in which the ants nest helps to eliminate these pests.

Odorous House Ant

The odorous house ant, so named because of its unpleasant odor when crushed, occurs throughout the United States, from sea level to 10,000 ft (about 3,000 m) elevation. It is second only to the Argentine ant as a household pest. However, in areas where the Argentine ant is also found, the odorous house ant tends to become less common or to disappear entirely. Nevertheless, it still frequently invades homes or apartments in southern California. The odorous house ant had a tendency to invade houses late in the year when the honeydew upon which it generally fed became less abundant. Also, rainfall and leaf fall, which reduce the numbers of aphids and other honeydew-excreting insects, may induce the ants to enter houses in search of food. It is most likely to invade houses during rainy weather. This ant does not possess a stinger.

Description: The odorous house ant workers are 2.4 to 3.25 mm long-slightly larger than the Argentine ant. The petiole is a single segment, with the node vestigial, flattened or inclined, and concealed from above by the base of the gaster. Like the Argentine ant, the odorous house ant travels in trails, and it also seeks the same kinds of food, preferring sweets. It can be distinguished from the Argentine ant by its darker colour, broader abdomen, by the fact that the petiole and its 1 vestigial node are hidden by the abdomen, and by its odor when crushed. Its pungent, "rotten-coconutlike" odor is detectable even when only a single individual is crushed.

Biology: The odorous house ant nests in many types of habitats - sandy beaches, pastures, woodlands, and bogs. It can also locate in buildings, and often nests in wall voids, particularly around hot-water pipes and heaters. Its nests may be under stones or logs, but also under loose bark, in plant cavities, insect galls, refuse piles, and bird or mammal nests. In the soil, the nests are shallow and indefinite in form. The number of ants in a colony may range from a few hundred to many thousands, and there may be many reproductive females. Mating usually takes place in the nest, but nuptial flights have been observed. Lone females have been observed to establish a colony, although colonies might also be formed when 1 or more fertile females left the parental colony accompanied by a number of workers. An average colony was found to have 345 eggs, 829 larvae, 533 pupae, 2,319 workers, and 1 to 6 dealated females.

Habits: As just mentioned, the workers travel in trails. They are very active, and have rapid movements. When alarmed, they dash about in an erratic manner, and the entire gaster is elevated. Odorous house ants are avid seekers of honeydew, and have been observed to transport honeydew-excreting insects such as aphids and mealybugs. They also feed on both live and dead insects. Although they will eat almost any type of household food, they prefer sweets.

Crazy Ant

This introduced species, possibly of Oriental or African origin, is now almost cosmopolitan, and is frequently intercepted in plant quarantine. It is common in the Gulf Coast region, especially in Florida, but is also well established in many towns and cities throughout the United States, including California. In the colder regions of the country, it occurs mainly in apartment buildings, hotels, and greenhouses, and can be a pest all year. Its common name axises from the worker's habit of running in a haphazard manner, as if lacking a sense of direction.

Description: The workers are 2.2 to 3.0 mm long. They have slender bodies, with very long antennae and legs. The antennae are 12-segmented, and do not have a club. The body is dark brown to black, with a "peculiar gray to violaceous luster or sheen," and has long, coarse hairs. The petiole has 1 node. The insect has no stinger.

Habits: Outdoors, the crazy ant nests in trash, cavities in trees, rotten wood, and in the soil under various objects in colonies of moderate to large size, but it can also nest in buildings. The workers run very rapidly, and can even jump, according to some observers. Their keen sense of smell enables them to find food quickly. They are almost omnivorous, feeding on live and dead insects, seeds, honeydew, fruits, plant exudates, and many household foods. In Puerto Rico, workers of *P. longicornis* have been reported to be predaceous on the larvae and adults of the house fly, flesh flies, and other Diptera.

They have been observed to attack the larvae of the oriental rat flea under laboratory conditions, indicating that they may be important in reducing flea infestations under natural conditions. On the debit side, the workers are also known to gather small seeds of such crops as lettuce and tobacco from seedbeds, and their fondness for sweets forced the proprietor of a soda fountain in Florida to discontinue business.

False Honey Ants

False honey ants occur throughout the United States. The workers are 3 to 4 mm long, light to dark brown, very shiny in appearance, and have a somewhat triangular abdomen. They seldom invade the home. Their name probably comes from their habit of feeding on honey, honeydew, and nectar until their gasters are greatly extended, almost as much as those of the honeypot ants (*Melophorus*) of Australia and of *Myrmecocystus* of the United States Southwest. Honey ants also feed on the exudates of oak galls, live and dead insects, the fluids of dead earthworms, and the young, tender tissues of plants.

Small Honey Ant

This species ranges throughout much of the United States, is common in California, and often replaces the Argentine ant where the latter has been exterminated. When these ants are found in houses, they have usually come from outdoors, but it is likely that they often nest inside, for on many occasions winged females and males have been collected inside houses from January to March. This ant lacks a stinger and seldom bites, for it is not aggressive. A striking characteristic of *Prenolepis imparis* is that it is most active outdoors at temperatures between 35 and 55 °F (2 and 13 °C), and has been observed to emerge from its subterranean galleries at 6 °F (–14.4 °C).

Ants feed on such household foods as "sweet corn, corned beef, meats, cakes, breads, sugar, honey, syrup, watermelon, and fruits, but are especially fond of sweets." Outdoors, the ants can damage flower buds. H. M. Armitage observed *Prenolepis imparis* feeding extensively on tender, young growth and the calyxes and unopened petals of orange buds. *Description: Prenolepis imparis* varies in colour from pale to dark brown or black, with the head and thorax often lighter in colour than the abdomen. The body is smooth and shiny. The thorax is small, slender, and almost cylindrical, and is divided into 2 parts by a strong constriction. The abdominal petiole is composed of a single segment (1 node). The antennae are 12-segmented, do not have a club, and the scape extends beyond the posterior border of the head by half the length of the scape.

The Nest

This ant normally constructs its nest in soil, preferably moist clay or loamy soil in shady woodlands, but seldom under stones or other objects. The nest has a central opening, surrounded by a crater of characteristic earthen pellets. A single gallery leads to the nest, which may be from 18 to 51 in. (45 to 130 cm) deep and have 6 to 40 lateral chambers leading from it. There are seldom more than a few thousand ants in a colony.

Life Cycle: In winter, an old colony contains a single reproductive female, winged males and females, and many workers. Usually in March to April, the overwintering males and females make their nuptial flight, although most of the mating appears to take place on the ground. The female then discards her wings and seeks a crevice in the ground in which to oviposit and start a new colony.

Species of Lasius

The Lasius species are relatively unimportant in California. The workers are monomorphic, active, emit a strong formic acid odor, but do not have a stinger. The antennae are 12-segmented and lack a club. The eyes are well developed, and are located closer to the posterior border of the head than to the base of the mandible. If ocelli (simple eyes) are present, they are small and indistinct. The abdominal petiole has only 1 segment, and the petiolar node is narrow in profile and either vertical or not strongly inclined.

Iridomyrmex Pruinosus (Roger)

This is a native species that ranges from Wisconsin to New York, south to New Mexico and Florida, and also occurs in Colorado, Arizona, Utah, Nevada, California, Oregon, and Idaho. It is common and widely distributed in the southern United States. It infests houses, particularly in the Gulf Coast states, generally invading them from outside, but is also believed to be able to nest within a house.

Description: This species can be distinguished from the Argentine ant by the difference in the shape of its head, the presence of erect hairs on the thorax, and by having a more densely pubescent body. The body colour is variable. In the eastern states, this species is commonly uniform dark brown or black, whereas the body of the Argentine ant is usually light to dark brown. In the western states, the body colour is even more variable, some specimens being even lighter in colour than *Iridomyrmex humilis*. The workers are 1.8 to 2.5 mm long. The odor of crushed specimens of *I. pruinosus* is like "rotten coconut," similar to the odor of the odorous house ant (*Tapinoma sessile*), whereas a mass of crushed specimens of *I. humilis* has a stale and musty or greasy odor.

Habits: Iridomyrmex pruinosus prefers open spaces, such as fields, meadows, pastures, or entirely bare areas, but will also nest in open woods away from dense and prolonged shade. The nests may be constructed in exposed soil or under cover of stones or the bark of logs and stumps, and contain small to moderate-sized colonies. Like the Argentine ant, this species feeds on live or dead insects, is particularly fond of honeydew, and forms foraging trails.

Western Bigheaded Ant

This species may be found throughout the San Joaquin and Sacramento valleys of California, as well as in most of the state below 3,000 ft (900 m) elevation, and east to Texas. These ants are 3 to 6 mm long, uniformly light to dark reddish, with the soldiers sometimes darker and more brownish than the workers. The soldier's head is deeply furrowed in the frontal view. The body has abundant long hairs. *Pheidole hyatti* is common in arid and semiarid areas, and makes nests in the soil, often beneath stones and boards. It is usually nocturnal, and is omnivorous, with a decided tendency to be predaceous. The ants are sometimes found in and around houses; in kitchens, they are usually found around the sink. They are especially fond of greasy meats.

Acrobat Ants

Acrobat ants are characterized by the heartshaped abdomen, the upper surface of which is flattened and the lower surface curved. They may be black, brown, or yellow, and the petiole has 2 nodes. Their common name is derived from their habit of raising the abdomen over the head and thorax. They are slow moving ants that tend aphids, and are known to construct "tents" or "cowsheds" of earth and vegetable debris over the aphids from which they obtain the sugary excrement known as "honeydew". They are generally found under stones or in old stumps, and have sometimes been found to tunnel and nest in wood. They occasionally become pests by invading houses.

Lined Acrobat Ant

This ant ranges from southern Canada southward along the Rocky Mountains and eastward to Florida, and has been found at altitudes up to 5,500 ft (1,700 m). It varies in colour from light brown through dark brown to black. The scape (first segment) of the antenna extends beyond the posterior border of the head. The workers are 2.5 to 3.5 mm long. Medium-sized to large colonies may nest in exposed soil, under stones or other objects, or in logs, stumps, and dead trees. The ants may also nest in the woodwork of houses and infest household foods, where they show a slight preference for sweets and meats or other high protein foods. In nature, they feed on honeydew and dead or live insects, including winged termites. The workers bite aggressively, and give off a repulsive odor when alarmed.

Harvester Ants

These ants are predominantly black, and have conspicuous thoracic spines. The petiole has 2 nodes. Their habits are somewhat similar to those of *Pogonomyrmex*, except that they do not sting, their stinger being vestigial and nonfunctional. Homeowners object to their nests in yards and lawns. They are very active, and harvest large quantities of seeds.

Longspined Harvester Ant, Veromessor Andrei

This species is 4.5 to 7 mm long, with a reddish-brown head and thorax, and in some specimens the posterior part of the abdomen is black. The remainder of the body is lighter in colour - in southern California, generally reddish brown. It is limited almost entirely to California in its distribution. Wheeler and Wheeler point out that while *V. andrei* does not sting, its bite is quite tenacious and annoying.

Jetblack Harvester Ant

This species is 2.5 to 6 mm long, and as the common name indicates, is a beautiful "jet-black, patent-leather-coloured" ant. It inhabits the deserts of southern California, northwestern Arizona, and Baja California. The nests of *V. pergandei* are usually surrounded by a crater of excavated soil, whereas those of *V. andrei* usually are not. *Veromessor pergandei* makes a neat pile of chaff with seed husks in the form of a semicircle or ring around the nest, whereas *V. andrei* makes a more ragged pile or scatters the chaff about. *Veromessor pergandei* can tolerate the most extreme heat of the deserts, but *V. andrei* cannot.

Bigheaded Ants

These ants are so named because of the exceptionally large heads of the soldiers in relation to their bodies. They live in situations somewhat similar to those inhabited by fire ants, have similar habits, and are distributed throughout the warmer areas of the United States. As in the case of the fire ant, the petiole has 2 nodes. They resemble fire ants superficially, but can be distinguished from them by their 12-segmented antenna, 3-segmented antennal club, and the large head, particularly of the soldier. The natural foods of *Pheidole* spp. are generally live and dead insects, seeds, and honeydew, but in houses they will feed on meats, grease, peanut butter, fruit juice, and other foods.

Pharaoh Ant

This species is native to Africa, but has been spread by commerce to nearly every part of the world. The specific name, from which the common name was derived, resulted from Linnaeus' mistaken belief that the ant was one of the plagues of Egypt in the time of the Pharaohs, mentioned in Scripture, whereas the only plagues mentioned were lice, flies, and locusts.

Although the Pharaoh ant occurs throughout the United States and Canada, it is rarely found in California, but appears to be established in at least 2 locations in the state. It has never been identified as a household pest in California, but is important in some other parts of the country. It is especially common in hotels, large apartment houses, grocery stores, or other places where food is commercially handled. In residential areas, nests may be found outdoors in the lawn or garden and indoors in wall voids, subfloor areas, attics, cracks, crevices, and behind wainscoting, baseboards, plaster, or mantles, under hearthstones, between flooring, or in furniture. The ants tend to seek warm and inaccessible locations.

They breed throughout the year, and produce enormous numbers of individuals. They feed on dead and live insects, and often damage the dried specimens in collections. In the home, they appear to prefer grease, fats, and meats, but are practically omnivorous. In the early years of American agriculture, they caused great damage to crops. For example, they injured young corn plants by gnawing the blades and drinking the sap.

Monomorium pharaonis has become particularly well adapted to existence in close association with man by abandoning the nuptial flight and territorial boundaries. Colonies multiply by "budding." If groups of these ants in a large building are joined by odor trails, they may constitute a widely dispersed colony that can contain up to millions of workers and thousands of queens.

Description: The Pharaoh ant is one of the smallest ants, 1.5 to 2 mm long, and is light yellow to reddish in colour. The petiole has 2 nodes. It is often confused with the thief ant, *Solenopsis molesta*, but averages slightly larger, and can be distinguished by 3 segments in the antennal club, compared with 2 in the thief ant.

Life Cycle: Although the usual sexual forms are winged, nuptial flights have never been seen. Mating takes place in the nest, and the formation of a new colony is not normally accomplished by a single female after a nuptial flight, as with many other ant species. Instead, a large part of a colony will migrate to a new location, the workers carrying with them eggs and young stages. In this way, new nests can be developed throughout a building, and such nests have been found to contain from 400 to 4,800 ants. In subsequent investigations, they found that the various life stages of the workers required the following numbers of days: egg, 5 to 6; larvae, 22 to 24; prepupal, 2 to 3; and pupal, 9 to 12; thus, the life cycle required 38 to 45 days at room temperature. Sexual forms required about 4 days longer. The maximum period of adult life of the workers was 9 to 10 weeks, females lived as long as 39 weeks, and males generally not more than 3 weeks. The females were able to lay from 350 to 450 eggs.

Habits: Pharaoh ants crawl great distances from their nests, and do not always follow definite paths. Therefore, the nests may be difficult to find, increasing the difficulty of control. A shipowner spent $4,000 in trying to eradicate Pharaoh ants from his vessel. They have been stated to be the most persistent and difficult of all our house-infesting ants to control. An interesting aspect of Pharaoh ant infestation is their tendency to nest in odd places, as between sheets of stationery, layers of linen, between clothing in trunks, in an electric iron, in a can of modeling clay, or, in piles of trash. Many investigators have noted the tendency of the Pharaoh ant to be attracted to water, such as from dripping faucets, and one observer noted that their trails to the earth beneath the house did not necessarily indicate the presence of a nest, for the ants might be seeking moisture in the soil.

Scientific classification

Kingdom: Animalia
Phylum: Arthropoda
Class: Insecta
Order: Hymenoptera
Suborder: Apocrita
Superfamily: Vespoidea
Family: Formicidae

Anatomy

The main body parts of ants, and the information below to learn about about the anatomy of ants.

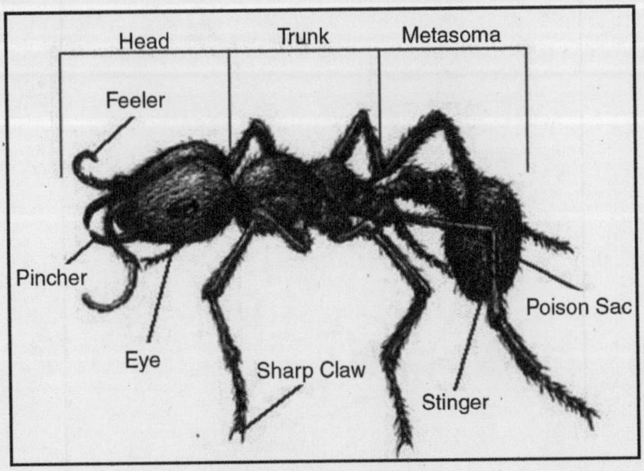

Fig. Ant Anatomy

The first thing you will notice in this diagram of an ant's anatomy is the three main divisions. The head, the trunk, and the metasoma. Ant bodies, like other insects have an exoskeleton. Their skeleton is on the outside - not covered by skin, muscles, and tissue like humans. The head of an ant has several imporatant parts. First the eyes. Ant eyes are a lot like flies. They have many small eyes connected together. This enables ants to see movement very well.

Also Attached to the head of the ant are two feelers. The feelers are special organs of smell that help ants communicate. If you have every seen an ant find a food source, it is not long before the ant communicates to his friends and their are many ants there. Ants release pheromones (chemicals that have different smells) to communicate with each other. The feelers pick up these smells as signals. The head also has two strong pinchers which are used to carry food, to dig, and to defend. Just inside the mouth is a small pocket where ants can store food. They can give this food to other ants in need. The trunk of the ant is where all 6 legs are attached. At the end of each leg is a sharp claw that helps ants climb and hang onto things. The metasoma of the ant is a poison sack. Ants are closely related to wasps and other stinging insects. Many types of ants have stingers and can inflict a very painful sting. This is a useful way to defend against the many predators ants have.

Inside Ants do not have lungs. Oxygen enters through tiny holes all over the body and Carbon Dioxide leaves through the same holes. There are no blood vessels. The heart is a long tube that pumps colorless blood from the head back to the rear and then back up to the head again. The blood kind of coats the insides of the ants and is then sucked into the tube and pumped up to the head again. The nervous system of ants consists of a long nerve cord that also runs from head to rear with branches leading to the parts of the body, kind of like a human spinal cord.

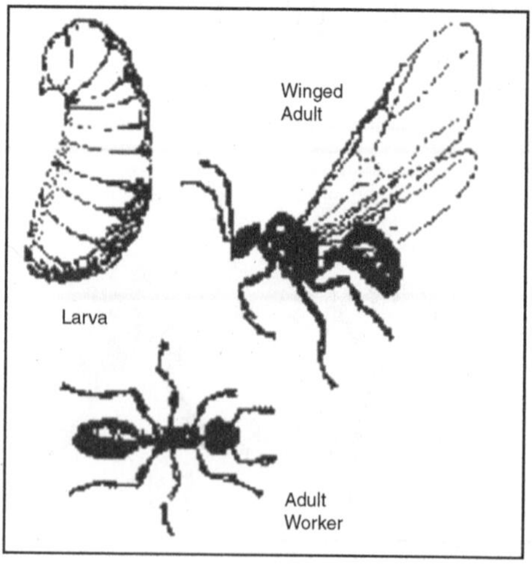

Fig. Stages of Ant

Adult Males and Females

When ant colonies reproduce, the new queens and males may be found in the colony. These are "flying ants" and have two pairs of wings. Males generally have small heads, large eyes, large thoraces, and a pair of claspers at the end of the gaster. Once they fly (and mate), males do not live very long.

After mating, new queens break off their wings and never fly again. Without wings, they can generally be distinguished from workers by their larger body size, larger thorax and larger abdomen. All workers are females.

Immatures (different stages)

Ant larvae are white and grub like. They have no legs and don't move about much on their own. You can generally see a large, dark stomach through their cuticle. Ant pupae look like white adult ants, with their legs and antennae pressed close to their bodies.In some species, larvae spin silk and the pupal stage is inside a cocoon. Newly emerged adult ants are often paler than older ones. It may take them several days to reach their final colour.

Natural History

Food

Most ants that are easy to keep in the classroom are generalists, eating a variety of small insects that they capture, dead insects they happen to find, nectar, or honey dew. They need a balance of carbohydrates and protein. Protein is especially needed for the queen to make eggs and for the larvae to grow.

Habitat

Most ant species live in the soil. Some, like the carpenter ants, also live in wood (they excavate, but do not actually eat the wood). Some ants live in cavities made inside plants, such as acorns, twigs, and galls.

Predators

A variety of reptiles or amphibians (particularly toads and lizards), spiders, other insects such as assassin bugs, and other ants may prey on workers. Bats, birds, and occasionally, people capture and kill or eat the flying males and females.

Interesting Behaviors

Since ants are social they display many behaviors that remind us of our families and society. For example, worker ants take care of larvae by feeding and washing them. Ants are able to communicate with each other. They are able to communicate, among other things, directions (to where the food is) and alarm.

Impact on the Ecosystem

Positive

World wide, ants are one of the most important predators on small invertebrates, including other insects. Leafcutting ants in the American tropics are the most important herbivores (plant consumers), outranking grazing mammals. In many ecosystems, ants are important dispersers of the seeds that they harvest. In desert regions, they are one of the principle consumers of seeds. Wherever ants live, they turn over and aerate the soil as much or more than earthworms (depending on the specific ecosystem).

Negative

A few ant species are considered pests, because they live in and protect territory that we consider ours or because they want to consume resources that we need. For example, leafcutting ants (see "Positive" section above) compete with us for crop plants in the American tropics. Fire ants colonize damp grasslands (including lawns!) with alarming ease. Carpenter ants, adapted for living in dead wood, consider the dead wood (lumber) in houses fair game, especially if it is damp. A number of opportunistic ant species can overrun kitchens, pantries, and pet food areas in search of suitable food items. Also, some ants (like their relatives the wasps and bees) have a potent sting. As with bees, some people can become hypersensitive to ant stings.

Taxonomy and Evolution

The family Formicidae belongs to the order Hymenoptera, which also includes sawflies, bees and wasps. Ants evolved from a lineage within the vespoid wasps. Phylogenetic analysis suggests that ants arose in the mid-Cretaceous period about 110 to 130 million years ago. After the rise of flowering plants about 100 million years ago they diversified and assumed ecological dominance around 60 million years ago.

In 1966, E. O. Wilson and his colleagues identified the fossil remains of an ant (*Sphecomyrma freyi*) that lived in the Cretaceous period. The specimen, trapped in amber dating back to more than 80 million years ago, has features of both ants and wasps. *Sphecomyrma* was probably a ground forager but some suggest on the basis of groups such as the Leptanillinae and Martialinae that primitive ants were likely to have been predators under the soil surface.

During the Cretaceous period, a few species of primitive ants ranged widely on the Laurasian super-continent (the northern hemisphere). They were scarce in comparison to other insects, representing about 1% of the insect population. Ants became dominant after adaptive radiation at the beginning of the Tertiary period. By the Oligocene and Miocene ants had come to represent 20–40% of all insects found in major fossil deposits. Of the species that lived in the Eocene epoch, approximately one in ten genera survive to the present. Genera surviving today comprise 56% of the genera in Baltic amber fossils (early Oligocene), and 92% of the genera in Dominican amber fossils (apparently early Miocene).

Termites, though sometimes called *white ants*, are not ants and belong to the order Isoptera. Termites are actually more closely related to cockroaches and mantids. The fact that ants and termites are both eusocial came about by Convergent evolution. Velvet ants look like large ants, but are wingless female wasps.

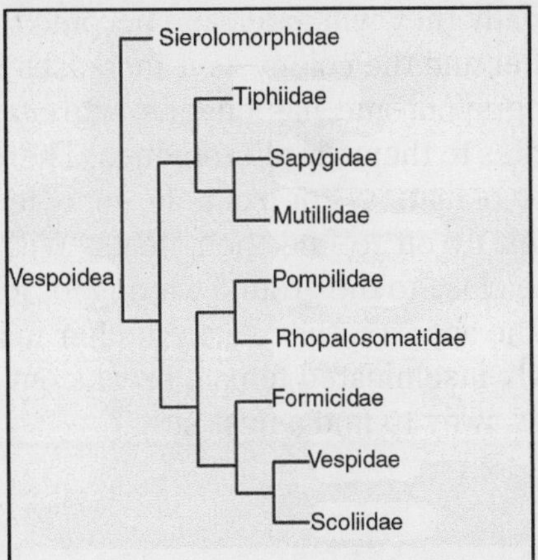

Phylogenetic position of the Formicidae

Origins

The oldest fossil remains of modern ants is preserved in amber and has been dated by radioactive isotopes to be approximately 80 million years old. It has many of the characteristics of modern ants while also presenting many characteristics of the ant's distant relative, the nonsocial aculeate wasp. Like modern ants, the fossil contains the metapleural gland, petiole turning down towards the gastor, and a relatively small thorax without wings.

The fossil contains the following wasp like characteristics: short mandibles with only two teeth, stinger that can be extruded, unconstricted gastor, and double tibial spurs. The fossil antennae have a short first segment and a long flexible segmented structure for the rest of the antennae. This antennae combines both wasp and ant features. An early Cretaceous fossil that may be an ant has been discovered. Unfortunately, the gastor has been folded back over the thorax, thereby restricting the view of the petiole.It is a less well preserved specimen, and, therefore, scientists are uncertain as to its' morphology.

Mating

Most all ants mate in one of two ways. The first is known as the male-aggregation syndrome. At a time usually triggered by environmental events

.such as a large rainfall, the males and unfertilized queens, which have been patiently waiting underground, suddenly take to the air and swarm in what is commonly called the nuptial flight. This swarming often appears as a column of flying ants rising off the ground. The queen is usually inseminated in mid-air and then flies off to a locale where she breaks off her wings and begins to dig a hole either in the ground or in plant material, which will become her nest. She lays her eggs and remains with them until they hatch, living off of her own body reserves. Eventually they will hatch and become the first of her workers that will take care of her and the colony as it increases in size and numbers.

The female-calling syndrome has queens aggregating at the surface of colonies and calling males to them via pheromones. There are many exceptions to this basic description of mating. For example, some females are inseminated on the ground and then fly off to find their nests. With another species, the males will be swarming close to the ground waiting for females to emerge from their colonies. When she appears, they surround her in a rolling, writhing ball of males until the newly inseminated female breaks out of the ball after about 20-30 seconds and flies away to find a nest site.

Ant Reproduction

Egg

The life of an ant starts with an egg. If the egg is fertilized, the ant will be female; if not, it will be male. Ants are holometabolous (a specific kind of insect development which includes four life stages) and develop by complete metamorphosis, passing through larval and pupal stages before they become adults.

Ant Eggs

Fig. Eggs of Ant

Larva

The larval stage is particularly helpless – for instance it lacks legs entirely – and cannot care for itself. The difference between queens and workers (which are both female), and between different castes of workers when they exist, is determined by feeding in the larval stage. Food is given to the larvae by a process called 'trophallaxis' in which an ant regurgitates food previously held in its crop for communal storage. This is also how adults distribute food amongst themselves.

Larvae

Pupa

Larvae and pupae need to be kept at fairly constant temperatures to ensure proper development, and so are often moved around various brood chambers within the colony.

Pupa

Adult

A new worker spends the first few days of its adult life caring for the queen and young. After that it graduates to digging and other nest work, and then to foraging and defence of the nest. These changes are fairly abrupt and define what are called temporal castes.

One theory of why this occurs is because foraging has a high death rate, so ants only participate in it when they are older and closer to death anyway. In a few ants there are also physical castes – workers come in a spectrum of sizes, called minor, median, and major workers, the latter beginning foraging sooner. Often the larger ants will have disproportionately larger heads and so will have stronger mandibles. Such individuals are sometimes called 'soldier' ants because their stronger mandibles make them more effective in fighting other creatures, although they are still in fact worker ants and their 'duties' typically do not vary greatly from the minor or median workers. In a few species the median workers have disappeared, creating a sharp divide and clear physical difference between the minors and majors.

Most of the common ant species breed in the same way. Only the Queen and breeding females have the ability to mate. Some ant nests have multiple queens. The male ants, called drones, along with the breeding females are born with wings and do nothing throughout their life except eat, until the time for mating comes.

At this time, all breeding ants, excluding the queen, are carried outside where other colonies of similar species are doing the same. Then, all the winged breeding ants take flight. Mating occurs in flight and the males die shortly afterward. The females that survive land and seek a suitable place to begin a colony. There, they break off their own wings and begin to lay eggs, which they care for.

The first workers to hatch are weak and smaller than later workers, but they begin to serve the colony immediately. They enlarge the nest, forage for food and care for the other eggs. This is how most new colonies start. Worker ants are sterile (do not breed), therefore their jobs are to look for food, protect the eggs, take care of the young, and defend the nest from unwanted visitors. During the night, the worker ants move the eggs and the larvae deep into the nest to protect them from the cold. During the daytime, the worker ants move the eggs and larvae of the colony to the top of the nest so that they can be warmer. If a worker ant finds a good source of food, it leaves a trail of scent so that the other ants in the colony can find the food. Army Ants are nomadic and they are always moving. They carry their larvae and their eggs with them in a long trail.

Gene Flow

It is often argued whether ants and other social insects are altruistic in their contribution to the colony rather than trying to reproduce and directly pass on their own genes to their offspring. Yes, they seem to be altruistic, in they are helping their sisters at their own expense without directly passing on their genes. No, they aren't altruistic, in that they are passing on their own genes by allowing the colony to survive and expand to new colonies. This type of natural selection is called Kin Selection. Genes are selected in nature according to their ability to contribute to the success of the species. If an altruistic gene helps a colony to survive, then that gene will be passed onto kin which will in turn have the altruistic gene that will help that colony survive. If the gene for altruism works against the success of the colony then the individuals with the gene will not be successful and the gene will be selected against.

One possible explanation for kin selection arises from the fact that all workers are females. It turns out that the workers are more closely related to each other than they would be to their own offspring. This means that they are more successful in making more genetically similar individuals by helping the colony than by having their own offspring.

Specifically, males arise from unfertilized eggs so they have only a half complement of genes, all from their mother. Female workers arise from fertilized eggs and therefore have a full complement of genes of which half come from their mother and half come from their father. Workers always get the exact same half of their gene complement from their father since he only has a half to begin with. Workers are, therefore, at least fifty percent related to each other since half of their genes always come from their father. Workers can end up with either half of their mother's genes which means that they will be on average 75 percent related to each other. If the workers did have the opportunity to reproduce, (which they don't), the best they can do is to contribute 50 percent of their genes to their offspring.

Communication

Ants need to be able to communicate for an array of reasons. The following is a list of the 12 major reasons for communication that researchers have identified.

- Alarm
- Simple attraction
- Recruitment, as to a new food source or nest site
- Grooming, including assistance at molting
- Trophallaxis (the exchange of oral and anal liquid)
- Exchange of solid food particles
- Group effect: either facilitating or inhibiting a given activity
- Recognition, of both nestmates and members of particular castes, including (broadly) discrimination of injured and dead individuals.
- Caste determination, either by inhibition or by stimulation.
- Control of competing reproductives
- Territorial and home range signals and nest markers
- Sexual communication, including species recognition, sex recognition, synchronization of sexual activity, and assessment during sexual competition.

The majority of communication seems to be chemical. Ants also tap each other, feel each other out with their antennae, straddle each other to give certain messages, and grasp and stroke each other as well. One gland, the pygidial gland on the gastor, is used to lay down trails for the same individual or for others to follow in the same track. It also seems to be used to warn colony members of danger and as a pheromone to attract fellow foragers to food sites. In fire ants, the Dufour's gland is the source of trail- laying chemicals. Some trail pheromones can last several days. In leaf cutter ants, they may create a main trunk trail leading away from the colony only to branch out in several directions a short distance later. They continue to branch out like arteries to capillaries until single ants are foraging for leaf material. When they obtain their leaf fragment, they turn around follow their pheromone trail back along the path to the main trunk and then the colony.

Division of Labour

Ant colonies are grossly divided into queens, males, and workers. The job of the queen is to lay eggs. The males generally do nothing for the colony. They wander around accepting food from the workers until the time comes for mating. They die almost immediately after mating. Workers are generally sterile females. They have a variety of tasks to perform for the colony ranging from foragers, defenders, brood feeders, food preparers, caretakers of the queen and nest construction, among other things. The size of the ant in the colony is often different according to the task the ant needs to perform. The queen is often the largest; brood keepers are the smallest. Soldiers are large, and foragers are smaller.

When ants specialize, they help the colony survive in a number of ways. For example, foragers are constantly exposed to danger while out on the hunt. A small percentage of the colony acts as foragers though, so the majority of the colony is safely tucked away in the nest. The foragers are sacrificing for the majority of the colony, which means that the colony will probably survive longer.

If every ant in the colony had to forage, then every individual would be exposed to great dangers. Interestingly, duties are often assigned according to the age of the ant. The younger ants are closer to the queen, taking care of her and the brood, while the older ants are usually delegated the more dangerous tasks of foraging and defence.

Foraging

The overriding goal of every foraging ant is to spend the least amount of energy to obtain and deliver the greatest food value to the colony. As long as enough foragers are successful, the colony can maintain the energy flow to continue reproducing indefinitely.

One major deciding factor as to foraging technique takes into account the external dangers presented to ants while foraging. They are constantly under attack by predators and competitors. When an ant is lost to predation or injury, the whole colony loses a small energy packet. Predators may influence whether a colony hunts with packs of ants going off at a time or if the colony uses individuals to hunt on their own. Will they hunt out in the open or will they hide under some sort of canopy?

If competition is a problem, ants may expend a larger amount of energy to retrieve a food source quickly, rather than to go slowly and allow other colonies to take the food.

Temperature and humidity are two important environmental factors effecting foraging. Desert ants are adapted to higher temperatures and lower humidity, while cold weather species do better in higher humidity, cooler temperatures. Some ants forage before sunrise and after sunset when the temperatures are not too hot. Others forage only after sunrise and before sunset to take advantage of the warmer temperatures. Some ants may forage at unseemly times of the day to avoid predators. Foraging may be controlled by the time of availability of the food source. High humidity seems to present a higher temperature window as to when ants will forage. Most ants will not forage during or shortly after a rainfall.

When foraging, ants seek a variety of materials depending on the needs of their colony. At different times, the colony may need protein or may need carbohydrates. Ants can be predators, omnivorous, seed specialists, or fungus gardeners. There are no known ants that actually eat leaves, but there are ants that use leaves for different purposes.

Symbiotic Relationships

Relationships between ants and other organisms are numerous and varied.

- *Ant/Ant:* Some species of ants are extreme in their dependence upon other ant species. For example, the ant *Tuleutomyrmex schneidere* spends almost its entire life riding on the backs of host ant species. They seem to contribute nothing to the hosts, but are tolerated and even fed. Slavemaker ants (*Formica subintegra*, for example) steal brood from other colonies and return the brood to develop and serve the Slavemaker colony. The slaves are absolutely dependent in that if they don't work, they don't get fed. Other ants work together as with the *Crematogaster limata parabiotica* and *Monacis debilis*. These ants have their nests close together and share the same foraging trails. *Camponotus* has also been seen giving food to the *Monacis* workers.
- *Ant/Other Insect:* These relationships are many and diverse, ranging from commensual to parasitic. Aphids and ants have many species relationships where both the ants and aphids benefit (mutualism). Aphids secrete honeydew and amino acids through their anus. The ants eat or store the honeydew. The ants sometimes incorporate the aphid territory into their own territory, which allows easier access to the aphids and affords the aphids protection by a greater number of ants. The honeydew sometimes contains chemicals that are

purposely directed at attracting ants. The aphids sometimes release chemical signals that warn other aphids of a predatorial attack and also alert the ants so they can attack the invader.

- *Ant/Plant:* These relationships are also known to be abundant. Some carnivorous plants allow ants to hunt herbivores on them. In turn, the ant protects the plant from the herbivores eating their plant tissue. Many plants have extrafloral nectaries on various parts of the plant. These are nectar- producing structures not associated with flowers. The ants are attracted to the plant where they can obtain small amounts of sugar and, in turn, defend the plant from other insects. Ants provide this same service of eliminating herbivores to many plants. Other ants confiscate plant parts to grow fungus on in fungus gardens deep with colonies. These leaf cutter ants process the leaves and use the fungus grown upon the leaf material for food. Sometimes ants live in tree hollows and have no effect on the plant at all. Harvester ants do a great service to plants by collecting and transporting seeds. In one case, the ants eat a small part of the seed and leave the rest of the still- viable seed to germinate.

CONTROL OF ANTS

Recognition of species is important in ant control. The odorous house ant and thief ant may nest in houses, and their trails will lead to the nesting areas, which can then be treated. When they nest in large, inaccessible areas, such as attics, wall voids, and voids under cabinets, dusting with a chlordane or silica aerogel dust may be the most convenient and also the most effective way to obtain complete coverage and good control. Although the Argentine ant nests outdoors, it infests houses more frequently than any other ant species. The housewife may wish to treat for Argentine ants indoors in order to get rid of them quickly, but for effective and long-term control of this and other outdoor-nesting species that may invade the house, the main reliance should be placed on outdoor treatment.

The Argentine, odorous house, thief, jetblack harvester, field, and velvety tree ants travel in trails, and the trails help to trace them to their nests. Repeated spraying or dusting of trails can result in good ant control. Some ants are attracted to sweets only, and the thief ant is usually attracted to fats only.

The other species are attracted to both sweets and fats, except for the harvester ants (*Veromessor*) which eat mainly seeds. A knowledge of the food habits of ants is necessary in order to decide on what type of bait to use when control is to be accomplished by the use of poison baits.

For control of ants within a house, sprays of 2 to 3 per cent chlordane applied with a sprayer or paintbrush to baseboards and cracks where ants are seen, or a 50 per cent chlordane dust applied behind baseboards or in out-of-sight areas, are effective. It may be most convenient to apply an aerosol containing chlordane, holding the canister close to the sprayed surface so as to wet it visibly. If ant trails are found outside the house, a spray or dust may be applied next to the foundation in the general area of infestation. Aphid-tending species of ants, like the Argentine ant, may be controlled in yards and gardens by spraying or dusting the ground under the infested shrubs or trees. The shrubs or trees themselves should not be sprayed, for they may be injured and natural enemies of the pests may be destroyed. This should not only help to control the honeydew-producing insect pests that are favoured by the presence of ants, but should also decrease the ant population in the general area of the home and thereby decrease the chance of infestation in the house.

A "hose-attachment" sprayer is a great convenience, and is a time-saver if large areas need to be sprayed. Best results can be obtained with power sprayers, such as those used by commercial pest control operators, to spray the entire area around a house where nests may be located as well as tree trunks and subfloor areas. Some operators control ants by periodically spraying the entire foundation and the adjacent soil.

The homeowner can effectively use baits, such as 0.125 per cent Kepone or 0.075 per cent mirex, placing them in ant-nest openings, around foundations, and in other areas frequented by ants. It is important to use the proper percentage of toxic ingredient when placing ant baits. For example, dieldrin at 0.05 per cent concentration in syrup gave complete control of the odorous house ant, but at 0.5 and 1 per cent, the baits were repellent and the ants would not feed on them. Dieldrin happens to have amazing toxicity against the odorous house ant - it required a 67 times greater concentration of chlordane to obtain an equal mortality level. However, dieldrin is no longer registered for such household uses.

Poisoned syrups are still commercially available for ant control, and are effective against species attracted to sweets, such as the Argentine ant. The receptacles containing the syrup should be placed at intervals of 10 to 20 ft (3 to 6 m) around the foundation of the house. The syrups contain sodium arsenite, and it is not advisable to use them if children or pets are in the vicinity.

For the control of Pharaoh ants in hospitals, where they are considered to be potential carriers of disease germs, baiting with arsenicals to be the only fully effective procedure. One successful bait had the following ingredients: ground beef, 1,000 g; arsenic, 200 g; and honey, 80 g; mixed by hand, using rubber gloves. Ground liver was the most attractive bait, and did not require the addition of honey. However, it dried out too rapidly in warm rooms, and was therefore more effective when mixed with beef. The bait was placed in locked wooden bait stations. Ground meat was left 3 days before poison bait was used, to accustom the ants to the new food station. Baits were renewed as often as required to avoid a dry bait surface. If an ant trail was disturbed during the baiting period, the insects fled to their nest, and it took several days for them to re-establish a trail to the bait station.

COCKROACH

Cockroaches (or simply "roaches") are insects of the order Blattaria. This name derives from the Latin word for "cockroach", blatta. There are about 4,000 species of cockroach, of which 30 species are associated with human habitations and about four species are well known as pests.

Among the best-known pest species are the American cockroach, *Periplaneta americana*, which is about 30 millimetres (1.2 in) long, the German cockroach, *Blattella germanica*, about 15 millimetres (½ in) long, the Asian cockroach, *Blattella asahinai*, also about 15 millimetres (½ in) in length, and the Oriental cockroach, *Blatta orientalis*, about 25 millimetres (1 in). Tropical cockroaches are often much bigger, and extinct cockroach relatives such as the Carboniferous *Archimylacris* and the Permian *Apthoroblattina* were several times as large as these.

World-Wide Distribution of Cockroaches

Cockroaches are tropical or subtropical in origin, and can maintain themselves outdoors in such areas. Species that became adapted to living in human habitations - "domestic" or "domiciliary" species - all originated in north or tropical Africa. Species names such as *americana*, *australasiae*, *germanica*, or *orientalis*, and the corresponding common names, are therefore misleading.

The distribution of domiciliary cockroaches throughout the world was made possible by their ability to infest the various means of intra- and intercontinental transport, particularly ships. Originally, and probably to this day, their principal means of transportation between continents has been by ships. However, transport of cockroaches in the baggage compartments and kitchens of aircraft has been abundantly documented.

For many years, the principal species infesting ships was the American cockroach, but in more recent times the German cockroach has become the dominant ship-infesting species.

To judge by the enormous numbers of cockroaches infesting ships before the advent of effective insecticides, these insects must have found abundant food and water. Even without food and water, however, cockroaches can live for considerable periods. Newly emerged adult females of the German cockroach, *Blattella germanica*, have been found to live at 27°C and 36 to 40 per cent relative humidity for an average of 20.1 days without food or water, 14.7 days with dry food (dog biscuits) and no water, 35.1 days with water but no food, and 82 days with both food and water. The corresponding figures for the American cockroach, *Periplaneta americana*, were 41.7, 40.1, 89.6, and 190. The apparent ability of the 2 species to survive slightly longer without food and water than with dry food alone is interesting, but in any case, cockroaches are well adapted to dispersal over great distances by man, even when food and water are lacking.

Domestic species of cockroaches never or rarely fly, but are readily carried about in and on such items as sacks, cartons, or packages (particularly corrugated cartons) of food, in laundry, or in kitchen appliances and furniture. Beverage cartons are particularly important means of distributing cockroaches. The cartons are often contaminated with spilled syrups or malts, which attract cockroaches. Empty and unrinsed softdrink or beer bottles in night clubs, restaurants, markets, and homes form a part of the infestation chain. More than 200 cockroach nymphs have been found in a single soft-drink bottle brought to a market for exchange. Cockroaches may be found by the thousands in insulation in the walls of refrigerators and gas or electric ranges, surviving for months if necessary without access to their usual foods and feeding only on cast skins and dead insects. Cockroaches may become established in basements or crawl spaces, particularly, if these places are dark and damp, and they may then enter the building around utility pipes, air ducts or ventilators, or under doors.

Economic Importance of Cockroaches

Cockroaches rank with termites as the most important insects that must be dealt with by the owners or occupants of homes or other buildings or by pest control operators hired to combat insects. Flies and mosquitoes may be more important as household pests in some areas, but they breed elsewhere and then invade dwellings. Therefore, in the United States and many other countries, fly and mosquito control is generally considered to be primarily a community or regional responsibility, and the homeowner is generally left with the job of controlling only minor or seasonal infestation.

The Cockroach as a Domestic Nuisance

Cockroaches breed within the home, sharing with man his food, water, shelter, and warmth, and are thus active throughout the year. They consume any human or animal food or beverage, as well as dead animals and plant materials, leather, glue, hair, wallpaper, fabrics, and the starch in bookbindings. They contaminate food, imparting an unpleasant odor and taste, and may transmit disease organisms.

Although cockroaches are nocturnal, if they are abundant a few of them may be seen during daylight hours, particularly if articles in pantries, cabinets, and closets are moved about to disturb their hiding places. Other evidences of a cockroach infestation may be dead cockroaches, cast skins of the various nymphal instars, empty egg cases, and fecal droppings. The droppings are variable in size, and range from as small as "flyspecks" to as large as mouse droppings. Intersections and corners of shelves and the hinges of cabinet, pantry, and even closet doors show fecal stains when there are heavy infestations. The musty odor left on objects cockroaches contact may also be evident in heavy infestations.

Cockroaches as Disease Carriers

Many of the organisms causing diseases of man have been found on the tarsi and other body parts, or in the fecal pellets, of cockroaches, but these insects are most commonly implicated in the transmission of *Salmonella*, the causal agent of food poisoning. When the feces of American cockroaches infected with *Salmonella oranienburg* were spread on human foods and on glass, the bacteria survived for the following periods: on corn flakes, 3.25 years; on crackers, over 4.25 years; and on glass slides, over 3.67 years. Mice placed in jars containing minute quantities of infected cockroach feces became themselves infected in as brief a period as 1 day.

Parasitic toxoplasmosis is a disease that is believed to infect more than a third of all adult Americans at some time during their lives. It is usually a mild disease, but can be extremely serious in pregnant women, causing congenital defects in an unborn child while remaining mild or symptomless in the mother. The protozoan parasite, *Toxoplasma*, can be acquired by eating infected raw or undercooked meat. If the blood serum of an infected person is added to cultures of the parasites, the differing colours of fluorescence of the latter will indicate either a positive or negative reaction and thereby serve as a test for the disease.

Cats that consume birds or rodents carrying the parasite can become carriers. Filth flies have been found capable of transmitting infectious *Toxoplasma* cysts to human food 1 or 2 days after consuming infected cat feces. In laboratory tests, *Toxoplasma* parasites were isolated from the digestive tracts and feces of cockroaches as long as 7 and 10 days, respectively, after the cockroaches had last been in contact with infected cat feces. Filth flies and cockroaches might consume infectious cat feces and contaminate human food. They might also serve as food for birds and small rodents that could in turn be eaten by cats.

Goodwin recorded the successful passage of hepatitis B antigen (AB-Ag) through the alimentary tracts of American cockroaches, and identified the antigen in the feces of the insects. The hepatitis antigen was found in the feces of the cockroaches for up to 9 days after their initial exposure to the antigen-positive test meal.

The Cockroach Allergen

Some people are allergic to cockroaches. In one investigation among 253 normal persons, 7.5 per cent showed positive skin tests with extracts of cockroaches compared with 28 per cent of an unselected group of 114 allergic patients. Skin-sensitizing antibodies were present in the blood sera of positive reactors. The allergen is thermostabile. There is evidence that in some instances symptoms attributable to food allergy may be caused by cockroach allergen with which the food is contaminated. In an investigation to test the age of onset of skin reactivity, 38 out of 102 allergic children, ranging from infants to 12 years old, gave positive cutaneous reactions to body extracts of the German cockroach, *Blattella germanica*, compared with only 5 out of 100 nonallergic children. A 4-year-old asthmatic child was the youngest to give a positive reaction. In a later investigation, it was found that an extract of *B. germanica* caused an attack of asthma in 10 asthmatic persons with skin hypersensitivity to the extract and other allergens, but not in asthmatics without such skin hypersensitivity.

An allergen in the feces of *B. germanica* acts as an ingestant when it contaminates food and as an inhalant when dried fecal particles become incorporated with house dust.

Natural Enemies of Cockroaches

Species from at least 6 families of Hymenoptera had been recorded as developing on cockroach eggs. They quoted Edmonds as stating that evaniid egg parasites were so abundant in a home that the occupants considered them a nuisance, although the oriental cockroaches in the basement did not annoy them. Roth and Willis also listed many predators, including scorpions, spiders, dragonflies, mantids, bugs, beetles, wasps, ants, toads, frogs, lizards, birds, poultry, and various mammals. Cannibalism among cockroaches, including the devouring of othecae, has been noted by many investigators even when food was adequate.

When humidity is too high in cockroach culture colonies, mites may become numerous, and have been known to cause German cockroaches to drop their othecae prematurely, resulting in a low percentage of eggs hatching. Control of mites with acaricides has been known to increase the vigour of cockroach colonie. A pterygosomid mite *Pimeliaphilus podapolipophagus* Trgrdh, must actually feed on live cockroaches to survive; it cannot subsist on feces, cast skins, or dead insects. It has been accused of biting people, and its presence in homes is an indication of cockroach infestation.

Rearing Cockroaches Experimentally

A continuous supply of several species of cockroaches has been reared for experimental purposes in our laboratory at UCLA. These species are, in the order of what we consider to be their importance in California, the German, oriental, brownbanded, and American cockroach. Later the Australian and the brown cockroach were added.

The insects were originally reared in 15 gal (57-L) garbage cans with an electric barrier, later replaced by 20-gal (76-L) plastic trash cans. A double layer of strips of aluminum foil with a strip of insulation tape between them, and with the terminals from a 22.5-volt battery connected to the aluminum strips, serves as an electrified barrier to the escape of the cockroaches. In each trash can there is a poultry-watering device, Purrina Dog Chow, and 3 mazes of concentrically rolled and closely approximate layers of corrugated paper to serve as the harborage areas that cockroaches require for optimum development. The trash cans are covered with their usual plastic lids, provided with screens for ventilation.

It is important to prevent contamination of a colony by stray individuals of other species. It is especially important to keep German cockroaches out of colonies of other species because of the ability of this insect to thrive and rapidly increase its numbers at the expense of the others.

In the cockroach rearing room, boric acid powder applied under benches, radiators, appliances, cabinets, etc., will serve to permanently prevent a buildup of the cockroach population, in the rearing room and adjacent rooms, from those insects which escape when they are being removed for experimental work.

Twenty cans of German cockroaches and a can or more of each of the other 5 species are continuously maintained in our laboratory. The temperature is kept near 80 F in the rearing room. Provided that a water supply is constantly maintained, the author and his colleagues have had no indication that the maintenance of relative humidity above ambient is of any advantage in the rearing of cockroaches, at least in the rearing cans just described.

With ambient humidity and monthly removal of the debris at the bottoms of the cans, infestation of German cockroaches by parasitic mites has been avoided. American cockroaches drop their egg cases to the bottoms of their rearing cans; therefore, the cans are vacuumed only about twice a year, but their colonies do not produce so much debris and do not seem to be so susceptible to mite infestation as those of the German cockroaches.

German Cockroach

The German cockroach is generally the most important cockroach species in the United States and many other areas of the world in homes, apartment buildings, restaurants, and in the food preparation and food-storage rooms of hotels, hospitals, and other private and public buildings in which food is prepared, stored, or served.

Description: Adult German cockroaches are 1.3 to 1.6 cm long, pale brown or tan, and have 2 parallel dark streaks on the pronotum. They have chewing mouthparts. Their movements are very rapid when they are disturbed. They are nocturnal. If a few are seen crawling about in open spaces during daylight hours, this indicates that the infestation is already severe.

The female is darker and has a broader abdomen than the male, and is rounded posteriorly. If German cockroaches are placed on their backs, the males are easily distinguished by the yellowish, slender abdomen, tapering gradually to the posterior end. The abdominal cerci of the males have 11 segments and those of the females have 12. In their investigation of the German cockroach, the ratio of females to males to be 1.12:1. The male of the German cockroach could not detect the female from a distance, even when in close proximity, but had to make physical contact, ordinarily via the antennae. The antennae of the female, as well as other body regions, contain a chloroform-soluble, nonvolatile substance that will stimulate the male sexually. Thus, sex discrimination by males is mainly owing to "contact chemoreception." It follows that the male German cockroach probably cannot be attracted from a distance by means of synthetic sex attractants (provided they are available) as some other insect species can be.

When male and female meet, their antennae touch and vibrate against each other. The male then turns around, raises his wings to expose the orifices of a pair of dorsal glands located on the seventh and eighth tergites, and extends his abdominal segments to expose the openings of the 2 pairs of glands. The glands are not normally visible, being covered by the wings and by the margins of the preceding abdominal sclerites. Both wing-raising and extension of the abdomen are required to uncover the glands. The female eats a secretion from these glands. After she has fed for a few seconds, the male-pushes his abdomen farther back, and connection of the genitalia is made. The male then moves out from under the female, and the pair remain attached in a linear position for an average of about 86 minutes.

These authors also pointed out that the fact that German cockroaches do not need to be attracted from great distances, as do many other insects for the meeting of sexes, is understandable in view of the habits of cockroaches in general. The chance meeting of opposite sexes is enhanced by the fact that they are negatively phototactic and positively thigmotactic (principally guided by contact), and that they are gregarious, with large numbers seeking the same environment. They seek secluded and particularly very narrow hiding places, such as cracks, crevices, and voids, especially those having optimum temperature and moisture conditions. Attraction to the odor of the species is another factor that favours aggregation. Once the sexes have been brought together by these different stimuli, more refined stimuli lead to sexual discrimination.

Life Cycle: The German cockroach lays more eggs and has more generations per year than the other common domiciliary species, and therefore tends to increase in numbers much more rapidly. There may be 3 or 4 generations per year. In one investigation at a temperature of 76 F, the developmental period varied from 54 to 215 days, with an average of 103. It has been found to be as little as 60 days for males and 65 days for females at 95 F and 90 to 95 per cent relative humidity.

An individual German cockroach, however, lives an average of 200 days at room temperature and may live as long as 10 months. A comparison of the developmental periods of the 4 principal domiciliary cockroaches at 82 F and at ordinary room temperature.

Scientific Classification

Kingdom: Animalia
Phylum: Arthropoda
Class: Insecta
Subclass: Pterygota
Infraclass: Neoptera
Superorder: Dictyoptera
Order: Blattaria

Anatomy

Size:

The American cockroach is a large cockroach, adults are approximately 1-1/2 inches long (38mm).

Colour:

The adult is a shiny reddish brown to dark brown and has a yellow margin on the pronotum (region directly behind the head). Immature American cockroaches are also reddish brown to dark brown in colour and often have yellow markings on the abdomen.

Description:

Adult American cockroaches have wings and will occasionally fly. However, they are awkward fliers and prefer to run when disturbed. Male and female American cockroaches are about the same size and look very similar. Both have a pair of cerci, finger-like appendages, at the tips of their abdomens.

The cerci are used to detect air currents in the cockroach's surroundings. Male cockroaches have an additional set of appendages called styli on their abdomens. The styli are located between the cerci but are smaller and more delicate. The presence of styli is the easiest way to distinguish male from female cockroaches. American cockroaches resemble adults, except they are wingless. The American cockroach egg capsules are mahogany brown and about 1/3 inch long.

EYE: Cockroaches have "compound eyes" which allow them to see simultaneously in almost all directions around its body. Each cockroach eye is made up of 2,000 individual lenses. In contrast, human eyes have only one lens. Weird fact: cockroaches cannot see in red light, but see in green light very well.

LEG: Hairs on the legs of a cockroach provide an extra sense of touch.

ANTENNAE: Sniff, sniff. Antennae, or feelers, provide roaches with a sense of smell. Male cockroaches have special sense receptors that allow them to smell the odor produced by a female when she is looking for a mate.

CERCI: Cerci are the two little hairs on the rear end of a cockroach. They act like a motion detector. When something tries to sneak up on a cockroach, the cerci detects the slightest breeze or movement, and alerts the cockroach to run in the opposite direction.

MOUTH: A cockroach mouth can smell as well as taste. (Imagine eating with your nose!) It moves from side to side, not up and down the way human mouths do.

SALIVARY GLANDS: Yup, cockroaches know how to and actually do spit thanks to their salivary glands. They can't blow bubbles, though.

REPRODUCTION: Cockroaches don't date. To attract mates, the female cockroach produces an odor (a pheromone) that excites male cockroaches. Males produce a package of sperm that may keep female cockroaches pregnant for most of their lives. Depending on the species, females can produce egg cases with anywhere from 16 to 64 embryos (young cockroaches).

SKELETON: The skeleton of a cockroach is on the outside of its body. The skeleton is shed when the cockroach outgrows it.

ESOPHAGUS: A cockroach esophagus (a fancy word for throat) isn't as long as a human's, but it gets the job done. Food travels from the mouth, through the esophagus to the stomach.

CROP: The crop is a section at the end of the esophagus used to store food for a while. If a cockroach doesn't chew its food properly, a second set of teeth in its digestive tract, also called a gizzard, gives it another chance.

GASTRIC CAECEA: If cockroaches couldn't digest their food, they'd explode! The gastric caecea are little sacks inside a cockroach that keep germs and enzymes available to help a cockroach digest its food.

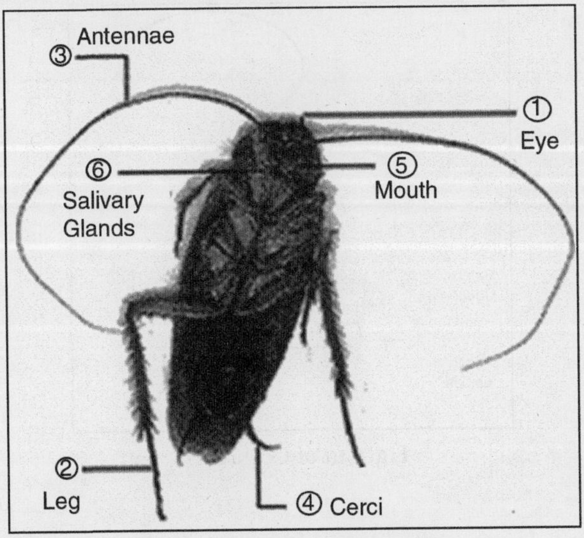

Fig. Anatomy of Cockroach

MIDGUT: The middle part of the intestine where the cockroach absorbs nutrients is called the midgut.

MALPIGHIAN TUBULES: You say po-tay-toe, I say po-tah-to. You say malpighian tubules, I say cockroach kidney. Either way, it's the same thing! Malpighian tubules clean waste out of the blood. And when they do their job they wiggle like snakes in the "blood" that sloshes around the cockroach's digestive tract.

COLON: Cockroaches not only have a colon, but a rectum, too! Hooray! Special glands in the rectum reabsorb water used during digestion, leaving behind excrement (poop).

HEART: A cockroach has a heart shaped like a tube. Its heart can even stop without the cockroach dying.

SPIRACLES: Cockroaches breath through a bunch of little holes called spiracles in the side of its body. Thin branching tubes carry oxygen from the air to all the parts inside.

FAT BODY: Ever crunch a cockroach? The white gunk that squirts out is the fat body. Like fat in humans, it stores energy. It also breaks down nutrients to provide energy, and helps treat or detoxify insecticides that may be sprayed on the roach (no wonder they're so tough to get rid of).

BRAIN: The cockroach brain is spread throughout its body. The head holds a bit of a nervous system, while the rest of the brain is scattered along the ventral (belly-side) part of its body. So, now you understand that if you cut off the head of a roach, it will still live for as long as one week; eventually dying from starvation or thirst.

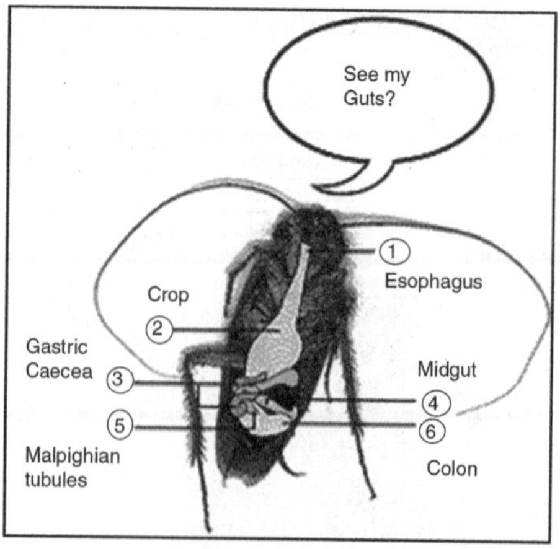

Fig. Gut of Cockroach

Habitat

American cockroaches are a "peridomestic species" this means that they generally live outdoors. However, populations can also move indoors and live in human structures. American cockroaches usually live in moist, humid environments but can survive in dry areas if they have access to water. The cockroaches prefer temperatures between 70°F and 85°F and will not survive 15°F. In structures, American cockroaches are common in areas where food is prepared or stored and moisture is plentiful. They are frequently found in restaurants, grocery stores, and bakeries. They are also commonly associated with boiler rooms, sewers, steam tunnels, and other warm, moist locations. In residential and commercial buildings, American cockroaches usually infest basements, crawl spaces, bathrooms, and decorative landscaping. Indoor populations tend to forage outdoors during warm weather. Likewise, during the winter months, populations established outdoors will venture inside seeking moisture and warmth.

Life Cycle

After mating, the female American cockroach will produce an egg case in three to seven days. She will carry the egg case protruding from the tip of her abdomen for another two days. The egg case is then deposited in a hidden location and glued to a surface with the female's saliva. Hiding the egg case helps to protect it from predators, parasites, and pest technicians. Each egg case contains an average of 15 embryos. The immature cockroaches will emerge in 24 to 38 days under warm conditions. They will complete their development and become reproductive in six to 12 months. Adult American cockroaches can live approximately a year to a year and a half. An adult female will produce between six and 14 egg cases during her lifetime.

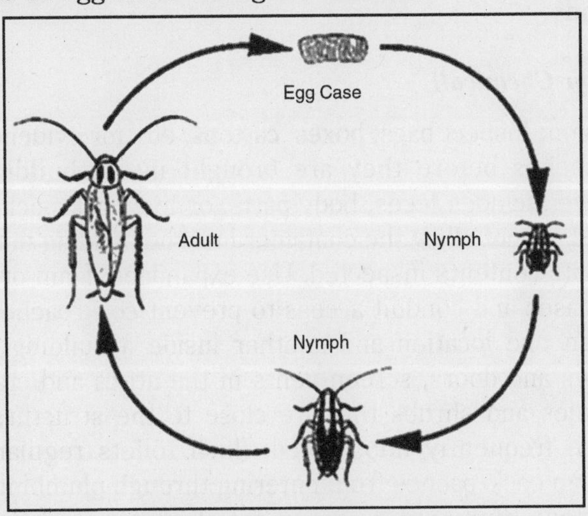

Fig. Life Cycle of Cockroach

Nymphs hatch from egg cases that the adult female cockroach deposits. As the nymph grows, it sheds its "skin" or exoskeleton so it can get larger. Each nymphal stage is larger than the previous one, and an adult cockroach emerges after the last nymphal stage. The figure shows a simplified image of the life cycle (egg stage, nymph stage and adult stage) of the German cockroach.

Type of Damage

American cockroaches feed on a wide variety of materials, including cosmetics, beer, potted plant shoots, wallpaper paste, soap, postage stamps, and fermenting fruit. They can foul human food, clothing, paper goods, and surfaces with their feces and body parts. American cockroaches also produce a strong unpleasant odor. This characteristic odor is not only detectable in infested buildings but is also transferred to items that the cockroaches crawl across when foraging. So a pest management professional can often detect an American cockroach infestation before he has actually seen any cockroaches.

Health Risks

When American cockroaches aggregate, their presence is primarily an aesthetic nuisance. However, members of this species are also known to carry infectious bacteria on their bodies and in their gut. These bacteria may be transferred to food and other items that the cockroaches contact.

Several bacteria commonly associated with American cockroaches are known to cause food poisoning, dysentery, and diarrhea in humans. American cockroaches also produce allergens on their bodies and in their fecal material. While American cockroaches are not considered to be major culprits of human respiratory problems like some other cockroach species, they have been implicated as a cause of allergic dermatitis and childhood asthma.

Control Methods

Prevention (Non-Chemical)

- Exclusion: Inspect bags, boxes, cartons, etc. for evidence of American cockroaches before they are brought into a building. Cockroach evidence includes feces, body parts, or live cockroaches. If evidence is found, do not allow the container into the building until it is emptied and all its contents inspected. Use expandable foam or caulk to block pipe chases and conduit access to prevent cockroaches from moving between one location and another inside a building. Weatherproof windows and doors, screen vents in the attics and crawlspaces, and trim trees and shrubs that are close to the structure. In buildings that are frequently unoccupied, flush toilets regularly to prevent American cockroaches from entering through plumbing traps.
- Sanitation: Eliminate as many moisture sources as possible. Fix leaking pipes, store recyclables (cans and bottles) outside the structure, insulate pipes to prevent condensation, and do not leave water standing in the sink. Eliminate all cockroach food sources. Clean

up spilled pet food every day. Use a vacuum attachment to remove fallen crumbs from behind the stove and between cabinets. Avoid leaving unwashed dishes on counter tops and keep food in tightly sealed containers. Rinse cans and bottles before putting them in the trash or recycling. Take the trash out every night and place it in dumpsters or receptacles with tight fitting lids. Remove clutter. American cockroaches can use boxes, bags, paper goods, old clothes, and magazines as places to hide. Eliminating cockroach harborage areas will make the environment much less conducive to cockroach infestation.

- Monitoring: Sticky traps are used to detect and monitor American cockroach infestations. Traps such as the Roach Motel can be placed at many locations throughout a structure. The traps should be left in place at least 24 hours so they are present when the cockroaches are most active, at night. Cockroaches caught in monitoring traps can let you know that there is a developing infestation in the structure. Although sticky traps can detect cockroach populations, they cannot control them. They usually will not catch enough cockroaches to control the population. Sticky traps are not recommended for outside use because they are not weather resistant and may capture non-target insects and small vertebrates.

Treatment (Chemical)

- Baits: Cockroach baits consist of a toxicant (active ingredient) formulated in a food source. American cockroach baits are usually packaged as dusts, gels, pastes, or granules. Dust baits are applied into cracks, crevices, and wall voids with a bulb duster or as an aerosol. Pastes and gels are usually purchased in a syringe, which can be used alone or inserted into a bait gun for more precise application. Most granular baits are applied outside in landscaping around the perimeter of the structure. However, some may be applied in wall voids using a bulb duster. The most common and effective active ingredients formulated in American cockroach baits include: hydramethylnon, fipronil, abamectin, and imidacloprid.
- Insect Growth Regulators (IGRs): IGRs for cockroach control are generally not available for homeowner purchase but are professional-use-only products. These compounds do not kill cockroaches. Instead, they disrupt the normal development of immature cockroaches making them functionally sterile as adults. The cockroach population then dies of attrition. Because of the slow-acting nature of IGRs, they are frequently used in combination with baits for faster control. While IGRs are certainly capable of controlling American cockroach populations, they are not commonly used for peridomestic cockroach

control because the cockroaches are not breeding indoors. However, to treat incipient populations in steam tunnels or boiler rooms, IGRs are available in spray formulations or point-source dispensers (the IGR is released on filter paper contained in a permeable plastic station and then transmigrates throughout the infested area). The most common IGRs (active ingredients) used in cockroach control products are hydroprene (Gentrol Point Source) and pryiproifen (Nylar).

- Inorganic Dusts: These dusts are normally applied for indoor cockroach control with a squeeze-bulb duster, which puffs the dust into cracks and crevices. Examples of these dusts are silica aerogel and boric acid. Silica aerogel is a finely ground silica (similar to glass) that adheres to the cuticle of the cockroach and absorbs the protective wax covering. This dehydrates the cockroach and eventually kills it. Boric acid is a stomach poison that also adheres to the cockroach when it walks across the dust. The cockroach will then ingest the dust while grooming itself.

BEETLE

Beetles are the group of insects with the largest number of known species. They are classified in the order Coleoptera which contains more described species than in any other order in the animal kingdom, constituting about 25% of all known life-forms. 40% of all described insect species are beetles (about 350,000 species), and new species are frequently discovered. Estimates put the total number of species, described and undescribed, at between 5 and 8 million.

Beetles can be found in almost all habitats, but are not known to occur in the sea or in the polar regions. They interact with their ecosystems in several ways. They often feed on plants and fungi, break down animal and plant debris, and eat other invertebrates.

Some species are prey of various animals including birds and mammals. Certain species are agricultural pests, such as the Colorado potato beetle Leptinotarsa decemlineata, the boll weevil Anthonomus grandis, the red flour beetle Tribolium castaneum, and the mungbean or cowpea beetle Callosobruchus maculatus, while other species of beetles are important controls of agricultural pests. For example, beetles in the family Coccinellidae ("ladybirds" or "ladybugs") consume aphids, scale insects, thrips, and other plant-sucking insects that damage crops.

Scientific Classification

Kingdom: Animalia
Phylum: Arthropoda
Class: Insecta
Subclass: Pterygota

Infraclass: Neoptera
Superorder: Endopterygota
Order: Coleoptera

Anatomy

Bearing in mind the wide diversity and number of species the anatomy of beetles is quite uniform. Beetles are generally characterised by a particularly hard exoskeleton, and the hard wing-cases (elytra) which tend to cover the hind part of the body and protect the second wings, the alae. The elytra are not used in flying, but generally must be raised in order to move the hindwings. In some cases the ability to fly has been lost, characteristically in families such as Carabidae and Curculionidae. After landing, the hindwings are folded below the elytra.

In a few families, both the ability to fly and the wing-cases have been lost, with the best known example being the "glowworms" of the family Phengodidae, in which the females are larviform throughout their lives.The bodies of beetles are divided into three sections, the head, the thorax, and the abdomen, and these in themselves may be composed of several further segments. The eyes are compound, and may display some remarkable adabtability, as in the case of the Whirligig beetles (family Gyrinidae), in which the eyes are split to allow a view both above and below the waterline. The dorsal appendage aids the beetle in stalking prey. Antennae can vary greatly and may be filiform, claviform, flabellate or genticulate.Oxygen is taken in via a tracheal system: this takes air in through a series of tubes along the body which is then taken into increasingly finer fibres. Pumping movements of the body force the air through the system. Although beetles have blood, it is not used for oxygen transference, although a heart is present. Adult beetle

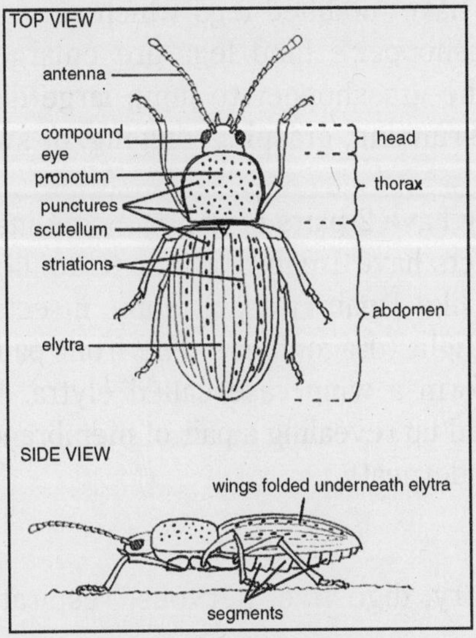

Fig. Anatomy of Beetle

HEAD

Contains brain, mouthparts eyes and antennae. The head functions mainly for food and sensory intake and information processing.

Mouth

- Insect mouthparts have evolved for chewing (beetles, caterpillars), piercing-sucking (aphids, bugs), sponging (flies), siphoning (moths), rasping-sucking (thrips), cutting-sponging (biting flies), and chewing-lapping (wasps).

Antenna

- Antennae are the insect's feelers. They are sensory organs that can smell, taste, detect sound and air movements. Some insects navigate solely using the antennae. Most insects use them to find and identify food.

Compound eye

- Allows the insects to see. Made up of many small lenses arranged in a globe which can give the insect a good field of view. Compound eyes do not produce good definition but they are very sensitive to movement. This is why flies cannot tell that there is glass in a window but can skillfully avoide being swatted by a newspaper.

THORAX

The thorax provides structural support for the legs (three pairs) and, if present, for one or two pairs of wings.

Legs

- Insects have six legs. Legs are segmented and vary in shape greatly. Some insects have modified legs which serve special purposes. For example, grasshopper's hind legs are enlarged and very muscular which allow the grasshopper to jump large disances. The legs may be adapted for running, grasping, digging, or swimming.

Wings

- Insects usually have 2 pairs of wings. Some insects have no wings at all. Many insects have 1 pair of wings (true flies) others have 2 pairs which behave like 1 pair (wasps). Some insects use all four wings to fast dynamic flight (dragonflies). The front pair of beetles wings are hardened to form a wing case called elytra. When beetles fly, the elytra are raised up revealing a pair of membraneous true wings which unfold from underneath.

ABDOMEN

Contains circulatory, digestive, nervous, respiratory and reproductive systems.

Circulatory system

- The internal anatomy of insects is characterized by an open circulatory system; with the exception of a heart and an aorta, there are few blood vessels – insect blood simply flows around inside the body cavity. The blood of the insect is rarely red, usually yellow or greenish. It's predominant function is the transportation of food, waste and hormones, and secondarily provides oxygen and carbon dioxide exchanges for remote body parts not directly served by the trachea.

Digestive system

- As with most animals, digestion starts at the mouth, which leads into the alimentary canal, a tube extending from the mouth to the anus. The alimentary canal is divided into three regions; the foregut (stomodaeum), the midgut (mesenteron) and hindgut (proctodaeum).

Nervous system

- The insect nervous system transports and processes information received from the sense organs (sight, smell, taste, hearing, and touch).

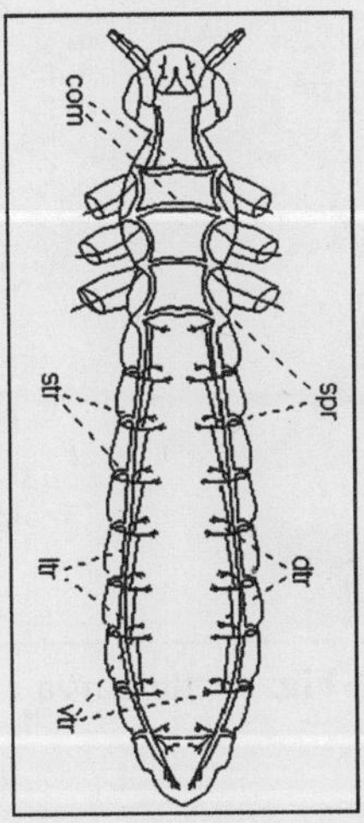

Fig. Beetle Abdomen

Respiratory system

- Gaseous oxygen is taken into the insects body through spiracles, travels via the trachea throughout the body, into the tracheoles which reach the tissues, is metabolized and leaves the body in the form of carbon dioxide in the opposite sequence although up to a fourth or so of the carbon dioxide may be eliminated through the body wall. The trachea are lined with cuticle, which is thickened in spiral rings for rigidity. The tracheoles are simple cells with thin walls sometimes filled with fluid. This process is usually accomplished by simple diffusion. The spiracles are located on the sides and number from 1 to 10 pairs.

com = commissural trachea; dt = dorsal trachea; lt = main longitudinal trachea; spr = spiracles; st = spiracular trachea; vt = ventral trachea.

Development

Beetles are endopterygotes with complete metamorphosis. The larva of a beetle is often called a grub and represents the principal feeding stage of the life-cycle. The eggs of beetles are minute but may be brightly coloured, they are laid in clumps and there may be from several dozen to several thousand eggs laid by a single female.

Once the egg hatches the larvae tend to feed voraciously, whether out in the open such as with Ladybird larvae, or within plants such as with leaf beetle larvae.

As with lepidoptera, beetle larvae pupate for a period, and from the pupa emerges a fully formed beetle or imago. In some cases there are several transitory larvae stages and this is known as hypermetamorphosis; examples include the blister beetles (family Meloidae).

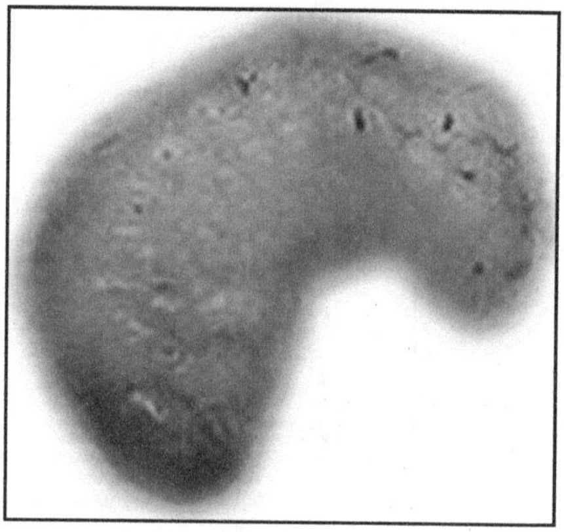

Fig. Beetle Larva

Reproduction

The larval period of beetles varies between species but can be as long as several years. Adults have an extremely variable lifespan, again, from weeks to years.

Beetles may display some extremely intricate behaviour when mating. Smell is thought to play significant importance in the location of a mate.

Conflict can play a part in the mating rituals for example in species such as burying beetles (genus Nicrophorus) where localised conflicts between males and females rage until only one of each is left, thus ensuring reproduction by the strongest and fittest. Many beetles are territorial and will fiercly defend their small patch of territory from intruding males.

Pairing is generally short but in some cases will last for several hours. During pairing sperm cells are transferred to the female to fertilise the egg.

Parental Care

As befitting such a large order, the parental care between species varies widely. It ranges from the simple laying of eggs under a leaf to scarab beetles, which construct impressive underground structures complete with a supply of dung to house and feed their young.

There are other notable ways of caring for the eggs and young, such as those employed by leaf rollers, who bite sections of leaf causing it to curl inwards and then lay the eggs, thus protected, inside. Generally the number of eggs laid is an indicator of the level of parental care subsequently employed, as they are inversely proportional.

Predation

Beetles and larvae have evolved to employ a variety of different strategies for avoiding being eaten. Many employ simple camoflage to avoid being spotted by predators. These include the leaf beetles (family Chysomelidae) that have a green colouring very similair to their habitat on tree leaves.

A number of longhorn beetles (family Cerambycidae) bear a striking resemblance to wasps, thus benefitting from a measure of protection. Large ground beetles by contrast will tend to go on the attack, using their strong mandibles to forcibly persuade a predator to seek out easier prey. Many species, including lady beetles and blister beetles, can secrete poisonous substances to make them unpalatable.

Evolutionary History and Classification

Beetles entered the fossil record during the Lower Permian, about 265 million years ago.

The four extant suborders of beetle are these:

- Polyphaga is the largest suborder, containing more than 300,000 described species in more than 170 families, including rove beetles (Staphylinidae), scarab beetles (Scarabaeidae), blister beetles (Meloidae), stag beetles (Lucanidae), and true weevils (Curculionidae).

These beetles can be identified by the cervical sclerites (hardened parts of the head used as points of attachment for muscles) absent in the other suborders.

- Adephaga contains about 10 families of predatory beetles, includes ground beetles (Carabidae), predacious diving beetles (Dytiscidae) and whirligig beetles (Gyrinidae). In these beetles the testes are tubular and the first abdominal sternum (a plate of the exoskeleton) is divided by the hind coxae (the basal joints of the beetle's legs).
- Archostemata contains four families of mainly wood-eating beetles, including reticulated beetles (Cupedidae) and telephone-pole beetles (Micromalthidae).
- Myxophaga contains about 100 described species in four families, mostly very small, including skiff beetles (Hydroscaphidae) and minute bog beetles (Sphaeriusidae).

These suborders diverged in the Permian and Triassic. Their phylogenetic relationship is uncertain, with the most popular hypothesis being that Polyphaga and Myxophaga are most closely related, with Adephaga an outgroup to those two, and Archostemata an outgroup to the other three. The extraordinary number of beetle species poses special problems for classification, with some families consisting of thousands of species and needing further division into subfamilies and tribes.

FLY

True flies are insects of the order Diptera (Greek: di = two, and *pteron* = wing), possessing a single pair of wings on the mesothorax and a pair of halteres, derived from the hind wings, on the metathorax. The presence of a single pair of wings distinguishes true flies from other insects with "fly" in their name, such as mayflies, dragonflies, damselflies, stoneflies, whiteflies, fireflies, alderflies, dobsonflies, snakeflies, sawflies, caddisflies, butterflies,calliphorid flys or scorpionflies. Some true flies have become secondarily wingless, especially in the superfamily Hippoboscoidea, or among those that are inquilines in social insect colonies. Diptera is a large order, containing an estimated 240,000 species of mosquitos, gnats, midges and others, although under half of these (about 120,000 species) have been described.[1] It is one of the major insect orders both in terms of ecological and human (medical and economic) importance. The Diptera, in particular the mosquitoes (Culicidae), are of great importance as disease transmitters, acting as vectors for malaria, dengue, West Nile virus, yellow fever, encephalitis and other infectious diseases.

Scientific Classification

Kingdom: Animalia
Phylum: Arthropoda
Class: Insecta

Subclass: Pterygota
Infraclass: Neoptera
Superorder: Endopterygota
Order: Diptera

Little House Fly

Males of this cosmopolitan fly are often found in houses; the females are rarely found indoors. They are more uniformly distributed than the house flies throughout the average home. In one investigation, 65.6 per cent of house flies, but only 40.9 per cent of little house flies, were found in the kitchen. Males of the little house fly tend to fly about aimlessly in the middle of the room or some shaded area outside, often hovering or making jerky, zigzag flights.

The little house fly appears earlier in the spring than the house fly, and in regions of high temperature it tends to disappear during the summer. In many areas, this difference is possibly explained by the fact that the little house fly becomes active at a lower temperature, and also suffers heat paralysis at a lower temperature. This difference in temperature requirements may account for the fact that in subarctic regions, Fannia canicularis often replaces Musca domestica, and is, for example, the common house fly of Iceland. This and other species of *Fannia* are attracted to honeydew, and are often found swarming under aphis-infested vines or trees.

Description: The little house fly resembles the house fly, *Musca domestica*, but is smaller (3 to 5 mm long) and more slender. The wings in repose are parallel and partly folded over each other, whereas those of the house fly are held in a diverging position. The abdomen of the female *F. canicularis* is ovoid and gray, while the male's is tapering, with pale yellow, semitransparent patches.

Life Cycle: The little house fly breeds in animal excrement or decaying vegetable material. Floatlike appendages on the eggs enable them to float on top of semiliquid material. The larva has spiny processes along the sides and top of the body. These have fine, whiplike hairs that are said to propel it through semiliquid media. The larvae change from translucent white to light brown as they mature.

Like those of the house fly, *F. canicularis* larvae move to a dryer environment near the surface of their food medium before pupating. The pupae resemble the third-instar larvae, but are darker and somewhat shorter. At 80 °F (27 °C) and 65 per cent relative humidity, the approximate durations of the immature stages were recorded as follows: egg, 1.5 to 2 days; larva, 8 to 10 days; pupa, 9 to 10 days. The period from egg to adult was 18.5 to 22 days, and from egg to egg, 22 to 27 days. About 50 per cent of the male fly population died within 14 days, and 50 per cent of the females died within 24 days. In another experiment, the life cycle took 24 to 29 days.

Latrine Fly

The latrine fly resembles the little house fly in appearance and life history, but its thorax is bluish black, contrasted to the thorax of *Fannia canicularis*, which is blackish gray. Also, a grounded protuberance of the median point of the middle pair of legs can be used to distinguish *F. scalaris*. It prefers human to animal excrement, and has been known to cause myiasis in man.

Coastal Fly

This species, of South American origin, is widely distributed throughout the United States. It is a little over half the size of *Fannia canicularis*, and is shiny black in colour. It sometimes occurs in large numbers, but is much less important than the little house fly as a pest.

Fannia Benjamini Malloch

This species is known from various areas of the West, particularly California. The adult is 3.5 to 4.5 mm long, with a brown-pollinose, unstriped thorax and a dark abdomen without a trace of the yellow translucence found in some species, such as the male of *Fannia canicularis*. The antennae and palpi are yellow, as are also the legs, with the exception of the black tarsi. The female is much lighter-coloured than the male.

The expansion of city suburbs into the foothills, as well as the increasing use of such areas for recreational purposes, is bringing *F. benjamini* into sharper focus as a pest. It is common in California along streams, particularly in groves of oak trees. It is a very annoying pest, and persists in its attempts to enter the ears, nostrils, and eyes, apparently attracted by perspiration and mucus. This species and *F. canicularis* have been implicated as vectors of a nematode eye worm, *Thelazia californiensis* Price. The known hosts of this nematode, besides man, are dogs, cats, horses, sheep, deer, coyotes, and bears.

House Fly

The house fly is cosmopolitan within certain latitudes, and is important not only because it annoys man and domestic animals, but also because its habit of crawling on garbage and excrement, as well as on humans and their food, makes it a potential carrier of various disease-provoking pathogens. Pathogens isolated from the house fly include those capable of causing diarrhea, dysentery, typhoid, cholera, poliomyelitis, intestinal worms, eye infections, yaws, anthrax, and tularemia. As with cockroaches, flies have been implicated as vectors of *Salmonella* bacteria, responsible for food poisoning and gastric infections. The house fly appears to be even more important as a potential vector of *Shigella* than of *Salmonella*.

It is not possible to incriminate the house fly as a transmitter of disease organisms with the same certainty with which, for example, *Anopheles* mosquitoes are identified beyond doubt in the transmission of malaria, or lice in the transmission of typhus. Flies are accused of spreading infections that can also be spread with the fingers or in contaminated food or water. They have been found to harbor over 100 different species of pathogenic organisms, and are charged with transmitting over 65 human and animal diseases, but only on circumstantial evidence. For example, outbreaks of diarrheal diseases are closely correlated with seasonal increases in fly abundance. Also, fly control has been closely correlated with a decline in the incidence of disease.

The scarcity of soundly documented cases of field transmission of disease by flies results in inadequate knowledge of how important they are in the natural transmission of the many pathogens commonly associated with them. Therefore, it is still principally the fact that they are annoying to have around domestic premises that motivates us to control them.

Description: The general size and appearance of the house fly are well known. It is 4 to 7.5 mm long, and may be distinguished from other house-frequenting flies by wing venation, the 4 dark stripes dorsally and lengthwise on the thorax, and the 2 velvety stripes, silver above and gold below, on its face.

The eyes of the female are more widely separated than those of the male. With its sponging-sucking mouthparts, the house fly can ingest only liquid material, but some solid foods can be liquefied by means of regurgitated saliva, and thus ingested. All the aspects of house fly morphology, as well as biology and bionomics, were discussed in great detail by West.

Life Cycle: House flies lay their eggs in warm, moist materials that are suitable for larval growth, such as animal excrement, garbage, piles of lawn clippings, decaying vegetables and fruits, or soil contaminated with such material. Immature stages are shown in plate IV, 1, and figure 158. During warm weather, the eggs, which are laid singly but piled together in small masses, can hatch within a day. The cream-coloured larvae (maggots), distinguished by their being narrow at the head end, burrow into the food material in which they hatched, and may complete their 3 instars and pupate in a week or less.

As the pupation period approaches, the larvae seek the drier outer regions of their manure pile or other breeding medium, or may leave it to pupate a considerable distance away in the ground, in loose material, or under boards or stones.

Under warm conditions, the pupal stage may last 4 to 6 days. Under the warmest natural conditions, a house fly may complete its life cycle in about 8 days, but this period is considerably longer at lower temperatures and when the larvae have had a less than optimum food supply. There may be as many as 10 or 12 generations in one summer. The adult flies generally live from 15 to 25 days, depending on temperature. In colder areas, the adults may overwinter. The potential increase in fly population in a single season is obviously enormous.

Stable Fly

Another fly that superficially resembles the house fly but rarely occurs in houses is the stable fly. Unlike the house fly, it has a stiff proboscis projecting forward from the lower part of the head, enabling both males and females to inflict painful bites and suck out blood. Stable flies normally are serious pests of domestic animals, but on occasion they also attack humans. They are most likely to bite in the region of the ankle. Socks and stockings offer no protection.

Description: The stable fly is more robust than the house fly and has a broader abdomen. It can also be differentiated by the difference in wing venation (A and B of the figure), by a prominent, light-coloured area between the longitudinal bands in the middle of the thorax, and by the presence of a number of nearly round, dark spots on the upper surface of the abdomen. When the stable fly is at rest, its wings are held widely apart instead of projecting backward on top of the abdomen, as with the house fly. The stable fly rests with the head cocked upward and abdomen held against the resting surface, whereas the house fly often rests in a crouching position, with head down and body parallel to the surface.

Life Cycle: Stable flies breed in decaying hay, straw, or grain; piles of lawn clippings or rotting fruit; and various manures mixed with grain or bedding. They also breed in the manure of dairy cattle and in poultry droppings when they are "stacked." They are often found in large numbers on beaches, where they lay their eggs in piles of seaweed, and are sometimes called "beach flies." Unlike *Fannia* species, the immature stages of the stable fly somewhat resemble those of the house fly. In warm weather, adults may emerge in 20 to 25 days after the eggs are laid, but this period can be extended to several months at lower temperatures.

Cluster Fly

Cluster flies derive their common name from their habit of entering a house in the fall and gathering together in a cluster that may emit a sickly sweetish odor. They tend to gather and hibernate in secluded areas, often in wall voids, and hibernate in them. Cluster flies and face flies (*Musca autumnalis*) often become trapped in large numbers in wall voids, attracting larder beetles (*Dermestes lardarius*). The beetles feed on dead flies and then migrate to other parts of the house, causing the occupants to believe they have an infestation of carpet beetles.

In the spring, cluster flies again become pests as they migrate from wall voids to the living space, via window pulleys, around baseboards, and through other small openings, and then crawl sluggishly over the walls. They are usually found in windows and other out-of-the-way places, and do not fly about noisily like other house-entering calliphorids. In this respect, they are not such nuisances.

Description: Cluster flies closely resemble house flies, but are larger, more robust, and slower in their movements. They are about 8 mm long, dark gray, with checkered black and silvery abdomens, and with many golden hairs on the thoraxes of newly emerged specimens which may be lost in older ones. The stripes on the thorax are not so prominent as those of the house fly. The wings overlap when at rest. *Pollenia rudis* can be distinguished from all other synanthropic North American calliphorids by its abdomen being nonmetallic in appearance and being covered by a pollenlike dust. A striking characteristic of the biology of cluster flies is that the larvae are parasitic on earthworms. Eggs are deposited in cracks in the soil, hatch in 3 days, and the larvae enter earthworms at almost any point on the body wall. The total developmental period of the fly varies from 27 to 39 days. Its breeding habits result in the cluster fly being less of a hazard than other calliphorids as a mechanical carrier of disease organisms.

Control of Cluster Flies

Lawns have been treated with 2.5 per cent heptachlor granules at 3 lb per 1,000 sq ft (1.36 kg per 93 sq m), or with a spray of heptachlor 25 per cent emulsifiable concentrate at 0.25 pt in 3 gal (0.12 L in 11 L) of water per 1,000 sq ft to kill earthworms, and thus prevent the development of cluster flies. The lawn should then be sprinkled to force the toxicant into the soil.

Blow Flies

The blow flies (greenbottle and bluebottle flies) are large scavenger flies, with a metallic blue or green sheen. They may become abundant in and around homes, particularly if they are located near such places as slaughterhouses, meat-processing plants, and garbage dumps. However, it should be borne in mind that some of these insects are stronger fliers than the house fly, and at least a small percentage may travel great distances from the point of dispersal. For example, in one experiment, 13 per cent of the black blow flies, *Phormia regina*, were trapped between 4 and 28 miles (6.4 and 46 km) from the point of origin, and more than 8 per cent at or beyond 11 miles (18 km) compared with only 3 per cent of house flies trapped at or beyond 4 miles.

Blow flies are most active during warm, sunny weather. When it is cool and cloudy, they tend to retreat to resting sites, usually on vegetation adjacent to areas where they oviposit. *Phormia regina* tends to disperse more widely than the other species, and in warm weather it often rests on sunlit walls. Some blow flies, such as *Phaenicia sericata* and *Eucalliphora lilaea*, freely enter houses, and because of their breeding habits and the fact that they feed and oviposit on food, mechanical transmission of disease becomes a distinct sanitary consideration. Meat or fish must be protected from oviposition by blow flies, not only for aesthetic and sanitary reasons, but also because the eggs and larvae of blow flies and some other species, when swallowed, may remain alive for some time, causing acute gastric or enteric disorders. Blow flies have been known to breed in the carcasses of dead birds or rodents in attics or wall voids of houses.

These may be sources, not only of adult flies, but even of the larvae, which are sometimes found crawling about in the living space of the house.

Because of their large size, buzzing sound, and tireless flying about throughout the house, blow flies are generally more annoying than house flies. They tend to fly to windows, for like all flies, except perhaps the males of *Fannia*, they are attracted to bright sunlight.

Although blow flies usually lay their eggs on meat or dead animals, when meat is not present they may oviposit on animal excrement or decaying vegetable matter. One female may lay as many as 600 eggs. Larvae first feed on the surface and then burrow down to less-decayed material. The full-grown larvae burrow into the ground to pupate, usually in the top 5 cm of soil. Tlere are several generations a year. The winter is normally spent in soil by full-grown hibernating larvae.

Anatomy

Antenna

Antennae are the insect's feelers. They are sensory organs that can smell, taste, detect sound and air movements. Some insects navigate solely using the antennae. Most insects use thhem to find and identify food.

Head

Contains brain, mouthparts eyes and antennae. Attatched to an insects head are its eyes, its antennae and its jaws. Some insects have biting mouthparts, others, such as bugs, have sucking mouthparts.

Compound Eye

Allows the insects to see. Made up of many small lenses arranged in a globe which can give the insect a good field of view. Compound eyes do not produce very good definition but they are very sensitive to movement. This is why flies cannot tell that there is glass in a window but can skillfully avoide being swatted by a newspaper.

Thorax

Contains muscles which drive the legs and wings which are attatched to it.

Abdomen

Contains digestive system and reproductive system. All organs are supplied with oxygen by a network of trachea. The trachea are connected to the surface of the insect's 'skin' where openings are called spiracles.

Legs

Insects have six legs. Legs are segmented and vary in shape greatly. Some insects have modified legs which serve special purposes. For example, grasshopper's hind legs are enlarged and very muscular which allow the grasshopper to jump large disances.

Wings

Insects usually have 2 pairs of wings. Some insects have no wings at all. Many insects have 1 pair of wings (true flies) others have 2 pairs which behave like 1 pair (wasps). Some insects use all four wings to fast dynamic flight (dragonflies).

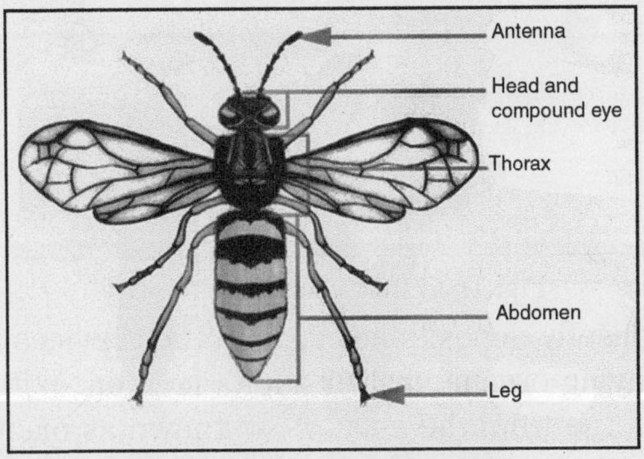

Fig. Anatomy of Housefly

ADULT MALES AND FEMALES ....

Males and females are hard to distinguish. Females are usually larger and can extend the tip of the abdomen to form an ovipositor which is used to lay eggs. Sometimes males have enlarged eyes which meet on top of the head.

Immatures (Different Dtages)

Flies are holometabolous, therefore they have four distinct morphological stages; egg, larva (maggot), pupa and adult. After hatching from the egg, larvae molt twice as they grow. Molting of maggots is difficult to see. Larvae are maggots with a legless soft body except for the dark mouth hooks. Pupae are dark, and look like a small barrel.

Development

The housefly life cycle closely mirrors that of most insects: a basic cycle that begins with an egg, then develops through a larva phase, a pupa phase, and finally, into an adult. During a warm summer — optimal conditions for a housefly — the cycle, from fertilized egg to adult, spans a mere seven to 10 days. After a male housefly chases down and fertilizes a female counterpart, she's ready to lay her eggs. Houseflies are solitary creatures. Like the rest of the insect world, males and females do not stick together after mating and, unlike nesting insects, females do not care for or protect eggs. Females simply leave the eggs where they will be safe from predators and have plenty to eat upon hatching.

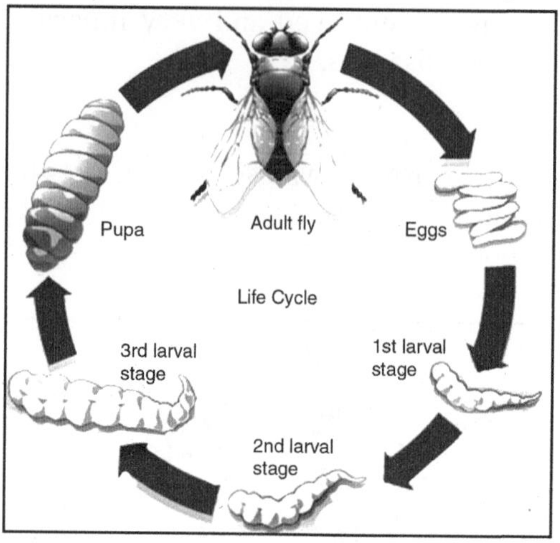

Fig. The Life Cycle of a Housefly

The female housefly deposits her eggs in the crevices and corners of the same kinds of decaying organic matter adults feed on. Within a day, the first larvae begin to emerge from the eggs. Also known as maggots, these worm-like creatures are little more than fleshy, sectionless tubes with hooked mouth parts used for feeding. The maggots grow rapidly. In less than two days they've doubled in size and therefore must molt. Molting is a process common to many invertebrates through which a growing insect sheds its former exoskeleton and grows a new one.

A maggot will molt twice more, emerging larger and more developed each time. Following its third molt, larvae will burrow deep into the substance they've been feeding on. Their skins will darken and harden as they enter the pupa stage. Inside this protective shell, the larva will fully develop the body segments and appendages of an adult housefly.

The only visible addition to the emerging housefly is a swollen bump on the fly's head, used to break through the shell. Since the housefly doesn't have teeth or jaws to chew its way out, it uses this fluid-filled pouch to break through the pupae shell.

Once fully emerged, the bump deflates back into the fly's head. A new adult housefly has, at most, three months to reproduce before it dies. With so many predators, a housefly's average lifespan is even shorter: 21 days. Luckily for the housefly, the phrase "breeding like flies" isn't just a figure of speech. Each female can lay up to 900 eggs during her brief life. The very thought of a housefly infestation may prove too disturbing for many homeowners. However, the next page will describe how a manageable amount of houseflies helps regulate the local ecosystem.

Food

Larvae feed on decaying meat and feces. Adult flies feed on sugary food of any kind, including nectar and rotting fruit.

Habitat

Flies live in garbage and wherever animal feces are available. Dead animals attract flies within hours after death. Most flies are diurnal.

Predators

Many birds, bats, spiders, and insects such as dragonflies eat the adults. Predatory and parasitic insects eat the larvae.

Interesting Behaviors

- The eyes of flies are among the most complex in the insect world. They are compound eyes with many individual facets, each representing a separate light-detecting unit. The light reflected from the eye of a horsefly can form a rainbow.
- Flies taste, smell, and feel with the hairs that cover their bodies. The hairs on the fly's mouth parts and feet are used for tasting. Flies taste what they walk on. If they walk onto something tasty, they put down their mouth and taste it again.
- Flies use other hairs to tell them when they touch something. These hairs bend when touched.
- The eyes of a fly do not have eyelids, so flies rub their eyes with their feet to keep them clean.
- A fly cleans itself constantly.
- Flies walk on smooth surfaces using sticky soft pads that act like glue. This allows them to walk on vertical glass surfaces and upside down.

Impact on the Ecosystem

Positive

Flies and other insects, such as burying beetles, are very important in consuming and eliminating dead bodies of animals. Flies are also essential in the conversion of feces and decaying vegetation to soil. Flies serve as prey to many other animals. Some flies aid in pollination.

Negative

Because of their habits of being attracted to feces and decaying meat, flies have been implicated in transmission of disease such as dysentery, typhoid fever, and cholera.

Bibliography

Alaka Pande and V.G. Rao: *A Compendium of Fungi on Legumes from India*, Scientific Publication, Delhi, 2001.

Anil Prakash: *Fungi in Biotechnology*, CBS Publication, Delhi, 2001.

Awani Kr. Singh: *Encyclopaedia of Infection Fungies*, Anmoln Publication, Delhi, 2010.

C.S. Chandoliya: *Fungi : Biological Diversity*, Cyber Tech Publication, Delhi, 2009.

D. Jayarama Bhat: *Fascinating Microfungi (Hyphomycetes) of Western Ghats - India*, Broadway Book Centre, Delhi, 2010.

E.J. Maria Florence: *Biodiversity Documentation for Kerala, Part 2 : Microorganisms (Fungi)*, Kerala Forest Research Institute, Kerala, 2004.

F.T. Bennett: *Fungi and Plant Diseases—for Students and Practitioners of Agriculture and Horticulture*, Shree Publication, Delhi, 2000.

H.C. Dube: *An Introduction to Fung*, Scientific Publication, Delhi, 2013.

H.C.I. Gwynne-Vaughan and B. Barnes: *Fungi : Their Structure and Development*, Biotech Books Publication, Delhi, 2004.

Kamal: *Cercosporoid Fungi of India*, Bishen Singh Mahendra Pal Singh Publication, Delhi, 2010.

M.C. Nair and S. Balakrishna:n*Beneficial Fungi and their Utilization*, Scientific Publication, Delhi, 2000.

M.K. Rai and S.K. Deshmukh: *Fungi : Diversity and Biotechnology*, Scientific Publication, Delhi, 2005.

P.P. Rajeshkumar and V.B. Hosagoudar: *Endomycorrhizal Fungi of Silent Valley National Park : Poochipara and Valakkadu Section*, BSMPS Publication, Delhi, 2014.

R K Verma; Nidhi Sharma; K K Soni and Jamaluddin: *Forest Fungi of Central India*, International Book Distributing Co.Delhi, 2008.

R.D. Khulbe: *A Manual of Aquatic Fungi (Chytridiomycetes and Oomycetes)*, Daya Publication, Delhi, 2001.

R.K.S. Kushwaha: *Fungi in Human and Animal Health*, Scientific Publication, Delhi, 2004.

R.M. Johri, Sneh Lata and Kavita Tyagi: *A Textbook of Fung*, Dominant Publication, Delhi, 2011.

Rajni Gupta: *A Text Book of Fung*, APH Publication, Delhi, 2004.

S K Singh and Seema Srivastava: *A Textbook of Fung*, Campus Books International Publication, Delhi, 2008.

Sanjay Kumar Singh: *Fungi*, Campus Books Publication, Delhi, 2008.

Sukhada Mohandas and P. Panneerselvam: *Arbuscular Mycorrhizal Fungi in Fruit Crop Production*, Daya Publication House, Delhi, 2016.

Udit Narain and Mukesh Srivastava: *Fundamentals of Fungi*, Advance Publication, Delhi, 2006.

V K Rastog: *Encyclopaedia of Fung*i, Anmol Publication, Delhi, 2007.

Vasant Rao, Manoharachary, G. Suresh Kumar and K. Subodh: *Fungi : Around Some Aquatic bodies in Andhra Pradesh, India*, BS Publication, Delhi, 2004.

Index